AF804387

# The Reliability of Generating Data

# The Reliability of Generating Data

Klaus Krippendorff

CRC Press
Taylor & Francis Group
Boca Raton London New York

CRC Press is an imprint of the
Taylor & Francis Group, an **informa** business

First edition published 2023
by CRC Press
6000 Broken Sound Parkway NW, Suite 300
Boca Raton, FL 33487-2742

and by CRC Press
2 Park Square, Milton Park, Abingdon, Oxon, OX14 4RN

© 2023 Klaus Krippendorff

CRC Press is an imprint of Taylor & Francis Group, LLC

Reasonable efforts have been made to publish reliable data and information, but the author and publisher cannot assume responsibility for the validity of all materials or the consequences of their use. The authors and publishers have attempted to trace the copyright holders of all material reproduced in this publication and apologize to copyright holders if permission to publish in this form has not been obtained. If any copyright material has not been acknowledged please write and let us know so we may rectify in any future reprint.

Except as permitted under U.S. Copyright Law, no part of this book may be reprinted, reproduced, transmitted, or utilized in any form by any electronic, mechanical, or other means, now known or hereafter invented, including photocopying, microfilming, and recording, or in any information storage or retrieval system, without written permission from the publishers.

For permission to photocopy or use material electronically from this work, access www.copyright.com or contact the Copyright Clearance Center, Inc. (CCC), 222 Rosewood Drive, Danvers, MA 01923, 978-750-8400. For works that are not available on CCC please contact mpkbookspermissions@tandf.co.uk

*Trademark notice*: Product or corporate names may be trademarks or registered trademarks, and are used only for identification and explanation without intent to infringe.

*Library of Congress Cataloging-in-Publication Data*
Names: Krippendorff, Klaus, author.
Title: The reliability of generating data/Klaus Krippendorff.
Description: Boca Raton : Chapman & Hall, CRC Press, 2023. |
Includes bibliographical references and index.
Identifiers: LCCN 2020030224 (print) | LCCN 2020030225 (ebook) |
ISBN 9780367630713 (hardback) | ISBN 9781003112020 (ebook)
Subjects: LCSH: Quantitative research–Reliability. | Data mining–Reliability.
Classification: LCC QA76.9.Q36 K75 2021 (print) |
LCC QA76.9.Q36 (ebook) | DDC 006.3/12–dc23
LC record available at https://lccn.loc.gov/2020030224
LC ebook record available at https://lccn.loc.gov/2020030225

# Contents

# *About the Author*

**Klaus Krippendorff**, PhD., PhD.hc, graduate of the Ulm School of Design and the University of Illinois, Urbana, was the Gregory Bateson Professor Emeritus for Cybernetics, Language, and Culture at the Annenberg School for Communication, University of Pennsylvania. He wrote his dissertation on content analysis at a time this method was quite underdeveloped. *Content Analysis, An Introduction to its Methodology* became a leading text now in its 4th Edition. The book earned the International Communication Association (ICA)'s 2004 recognition as the most influential work. The issue of reliability followed him into numerous empirical ventures. It taught him that data cannot be taken as the starting point of scientific research without knowing how they came about, what they mean, and for whom besides their analyst. In 2012, the Methodology Division of the Association for Education in Journalism and Mass Communication (AEJMC) recognized his "Agreement and Information in the Reliability of Coding" as their "Article of the Year." He has written over a hundred frequently cited publications, not only on reliability but also *on Communicating, Otherness, Meaning, and Information*. In the area of design, he advocated a human interactive approach to understanding technological artifacts as well as of the design discourse that guide them: *The Semantic Turn, A new Foundation for Design*. In the area of Cybernetics, he has received several awards for his contributions in the form of books and articles in academic journals. For some time he wrote and taught seminars on the discursive construction of realities, not just of the social world but also of what natural scientists say they describe. He co-edited *Discourses in Action; What Language Enables Us to Do*. He critically explored how such constructions impact everyday life – for example, the algorithms we live with. His focus was on the possibilities of emancipation from oppression due to widely shared but burdensome realities, mistaken as unalterable facts. This frames the first chapter of the book, starting with answering the question "When are Data?" Klaus Kripendorff passed away in October 2022.

# 0

# *How I Became Interested in Reliability Issues*

At a reception during an International Communication Association (ICA) conference, a colleague of mine asked me an initially puzzling question: How did you as a critical and qualitative communication scholar writing about discourse and analyzing the misuse of power metaphors get to develop what is widely known as Krippendorff's Alpha. I gave him an answer that was adequate for the situation in which I was asked, but it came to me that a longer answer to this question might serve as introduction to a book I had been thinking of writing on the Alpha-bet of measuring the reliability of data. At that occasion, the idea of this book was born. Here is my longer account of how it happened:

I wrote my (1967) dissertation on content analysis, largely because I felt it was a research technique that had a literature consisting of many studies but was methodologically underdeveloped. To me, content analysis is one of two research methods that are truly indigenous to the field of communication research, the other being network analysis, which developed out of studies of communication in social organizations. As a student, I had become familiar with Shannon's (Shannon & Weaver, 1949) mathematical theory of communication and thought I could devise a statistic that combined notions of reliability with the informational requirements to answer research questions. In 1966, I wrote a first paper on the subject (Krippendorff, 1971a) but I had little practical experience then.

## 0.1 The Initial Challenge

It was in the Fall of 1968. I was an assistant professor at the Annenberg School of Communications (as it was then known) when George Gerbner, Dean of the School, asked Martin Brouwer, a visiting scholar from The Netherlands, Cedric Clark, a post-doctoral fellow at the School, and me whether we would be willing to work with him on an urgent request from the United States Surgeon General's Office to provide data on the portrayal of violence on U.S. television. He made clear to us that the timeline was too short and the volume of television shows to be analyzed too big for any one researcher to handle. Only a team would be able to conduct such a study and if we would not do it jointly, it might not be done. Brouwer was an experienced public opinion researcher from The Netherlands and brought expertise in data analysis to the team. In his dissertation, Clark had analyzed the portrayal of African Americans on network television and this was pretty close to the challenge we were offered. I already taught a course on content analysis at the School, and working on such a project was an unquestionable yes. We agreed and spent all our free time during the following months on the project.

We quickly realized that violence is a concept with fluid boundaries. Is blackmail, visibly wearing a gun to a social event, beating a child for disobeying a parent, or putting someone in jail violence? Only the extremes of people killing each other seemed obvious. Yet, there is a difference between fighting a war and strangling a spouse. Most definitions of violence are promoted by institutions with diverse interest in its occurrence. Some psychologists considered watching violence to have cathartic effects, releasing emotions otherwise enacted. Sociologists were sure that the increase in reported violence resulted from watching unpunished violence on TV. There was evidence that several prison breaks and bank robberies were inspired by detailed Hollywood fictions. Broadcasting and the movie industry saw a benefit from the correlation of fictional violence and audience sizes, and newspapers lived from reporting sensationalist crimes.

The U.S. Congress wanted to conduct hearings to sort out claims and counterclaims regarding the conduct of the mass media and its effect on public and private life.

Our work started by cutting through the multiplicity of interests by conceptualizing TV violence in analyzable form. Its occurrence had to be identifiable instances to get us to our units of analysis. We were cognizant of how to describe instances of violence in terms of perpetrators, victims, and motivations. The latter could come from ethnic, gender, and religious prejudices, from greed, righteousness, the role of police, and legal considerations in particular. We brainstormed to define variables that could account for what everyone was exposed to. None of us had experiences in turning such conceptual tasks into an efficient research organization. All of us had undertaken research projects small enough for one principal investigator to manage. Our TV studio at Annenberg had facilities to tape TV programs, but by contemporary standards very primitive ones, limited to one show at a time. We developed coding instructions, employed students as coders, and expected them to follow these instructions when describing what they saw in TV shows. The records they created filled reams of data sheets which we transferred to Hollerith cards for subsequent processing. The University of Pennsylvania computer center had just made the transitions from IBM-card sorters to electronic processing of Hollerith cards.

We could not abandon our teaching responsibilities and students dropped in and out of the project, having to juggle educational obligations at the same time.

After our first round of generating data, we did not quite anticipate the considerable intercoder disagreements we faced and became worried that we would not be able to report findings based on data that were as unreliable as ours. We searched for sources of this variance. We reconceptualized our units of analysis, which were not as natural as subjects in psychological experiments, answers to predefined interview questions, or grammatically defined units of text. Television shows are continuous phenomena and selecting what we decided to be violent encounters was not easy. Cinematographic units, such as shifts in scenes, or dramaturgical changes of scenes with actors entering or leaving, did not suffice as violence is always prompted by preceding conditions, justified or not, and incidences of violence had consequences beyond the actions we observed on the screen. Counting the number of violent acts and victims failed to capture the brutality with which they were committed so we had to find new variables. Even the distinction between perpetrators and victims was far from easy; for example, when violent acts were committed in response to past violence mentioned but not depicted, or legally justified. We formulated distinctions and categories as well as we could, tried them out ourselves, and then gave them to coders.

To select coders for the task, we developed a training program that instructed prospective coders and tested their abilities to do what we expected them to do. It consisted of a collection of prototypical television shows, not to be included in our analysis, which the three of us had worked through and coded in all details. Students viewed the shows, one at a time, identified instances of violence, interpreted their narratives according to our written instructions, and marked the categories we had provided. After they completed the data sheets for one show, they were able to check for themselves whether their judgments conformed to or deviated from the standard we had set and went to a next show until they felt sufficiently confident and we let them "graduate" to work on their own.

However, while the standardized training program focused on calibrating the students' interpretation of what we needed, the criteria to judge the quality of the data they subsequently generated were far from satisfactory, yet very important. TV violence had become a public concern with divergent stakeholders articulating their competing perspectives. We needed to make sure that our results would withstand the scrutiny by the U.S. Surgeon General's Office, the U.S. Congress where legislation was being debated and hearings conducted, by critical scholars within the academic community, and last but not least vested interests by TV networks which thrived on producing violent fictions.

## 0.2 Coming to Understand the Issues

The practice of testing reliability in social research was far from settled at that time. In the first text on content analysis, Bernard Berelson (1952), (initially Berelson and Lazarsfeld, 1948) had discussed the importance of achieving reliability, but did not provide a measure. He reported how several researchers had addressed the reliability of their content analyses data by reporting percent agreements and correlations between coders. We considered both measures inadequate for different reasons: percent agreement because it does not account for agreement due to chance, and correlation because it does not measure agreement among coders. Instead it establishes whether coders are statistically independent of each other. Also, both measures applied to pairs of coders; we employed more than two, and because students had other obligations, they made unequal contributions which is to say that the number of coders that coded any one show varied.

We had internal epistemological disagreements about the purpose of assessing reliability. Gerbner wrote a memo declaring that the very unreliability we experienced should be considered a substantial finding. It is, so he argued, consistent with the strategy of the TV industry to make violence not too easily definable, using it to attract large audiences while avoiding being publicly criticized for doing so. We, the three other members of the team could not quite convince him of this misconception. We argued that he was certainly correct in his assessment of the industry's intentions but if the industry's motivation were to become part of our violence report, it would have to be assessed reliably as well and tested as a hypothesis.

We did not know of Bennett, Alpert and Goldstein's (1954) agreement coefficient *S*, Scott's (1955) $\pi$ (*pi*), introduced into the public opinion research literature but rarely used there, Cohen's (1960) $\kappa$ (*kappa*) at home in biometric research, and we had never heard of Cronbach's (1951) *alpha*, known largely among psychometricians and used to measure a battery of psychological tests aimed at establishing whether a psychological variable was underlying various responses to a stimulus. We rejected the suggestion of accepting data as reliable when the null hypothesis of no agreement failed. For data to be taken as reliable, it was not enough to prove that they are not entirely worthless but that they do not deviate too much from being perfect. In the content analysis literature, the most common way to claim data to be reliable was by calculating percent agreement between two coders, which is a measure that is not only limited to two coders but varies with the number of categories coded and includes the case of the total absence of reliability while measuring 100%. In retrospect, our ignorance of the established measure was a blessing in disguise because adopting any one of these measures would have meant adopting the epistemological assumptions of research traditions inappropriate to content analysis research which would have seriously constrained what we were hoping to be able to report in the end.

What we faced were tables of how several observers, coders, and interpreters selected incidences of violence in TV programs, our recording units, and judged them in terms of categories or scale values of several variables we expected would answer our research questions. Figure 0.1 represents the canonical form of such reliability data for one variable, an observers-by-units matrix whose different entries are visually distinguished by dots of unequal size:

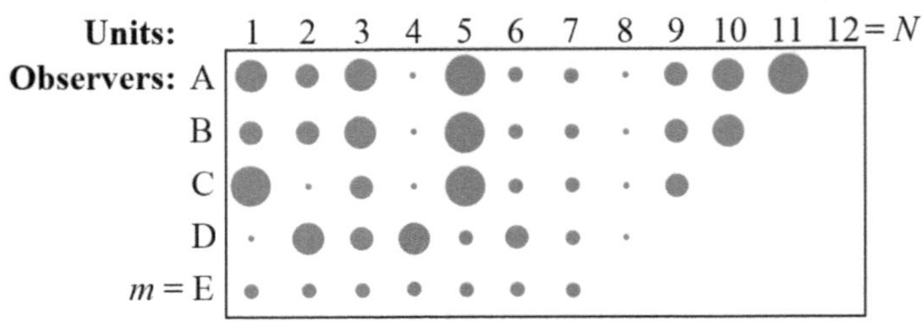


**FIGURE 0.1**
Reliability Data Generated by $m = 5$ Observers Coding $N = 12$ Units in Terms of 5 Values

To be clear, Figure 0.1 does not depict realistic reliability data. It is constructed to include the most typical disagreements among $m = 5$ observers coding $N = 12$ units with at least five values. The first six units in columns 1 to 6 exhibit all possible patterns of disagreements among five observers; the remaining units 7 to 12 are coded by a declining number of perfectly agreeing observers.

For data to be reliable, all units in Figure 0.1 would have to be valued identically as in unit 7. However,

- Unit 1 exhibits the maximum possible disagreement. The five observers do not agree on any one of the five values they have available to describe what they have seen, read, or experienced. Their disagreement is maximum, a condition rarely obtainable by chance.
- In unit 2, observers A and B agree but none of the others agree with them.
- Unit 3 shows A and B to agree on one value, C and D on another, with E disagreeing with both.
- Regarding unit 4, three observers A, B, and C agree but D and E disagree with each other and with the majority.
- In unit 5, A, B, and C agree on one value while C and D agree on another.
- In unit 6, coders A, B, C, and E agree among each other from which D deviates.
- In units 7 through 10, disagreements are absent, but the number of observers coding these units declines from five to zero.
- Unit 11 is coded by only one coder. Identifying agreement or disagreement requires at least one pair of values. A lone code cannot be included in any reliability test, as is the case for unit 12, which is not coded by anyone.

Figure 0.1 reveals other noteworthy properties.

- Observers A and B agree on eight out of ten units. If this were true for the remaining pairs of observers, their reliability may not be too bad.
- The other three observers join the first two only in regarding units 7 to 9 but to a decreasing extent.
- Observer E displays no variation in coding all seven units coded. This complete absence of variability across units is as uninformative about the nature of phenomena that a researcher wants to analyze as is coding by throwing a dice. Whether the observer is asleep or codes randomly, neither condition generates reliable data.
- As already mentioned, the number of observers coding given units declines starting with unit 8, which also contributes a declining amount of information to a project.

Conceptually, for a diversity of units to be reliably represented in data about the phenomena of interest, an ideally 1-to-1 relationship between the coded units and the nature of the phenomena of interest has to be achieved. For any one observer that 1-to-1 relationship may well be experienced as obvious although always only in their own terms. Yet, analysts who rely on data generated by employed observers cannot possibly check what these observers had observed, read, or experienced and any one observer's verbal assurances of the accuracy of this relationship become questionable when they do not agree. Unable to check that relationship yet required to rely on the given data poses an epistemological conundrum to the analyst who is well advised not to proceed without adequate reasons to trust the data. This trust could be justified when several independent observers, known to have faced the same set of phenomena, provide identical reports of what they saw. The extent to which this ideal is not achieved would be reasonable grounds to abandon our categories. Without direct access to the 1-to-1 relationship, we had to measure this condition indirectly by the extent to which independent observers agreed on the coding of units. We quickly recognized that the reproducibility of our coding instructions was the condition that reliable data need to satisfy. Measuring the extent to which data are reproducible is the key to all reliability concerns. The question was how to assess observed disagreement to infer whether data are sufficiently reliable to warrant proceeding to their analysis.

Visually detecting unwanted diversity of codes across observers is limited to very small data. We had reams of data and little time to examine them for all the variables we pursued. Spiegelman, Terwilliger and Fearing (1953) had encountered such bewildering complexities before we did. They solved them by creating all possible patterns of matching and mismatching values, much as in Figure 0.1, asked subjects to rank them according the degree of their (dis)agreement, matched these judged ranks with the reliability data obtained, and took the average of these ranks as a measure of the reliability of their data. This was a practical idea, but it required subjective judgments for all possible pattern of (dis)agreement and could not easily be generalized to varying numbers of participating coders who generated our data.

In my dissertation on content analysis, I had proposed to measure the amount of information in data and to trace this quantity through their analyses to the answers to given research questions. I argued that noise in the data could lead to invalid conclusions. With this conception in mind, I began to define an information measure that equated the observed disagreement with noise and expressed the reliability of data as the extent to which data provide information about unitized phenomena of interest as opposed to noise. Formally:

$$\frac{\text{Information between unitized phenomena } P \text{ and coded data } D}{\text{Entropy in } P \times D\text{, the whole matrix of relibility data}} = \frac{T(P:D)}{H(P \times D)} = 1 - \frac{H_P(D)}{H(P \times D)} \quad (0.1)$$

I was encouraged to pursue this path by the recognition that Shannon's measure of noise, $H_P(D)$, equal to the sum of the entropies associated with the coded units of analysis, correlated perfectly with Spiegelman et al.'s subjective rankings of different degrees of disagreements. The expression in the denominator of this fraction $H(P \times D)$ is the entropy in an observers-by-units matrix of assigned categories and equals the amounts of information plus noise $H(P \times D) = T(P:D) + H_P(D)$. It follows that the proportion in the preceding equation (0.1) yields 1 when noise is absent and 0 when information is absent. The ability to decompose entropies into quantities of information and noise was made even more attractive as it provided analytical flexibility. It enabled dealing with any number of observers variously categorizing any numbers of distinct phenomena.

We hand-calculated this measure for variables we could see as lacking reliability. Most of the time it produced plausible assessments, but there were occasions when it did not, which taught me an important lesson. Shannon's entropy measures derived their analytical flexibility by defining entropies in terms of the logarithm of probabilities. Its granularity was inextricably related to the number of probabilities assessed. Naturally, the granularity of the measure of noise, which is a function of the number of observers coding them, is smaller than the granularity of the reliability data generated by these observers. Moreover, as the number of observers varies, as in units 7 to 11 of Figure 0.1, entropies became numerically incompatible. Although the information measure promised several interesting analytical possibilities, which I had explored earlier and published much later (Krippendorff, 1971a), the agreement measure for inferring the reliability of data based on entropies proved unworkable and I had to abandon this effort. The lesson I learned from working through many situations is that all indices have internal properties that need to be compatible to the measured phenomena for their quantitative results to make sense.

Because the algebraic properties of entropies in the information calculus resembled that of variance analysis, I followed Rajaratnam's (1960) analysis of variance to conceptualize an agreement coefficient α (alpha – for "agreement") (Krippendorff, 1970a). In that paper I was interested in separating systematic from random variances and came up with a simple expression. Its analogy to the above information theoretical measure in (0.1) should be obvious:

$$\alpha_{\text{interval}} = 1 - \frac{\text{Variances within all units } u}{\text{Variance in all reliability data}} = 1 - \frac{\frac{1}{N_u}\sum_{u=1}^{N_u} V_u}{V_{\text{total}}} = 1 - \frac{D_o}{D_e} \qquad (0.2)$$

Where $V_u = \dfrac{\sum_{i=1}^{m}(x_{iu} - \bar{x}_u)^2}{m-1}$ = the variance of $m$ coders' judgments of one unit $u$

$x_{iu}$ = The value $x$ assigned by coder to unit $u$

$\bar{x}_u = \dfrac{1}{m}\sum_{i=1}^{m} x_{iu}$ = The average of the $m$ values $x$ in unit $u$

$$V_{\text{total}} = \frac{\sum_{u=1}^{N_u}\sum_{i=1}^{m_u}(x_{iu} - \bar{x})^2}{mN_u - 1} = \begin{cases}\text{The total variance in } m\text{-by-}N_u \text{ reliability data,} \\ \text{presuming all } m \text{ observers code all } N_u \text{ units}\end{cases}$$

$\bar{x} = \dfrac{1}{nN_u}\sum_{u=1}^{N_u}\sum_{i=1}^{m} x_{iu}$ = The average of the $mN_u$ values $x_{iu}$ in the reliability data

$D_o$ = The observed disagreement (variance within coded units)

$D_e$ = The expected disagreement (total variance within and between coded units)

Equation (0.2) ensures that whenever agreement among values within all units is perfect, the columns in Figure 0.1, are without exception, i.e., their variances $V_u$ are absent, $D_o = 0$ and $\alpha_{\text{interval}} = 1$. This condition is indicative of perfectly reliable data. And when the variance within units is indistinguishable from the variance in the whole table (within and between units), $D_o = D_e$ and $\alpha_{\text{interval}} = 0$, signaling the absence of any statistical relationship between the predefined units and the values assigned to them. Under these conditions, the data in hand could not possibly serve as a surrogate for any phenomena of analytical interest – in our case for the TV violence we hoped to analyze through the data we had to generate.

While $\alpha_{\text{interval}}$ could assume negative values, its range between 0 and 1, between the absence of any relationship between data and the phenomena of interest and perfect agreement, was clearly interpretable as a reliability scale. We decided that these two reference points should be constitutive of any valid agreement measure capable of evaluating the reliability of data.

(0.2) solved the granularity problem of (0.1) but created two others. The inconvenient one was that it required that all $N_u$ units be coded by all $m$ or at least the same number of observers. While we could force our coders to stay on for a whole reliability test, this strictly computational constraint was quite impractical as students had course work to do and good reasons to be inconsistently available.

More important was the unmitigable fact that variance required interval data. Most of our variables consisted of unordered categories. To these reliability data, the analysis of variance is inapplicable. Brouwer suggested a decomposition of nominal variables into a series of binary distinctions which an interval measure could handle as a special case, but this was awkward, extremely time consuming, and gave us only "atomic" answers.

On a long flight to Bangladesh, then East Pakistan, with nothing else to do, I tried numerous ways to generalize what I had found for interval data to a form suitable for unordered nominal data. I had almost given up when I realized that the standard deviation $SS$ in the definition of variance $V$ is a computational shortcut for averaging the algebraic differences among all possible pairs of values within a set of interval values:

$$V = \frac{SS}{df.} = \frac{\sum_{i=1}^{n}(x_i - \bar{x})^2}{n-1} = \frac{\frac{1}{n}\sum_{i=1}^{n}\sum_{j>i}^{n}(x_i - x_j)^2}{n-1} \tag{0.3}$$

I submit, for a statistician the recognition of the equivalence (0.3) of summing $n$ squared deviations from a mean $(x_i - \bar{x})^2$ and $n^{-1}$ of the sum of all $n^2/2$ squared differences $(x_i - x_j)^2$ among the $n$ values may be obvious. I didn't have that background. Yet, it allowed me to define an alternative difference for unordered categories as replacement of the squared difference in expressions of variance. It recognized that pairs of unordered categories are same or different without any degrees between these two alternatives:

$$_{\text{nominal}}\delta^2_{x_i x_j} = \begin{cases} 1 \text{ iff } x_i \neq x_j \\ 0 \text{ iff } \mathrm{x_i} = x_j \end{cases} \tag{0.4}$$

To make the interval differences compatible in its range with (0.4), I had to standardize the differences in (0.3) by the largest range which is a constant of a given set of data, and obtained the difference function for interval data:

$$0 \leq {}_{\text{interval}}\delta^2_{x_i x_j} = \left( \frac{x_i - x_j}{x_{max} - x_{min}} \right)^2 \leq 1 \qquad (0.5)$$

Their tabulation among the $n(n-1)$ possible pairs of $n$ values is depicted in Figure 0.2. If disagreement $D$ is defined as the sum of all differences ${}_{\text{interval}}\delta^2_{x_i x_j}$ among $n$ values as in (0.6), this results in a very simple linear relationship to the variance $V$ in (0.3) due to the fact that the variance is not standardized by $(xz_{max} - x_{min})^2$ and sums only one half of the $n(n-1)$ differences that constitute the disagreement $D$. The relationship between the interval disagreement $D$ and the variance $V$ is expressed by the fraction in (0.6) which is a constant for a particular analysis. It follows that the analytical results with either statistics would be the same.

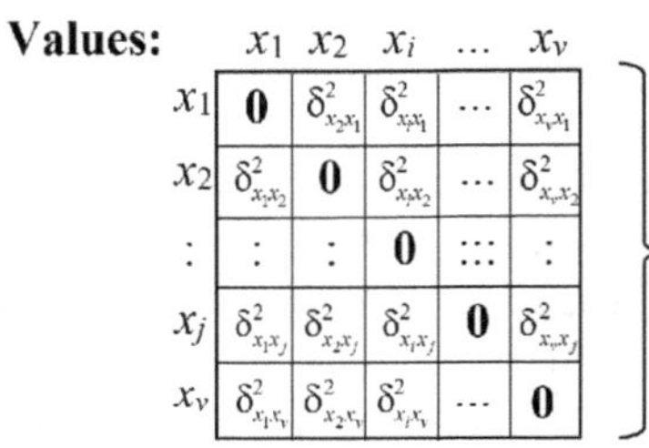

$$D = \sum_{i=1}^{n} \sum_{j=1}^{n} {}_{\text{interval}}\delta^2_{x_i x_j} = \frac{2n(n-1)}{(x_{max} - x_{min})^2} V \qquad (0.6)$$

**FIGURE 0.2**
The $n(n-1)$ Differences Among $n$ Values, Not Pairing Values With Themselves

The point of relying on difference functions ${}_{\text{metric}}\delta^2_{ck}$ is of course their generality which will be taken up in Chapter 5.

While Shannon's entropies had the undesirable properties of varying granularities mentioned above, the algebra of its calculus turned out to be isomorphic with Fisher's calculus of variance, save for the differences they respectively summed. Inserting either (0.4) or (0.5) into (0.2) resulted in unified measures of agreement, $\alpha_{\text{interval}}$ and $\alpha_{\text{nominal}}$, isomorphic to the variance-based computations.

This possibility also gave rise to a variance analysis for nominal differences (Krippendorff, 1971b), which will be the further generalized in Chapter 16.

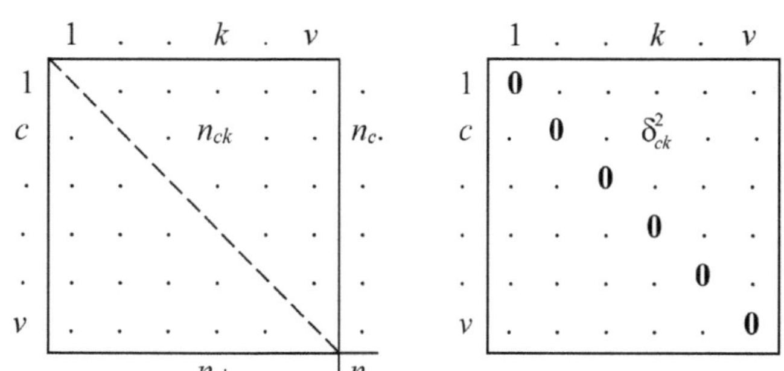

**FIGURE 0.3**
The Matrix of Coincidences $n_{ck}$ and Their Applicable Differences $\delta^2_{ck}$

Computing the newly developed reliabilities by hand was time consuming and prone to human error. It was helpful to separate computing the frequencies $n_{ck}$ from the corresponding difference functions ${}_{\text{nominal}}\delta^2_{x_i x_j}$ or ${}_{\text{interval}}\delta^2_{x_i x_j}$. We called the resulting $n_v$-by-$n_v$ tables in Figure 0.3 coincidence and difference matrices respectively and could take advantage of their symmetry around their diagonal cells which contained the matching coincidences and zero differences respectively.

As time was precious, we started hand-calculating reliabilities using students in assembly-line fashion for different steps, until I had learned the then state of the art programming language Fortran IV, which enabled me to write software for computing α and obtain reliabilities for our data from our computer center overnight.

In the end, we could report satisfactory reliabilities (Brouwer, Clark, Gerbner & Krippendorff, 1969). To my knowledge the reliability of our data was never challenged although our conception of violence triggered public debates (Krippendorff & Bock, 2009: 383–403).

Our practically motivated development plus the meanwhile discovered proposal by Scott (1955) and Cohen (1960), but especially Pearson's (1926) brilliant suggestion to turn his product-moment correlation coefficient into a coefficient of racial likeness, as he called it (Tildesley, 1921), brought me to appreciate the difference between our coincidence matrix representations of reliability data and the more familiar contingency matrices, and compare α with other agreement coefficients, which will be discussed in Chapter 4. Further extensions of α to other kinds of metrics, ordinal and ratio metrics at that time, heretofore unavailable, also brought several other agreement coefficients under the same roof (Krippendorff, 1980, 2019). This book presents numerous elaborations and further generalizations of these basic notions.

Subsequent to our TV violence study, I continued to teach content analysis at the Annenberg School for Communication, consulted on various research projects with reliability problems and was invited to publish further developments of α. This agreement coefficient has since found applications in medical research, psychotherapy, computational linguistics, library science, political discourse analysis, history, educational testing, comparing human coders to computer applications, and the reliability of crowd searching. It proved its generality far beyond its initial expectations. Over time, α has also grown into a whole system. It is applicable not just to coding predefined units with diverse metrics (levels of measurement) and incomplete data, as originally defined, but also to multiple descriptions and unitizing continuous phenomena, whose reliability we could not conceptualize at that time.

Already during the TV violence study, we used α not just to obtain publishable measures of the reliability of our data, but also to identify various sources of unreliability; for example, unreliable coders, ambiguously defined categories, and systematic coding errors, all of which could direct our attention to possible improvements of the reliability of data once coded. Today, α governs a wide territory. It embraces several specialized coefficients and has led to discussions of general reliability standards. This book will give a flavor of its scope.

I should mention that I also became interested in the more general issue of the grammar of information in data, for example in Clyde Coombs' (1967) *A Theory of Data*. Coombs was not concerned with the reliability of data and how well data represented the phenomena of analytical interest but informed my notion of data structures. With this interest in mind I wrote an award-winning paper "On generating data in communication research" (Krippendorff, 1970). It criticized Harold Lasswell's (1948) then widely celebrated model of communication which defined the field of communication research as answering five separate research questions. I argued that the structure of the data these questions generated could not possibly shed light on interactive processes of communication, or reflect on itself, and would leave our discourse unable to capture the emerging realities it was dedicated to explore. Therefore, my interest in the problem of analyzing data that fail to serve as a surrogate for phenomena of scholarly and practical interest goes beyond the narrow scope of measuring the reliability of generating data, variable by variable.

This personal history may explain why I pursued the development of α, not as a single reliability measure but as a vocabulary, an Alpha-bet for assessing data reliability, which generalized the original α to a huge family of reliability measures for different kinds of data, easily extendable to as-yet-unknown domains; maintaining the basic form in the last expression of (0.2) throughout all applications and assuring the comparability of reliability measures across many empirical domains.

I wish to add that the development of α, usually in response to other researchers' reliability problems, has always remained a sideline to several other personally more fascinating scholarly topics on human communication, cybernetics, language, and culture. This book presents only a small piece of my far more general scholarly concerns. It is meant to be a complementary text book for courses in content analysis and other qualitative/quantitative courses of study and a handbook for social researchers to guide them to solutions for assessing the reliability of their data, a vocabulary for software developers to provide algorithmic aids for empirical research, and an invitation to methodologists to continue α's development to whatever reliability issues may arise in the future.

## 0.3 Outline of the Chapters of This Book

Chapter 1 challenges the common conception of data as unquestionable facts. The etymology of facts is related to the word 'manu*fact*uring', producing arti*fact*s that do not come naturally, including data. Data are created for diverse purposes. The chapter starts by discussing four constitutive conditions for being data. All are related to human actions, not only by scientific researchers, although theirs is my main concern. Data have to be durable to be retrievable. They have to be comparable, computable, analyzable, or be able to be part of arguments. They have to serve as surrogates for analytical concerns, and represent phenomena other than themselves. And they have to inform current scholarly concerns, or answer given or emerging research questions. To these conditions I add two often-overlooked features: Data is the plural of datum and implies diversity. Also, data always link contemporary readings to their past origins.

Grounded in the possibilities that data can fail researchers, theorists, and decisionmakers, this chapter does not take data as unquestionable givens. It reviews how a variety of scholarly and practical disciplines have coped with assuring that their data can be trusted. These explorations lead to my own synthesis and the distinction of five kinds of reliabilities. Stability, replicability, accuracy, surrogacy, and decisiveness are formally related but respond to different deviations from perfect agreements and therefore serve different purposes. The chapter concludes with a discussion of how reliability and validity are related.

Chapter 2 serves as an introduction to measuring agreements among independent replications in order to infer the reliabilities of data. It relies on the most common and very simple form of reliability data generated by two independent observers categorizing a set of predefined units of analysis. After illustrating what is involved by means of an easily traceable numerical example of this situation, the chapter discusses the simplest possible case of making binary either/or distinctions and extends this common form of reliability data to a larger number of observers categorizing or coding a set of phenomena. The purpose of leaving the more exciting developments of the $_c\alpha$-measure for coding predefined units to subsequent chapters is to give readers just enough details to make sense of the following two chapters.

Chapter 3 highlights some of the mathematical properties of $\alpha$ generally and of the $_c\alpha$ of Chapter 2 specifically. It discusses $\alpha$'s form, range, scale, and behavior. Readers will get a sense of why $\alpha$ has the properties it has and how they manifest themselves numerically. Although comparisons of $_c\alpha$ with other coefficients will be discussed in the next chapter, this chapter presents several conceptions that readers may have to keep in mind when evaluating, criticizing or debating reliability measures found in the literature.

Chapter 4 begins with a rough sketch of $\alpha$'s lineages. It dismisses the common measure of percent agreement as unsuitable for inferring reliability and focuses on five prototypical so-called chance-corrected percent agreement measures. It uses a numerical example and visualizations of their respective expected agreements to reveal their differences, using criteria developed in Chapter 1. The properties of these measures are compared with the ones of $_c\alpha$ discussed in Chapter 3. All five of these popular agreement coefficients are applicable to the simple reliability data defined in Chapter 2. This chapter ends with a table of twelve agreement coefficients including the five coefficients examined in detail. It comparatively lists the kind of reliability data they are capable of handling, but most importantly, the conditions under which they could possibly serve as unbiased indicators of the reliability of data.

Chapter 5 provides the first of several generalizations of the simple $_c\alpha$ of Chapter 2 for assessing the reliability of categorizing predefined units to more powerful kinds of ordered data. The agreement coefficients discussed so far were limited to unordered data. The literature does contain a few agreement measures for other than nominal categories. However, just as the measures reviewed in the previous chapter, they have different origins, are the result of different assumptions, have different mathematical structures, and tend to be developed in different domains of application which makes their numerical results not comparable across data with unlike metrical qualities. To maintain the comparability of $\alpha$ across different metrics of data, its extension to other kinds of data is accomplished by defining metric-specific difference functions that weigh only the coincidences of $\alpha$ preserving the way $\alpha$ accounts for disagreements within the reliability data. The birth of this extension was already described in Chapter 0 when the possibility arose of replacing the familiar differences of variance analysis by nominal difference functions, creating the $_c\alpha$ discussed so far. This chapter defines six difference functions, one for each kind of data: unordered categories, rank ordered items, interval scales, ratio measures, e.g., proportions, polar scales, and circular data from modular arithmetic. Inserting them as needed creates a diversity of numerically comparable $_c\alpha$-agreement measures. The manner with which $_c\alpha$ can be made to respond to a variety of metric qualities of data leaves $_c\alpha$ open to currently unanticipated data structures.

Chapter 6 continues to expand $_c\alpha$, previously limited to evaluating the replicability of data, to other kinds of reliability already outlined in Chapter 1. The idea of a quadrilogy of reliability coefficients was born in the course of a challenge to apply $_c\alpha$ to very big data. This chapter begins by sharing the difficulties experienced in measuring the reliability of such data and operationalizes the quadrilogy of replicability, accuracy, surrogacy, and decisiveness for binary data. Such simple data have the advantage of being metric-independent and reveal most clearly the fundamentally different computations for these four reliabilities. In the literature, reliability and accuracy are often considered synonymous. However, this chapter considers accuracy as a measure of the correspondence of largely human coding with a trusted standard. Surrogacy as a measure of the extent to which a preferred observer, measure, or algorithm can take the place of what replications of human judgments have in common, and decisiveness as the extent to which coders conform to their majorities, medians, means or averages. These four reliabilities serve fundamentally different purposes, all under the roof of $\alpha$. The chapter ends by extending the binary versions of this quadrilogy to variables with larger numbers of values of diverse metric qualities.

Chapter 7 responds to the frequent need to obtain reliabilities for situations in which units of analysis are characterized by more than one category, value, or description. We know that words tend to have several meanings, individuals may maintain several friends, and artifacts may serve multiple and equally justified functions. Chapter 7 expands previous limitations of single-valued coding to multi-valued characterizations of predefined units of analysis. This is accomplished by a difference function capable of embracing all single-valued metric differences defined in Chapter 5. This extends ${}_c\alpha$ into heretofore unavailable territory of describing units in terms of sets of values signified by ${}_C\alpha$.

Chapter 8 seeks to overcome the computationally convenient but not always appropriate use of agreement measures for predefined units of analysis by focusing on continuous phenomena in which the units of analysis are not definable in advance. Texts are created and read linearly. Narratives have natural beginnings and ends. Conversations consist of conversational moves. Movies visualize flowing narratives. Chopping them into arbitrary units – words or pages, 30-second intervals, or scenes – just to apply widely available statistical methods to them, including agreement measures – hides the underlying unreliability of such data. This chapter develops a family of four ${}_u\alpha$-coefficients that respond to disagreements among freely identified segments of partitioned continua. Its agreement coefficients are concerned with the volumes of coded and relevant and uncoded and irrelevant intersections.

Chapter 9 presents an agreement coefficient ${}_U\alpha$ which also addresses the reliability of unitizations but acknowledges the integrity of identified segments. It is not concerned with the volume of coded intersections as is ${}_u\alpha$, but with the extent to which identified and coded segments coincide, in effect preserving their coherence or contiguity.

Chapter 10 represents the most basic form of dealing with continua. What are binary distinctions among predefined units here become simple distinctions drawn in a continuum. *D*-Alpha assesses the extent to which distinctions drawn within a continuum by several observers, coders, or unitizers coincide without regard to what they distinguish.

Chapter 11 ventures into the reliability of text mining and information retrieval. It questions the two widely used measures of the quality of search engines on several grounds. 'Precision' and 'recall', being percentages, have all the flaws of percent agreement measures discussed in Chapter 4, and 'recall', billed as the percent of relevant items retrieved from those contained in a data base, paradoxically requires knowledge of the content of a data base independent of retrieving data from it. In addition, these measures fail to separate what a search engine does with texts and how researchers interpret retrieved text as informing their research questions. This chapter cannot resolve all the epistemological problems of this kind of generating data, but it paves the way to comparing the results of different search engines separately from the reliability of researchers' interpretations.

Chapter 12 turns its lens to diagnostic uses of agreement measures. The original purpose of measuring the agreement among replications of processes of generating data was to give researchers the needed assurances that their data represent the phenomena of interest with a minimum irrelevant variation. However, there is a place for identifying the causes of unreliable data which can potentially inform corrective actions. This chapter proposes several diagnostic aids in the form of specialized agreement coefficients for three common phases of research projects. While developing or refining coding instruments, specialized agreement measures may inform researchers about ambiguously defined values, unreliable observers, and units of analysis that are difficult to code. For using machine learning algorithms, surrogacy and decisiveness measures are critical. While generating data, researchers may have to check the accuracy of the process to ensure the continuity of data quality. After data have been generated, agreement measures are provided that can support decisions not only about the omission of whole variables but more importantly about whether unreliable variables can be salvaged by lumping values or using several other transformations of trading dispensable information for improved reliability of data for subsequent analysis.

Chapter 13 presents several solutions to highly specialized problems not only to establishing the reliability of data but also to using agreement measures for other analytical purposes. For example to measure the agreement between groups' diverging assessments of an issue. It ends with a measure that generalizes percent agreement to ordered data of any of the metrics defined in Chapter 5. These examples are meant to inspire researchers to think outside of the box of α's definitions but the solutions offered may well resonate with readers' concerns.

Chapter 14 develops the statistical properties of α. It elaborates some issues mentioned in conjunction with the very simple $_c\alpha$ in Chapter 3. It highlights the importance of variance in data, and differentiates information about phenomena of interest and noise, both entering data-analytical processes indistinguishably. It exemplifies how noise can cause research results to vary even when reliability standards seem to be met. Based on the important distinction between reliability data and the data whose reliability is in question, this chapter addresses the drawing of representative samples from available observers, coders, or instruments; among phenomena of interest; and seeing to it that data provide adequate amounts of information for reaching the intended analytical goals. It also provides two methods of estimating the confidence intervals of observed αs at chosen levels of significance and the probability of mistakenly accepting data as reliable when they could measure below a chosen minimum $\alpha_{ok}$. For nominal data, the chapter recommends a computational approximation to the normal curve. For all other data, it suggests an algorithm for bootstrapping α's distribution from which these two statistical parameters of alpha are obtainable.

Chapter 15 discusses how to decide on standards for making decisions on the trustworthiness of data. It relates α to the demonstrable ability of researchers to decode the phenomena of their concern from the data they are analyzing, and not merely claim to know their subject matter. While readers may hope to find a magical cutoff value for α-agreements, this chapter gives three reasons not to support a universal baseline. For one, as developed in the previous chapter, most measured αs have confidence intervals whose width varies with the size of the reliability data. For another, biased or wrong conclusions derived from imperfect data may have unequal social consequences. When the lives of people are at stake, an acceptable α-agreement should be nearly perfect while a lower one may still be acceptable in mere exploratory academic research. Finally, this chapter acknowledges that some analytical techniques applied to data suppress evident unreliabilities while others magnify them. It follows that the relationship between observed unreliabilities and the validity of research results is not as obvious as some researchers claim but is certainly important.

The chapter ends with a proposal to measure the reliability of the results obtained by analyzing all relevant variables. This shifts the emphasis of the preceding chapter on the variable-specific reliability of generating data to the consequences of analyzing them. It recognizes that variables may have unequal impact on the analytical results; some severely unreliable data may make no difference, whereas small disagreements in other variables may cause one result to flip into its opposite. If the analyses of several possible data result in one result or an unambiguous answer to the guiding research question this would be very strong evidence for the reliability of research. This standard concerns the replicability of research and is limited to relatively formalized analytical techniques and sufficient number of replications. However, it provides the ultimate reliability standard which is difficult to imagine at the point of generating data.

The concluding Chapter 16 takes a birds-eye view of the logic that underlies the two kinds of disagreement which go into the ${}_c\alpha$-coefficient. It proposes a general calculus of differences and of agreements of which ${}_c\alpha$ is merely a special case. This calculus focuses on various differences and agreements in a chosen variable, as ${}_c\alpha$ does on coded phenomena. It expands ${}_c\alpha$'s reliance on replications and coded units to any number of explanatory variables for that variable and relaxes any restrictions on the number of items distributed in their multivariate space. This chapter proposes several accounting equations of the differences, decisivenesses, and agreements within the partitions of the focus variable by chosen explanatory variables. In the end, this calculus liberates Fisher's variance analysis from its limitation to interval data by being applicable to variables with any metrics for which difference functions are available or can be defined.

Following the last chapter 16, the reader will find an appendix of all $\alpha$-coefficients this book names, distinguishes, elaborates, and exemplifies. It lists these $\alpha$-coefficients in the contexts of their notational relatives, offers brief descriptions of what their notations mean, what they intend to measure, and the sections in which the reader can follow their development.

# 1

# *On the Epistemology of Reliable Data*

## 1.1 When Are Data?

All scientific pursuits I know of take data to be the trusted ground for reasoning, discussion, calculation, and as evidence of scientific claims.

The word *data*, the plural of *datum*, comes to English from the Latin *dare*, 'to give,' hence 'something given,' not open to be questioned. The *Oxford English Dictionary* dates the use of data to a 1646 theological tract that refers to "a heap of data." A contemporary dictionary defines a *datum* as a 'piece of information.' When researchers challenge each other and ask to see the data that support their claims, tested hypotheses, theoretical explanations, answers to research questions, or practical recommendations, data are equated with evidence on the lowest level of abstractions.

In such arguments, data are assumed unquestionably given. This assumption surfaces in answers to questions about their origin. When data are said to be 'found,' 'discovered,' 'uncovered,' or 'gathered,' the presumption is that they were already there to begin with. When data are said to have been 'sampled,' 'weighted,' 'counted,' and 'entered' in a computer, they are treated as if they were collections of 'objects' that exist as such in the world, waiting to be taken into the custody of researchers. Data-based research yields 'findings,' as if both the data and the results of analyzing them preexisted scientific interest in them. It is such uses of language that encourage researchers and critics to take the objectivity of data for granted, rendering the conditions under which they come to be data undisputable. This language conveniently frees researchers from having to justify who created their data or what brought them into their possession and focuses their attention instead on the analytical or computational methods employed to extract pattern underlying them. The possibility that data might be tailored to fit analytical purposes, could be biased, unreliable, and thereby introduce uncertainties in the conclusions drawn from them becomes conveniently sidelined.

However, I shall argue that all data are human artifacts. They are made into facts; if not physically so at least by declaration. The word *fact* derives from the Latin as well, from the verb *facere* and the noun *factum*, something done or manufactured by a factory. Regrettably, common usage of 'fact' ignores its etymology and renders invisible the role of scientific discourse in which facts are established as such. It is an epistemological mistake to confuse data and facts with truths as independent of scientific attention, observation, action and constitution. We should remind ourselves that 'true' and 'false' are properties of propositions, not of the world independent of language. Propositions are always someone's propositions, someone who formulated them in the expectation of being held accountable for them. The 'real' happenings that propositions claim to refer to are not knowable independent of experiencing, measuring, recording, and articulating if not creating them.

As Ruth Hubbard (1989) reminds us, "facts aren't just out there. Every fact has a factor, a maker." Interestingly, proposition 1.1 of Wittgenstein's (1933; 30–31) Tractatus reads:

1.1 *Die Welt ist die Gesamtheit der Tatsachen, nicht der Dinge.*
1.1 *The world is the totality of facts, not of things.*

In the German version of this proposition, "Tatsachen" are properties or circumstances (Sachen) brought about by actions or deeds (Taten). I am suggesting that Wittgenstein had it right by shifting our attention from conceptualizing the world as existing in the form of observer-independent objects or things to what we do to bring them forth, why they matter to us, and how our use of language relates them to each other. Reading "facts" as truths would derail his conception.

Along the same line, I am reminded of Werner Heisenberg suggesting: "What we observe is not nature itself, but nature exposed to our method of questioning." Or from a more recent theoretical physicist Sean Carol (2016) who summarized his account of the physical universe by concluding "What matters is what matters to people," and "It takes all kinds." I would say that data always are the product of discursive actions, texts, talk, and deeds by members of discourse communities.

For example, scientific experiments generate facts that do not exist in an observer-independent nature. Consider the Large Hadron Collider, built by CERN. It lies in a tunnel 17 miles in circumference crossing the French–Swiss border. It was designed to produce a hypothesized particle, the Higgs boson or God particle. Quantum theory had it that this particle, if it could be produced, and it was, would last only a fraction of a second, and it did, hence could never be 'found' on earth without the Collider. By all measures, this particle is an artifact of physics and created at a cost of $13.25 billion. Physicists deserve credit for manufacturing it, and nature for not objecting to their creative efforts.

Psychological experiments too are justified as a way of reducing the complexities of individual humans to what matters to experimenters in pursuit of particular theories. But in the social sciences, the recognition of data as artifacts is even more obviously because they almost always relate to the language used by those experimented with, observed, and interviewed, even with all the precautions that ethnographers have developed. For example, the voices of American Indians are channeled largely by anthropologists who have visited them for a short while. The accounts they bring back to their universities become certified facts when they are successfully defended in the form of dissertations or publications in peer-reviewed articles of academic journals. Similarly, public opinion research elicits answers to questions of interest to a polling agency's clients. There is no check of how interviewees understand the questions they answer, and no guarantee that the question-answer pairs elicited ever occurred in public conversations among interviewees. Most social research is intrusive by generating data in the process of changing the object of their investigation. Surely, interview data are the product of how the formal interviewer-interviewee relationship is enacted and interviewees' willingness to answer questions even if they do not understand them as the pollster does or are not sure how to answer them but pick one alternative anyhow. However, such uncertainties are not the data that public opinion researchers are reporting.

Even pebbles on the beach, while unquestionably experienced when walking on them, are not data until someone decides to collect them and put them in boxes by their shape, size, composition, origin, or utility for particular purposes, minimally: drawing distinctions among them. All data are the results of human actions, including declaring them as such, which makes them discursive constructions. By declaring some entities, records, or descriptions to be data, they acquire particular meanings for particular communities, particularly for scientific researchers who may use them in support of publishable conclusions. Talking of observing facts, of discovering data, or claiming to have found laws of nature perpetuates an epistemological myth in need of retirement.

I recognize that etymology is not the only way to conceptualize a word. Currently, 'facts' are not conceived of as manufactured. Claiming something to be a fact is a very convenient rhetorical device to evade accountability for their existence. "X is a fact" tends to mean that X is obviously real, unquestionable true, and free of any bias. Speaking of the difference between facts and data, Daniel Rosenberg (2013: 18) argues that statements can be true or false, but facts and data can never be false. Evidence, by contrast is defined as in support of particular propositions and articulated beliefs. Without that support, evidence is no longer evidence while data remain data whether they are true or not and whether they support a conclusion or deny it. Accordingly, data are ubiquitous, whether they are misleading or not.

Recognizing the actions required to tame phenomena to become data, Peter Checkland and Sue Holwell (1998: 86–89) suggest replacing the word data by '*capta*,' emphasizing that all observational data are selectively captured. Capta presumes that there are wild things out there that need to be tamed. More appropriately, Herminia Alfonso (1999: 132–153) suggests replacing 'data' by '*poieta*,' not only for being made in Richard Rorty's (1989: 3ff) sense but also for being 'consensually authenticated' as legitimate grounds for inquiry.

The concept of operational definitions of scientific concepts presents an important challenge to the idea of data as preexisting scientific interest in them. For example, we know 'time' is not observable as such. We live, talk, listen, read and write always in the present. It is the grammatical resources in the language we use that enable us to talk of past experiences and not yet observable futures. In the scientific discourses, time has become operationalized in the form of instruments designed to quantify change independently of individual actions – originally shadows on sundials, then autonomously working clocks, now atomic devices. They all operationalize time to the extent they are correlated with, consensually interpretable as and able to explain a host of observable phenomena and coordinate numerous institutionalized social activities. Obviously, time does not reside in an unattended nature. It has become real by talking of designed instruments for clocking it.

Although professional pollsters might not agree when I claim that 'public opinion' is operationally defined as well, to me, it is an artificially aggregated account of a sample of question-answer pairs generated during interviews with people drawn from a population of institutional interest. In the reality of everyday life, public concerns are articulated in places where people meet. A public may surface in joint actions ranging from cheering at sports events, collective responses to natural disasters, resistances to imposed policies, demonstrations of common concerns, unrests, revolts, or revolutions. But publics do not have opinions and do not speak; individuals do. Even when people find themselves in large crowds, they experience their fellow citizen's opinion mainly in conversations among a few in close proximity. By contrast, computed polling results enable institutional actors to bring their goods to where they are appreciated or design political messages for desirable effects. Their actions affect what people may talk of. Especially when polling results are published and reach sections of populations whose members have a stake in knowing where they stand relative to others. Whether they find themselves in a majority, minority, or ignored, they can and often do act on what is reported. Reports of public opinion can be opposed, invalidated or become reified. Without pollsters 'public opinion' could not exist (Krippendorff, 2005).

The essence of operational definitions of facts is that the data that give rise to them – public opinion, ethnicities, social institutions, or nations – participate in the very process that constructs them.

## 1.2 Four Conditions for Being Data

Whether facts are operationally defined or the product of analyzing data, none exist without researchers, analysts, or observers distinguishing phenomena of interest from those that do not mean anything to them. Texts have no meanings independent of their readers. Authors always are the first readers of their own writings. Measuring instruments never read themselves. For research results to be informed by data, presupposes reliable knowledge of how data came to be. The assumption that the world consists of data, ready to be picked up and analyzed for what they are amounts to epistemological blindness. I am suggesting that the ability to rely on particular data calls for data to satisfy the following four conditions:

### 1.2.1 Durability and Retrievability

Data must be sufficiently durable to survive at least their analysis and reexamination by critical scholars. Most natural phenomena that matter are transitory: the weather, earthquakes, speech, and cultures. Even the physical traces they leave behind are subject to inevitable entropy or disintegration. Heraclitus' famous quote that "No man ever steps in the same river twice," applies to the role of data as well. Indeed, the water of a river is always on the move. Yesterday's water, out of sight today, has gone down the river. One cannot compare two earthquakes centuries apart, unless traces, records, or recollections of them persist in time, are retrievable and subsequently examinable. To observe similarities or measure differences requires putting data side by side. Anyone's verbal promise disappears immediately after it is articulated. We heard it, may recall it after some time has elapsed, hold the speaker accountable for it, but its analysis requires some kind of record, for example, in the form of the written witness testimony, a video recording, or signed contract. This is why conversation analysts work from transcripts, economists from reported transactions, historians from documents, and archeologists from ruins and physical artifacts left behind by extinct cultures.

Research fundamentally is re-search, repeatedly searching for patterns among available records of phenomena of interest to be sure everything is accounted for and nothing is overlooked. Transitory phenomena escape repeated analysis, unless they are physically manifest in the form of durable records. To remain data, their manifestations need to survive into the present and withstand repeated examinations. Without fossil records of dinosaurs, we would not have a clue that they existed. It is the retrievability of durable traces or records from the past that makes their systematic examination possible and enables researchers to ensure that their claims can be substantiated by others. Re-search cannot happen without durable data.

Physical durability is essential. Left alone, all physical phenomena degrade into unstructured entropy. Established rates of decay may allow us to date its origin. Physical durability can be prolonged or eliminated by social institutions. Institutions tend to have their own reasons to preserve and store some records of their past and let go others. The records of wars tend to be kept by their victors. The voices of their victims have less of a chance to survive into the presence. The artifacts of ancient Greece we know today are made of durable stone. But before the 17th century few people cared about them. They were recycled as building material or trashed and would not have become data had European archeologists in the 17th and 18th century not developed an interest in the physical manifestations of European history, unearthed its specimens, organized them in conceptual categories, displayed them in museums, or encouraged their hosts to preserve them in the form of tourist attractions. Orally transmitted accounts of happenings reside in their retelling and are subject to modifications that serve their narrators' and listeners' needs. Until the Brothers Grimm transcribed German fairytales in the 19th century, they were not data. Now we can read them as records of folk life at the time of their transcription. Hospital records, purportedly of patients' problems and treatments are limited to what serves medical professionals, for example, to be reimbursed for their services or to protect themselves from possible malpractice suits (Garfinkel, 1967). The idea that scientific knowledge is progressing and systematically accumulating has been debunked since Ludwig Fleck's (1934/1979) and Thomas Kuhn's (1962) works. What is unquestionably accumulating is the wealth of durable data, the increased sophistication of analytical devices, and the retrievability of published research and theories generated for a variety of purposes. It is these largely institutionalized records of generations of scientists that allow us to examine how scientific facts emerged and how we construct and reconstruct our universe. While the increased availability of data might reduce the probability of reinvention, its institutionalization can also stifle developments.

Archives as the depositories of data have existed for centuries. Obviously not everything physically durable ends up there; only what serves the missions of their archivists. In our information society, durability has acquired new meanings. The physical location of data has become less important than their circulation, their ability to be communicated from one computer to another without or with only insignificant losses. The contents of computer memories are constantly copied, data bases are repeatedly updated, blogs are refreshed, and information is disseminated widely. Data have become fluid and are in constant transition. A contemporary form of durability is redundancy. As electronic records are stored in different data bases, and communicated through multiple channels, losses of data are more easily detected and restored. Leaking supposedly secret information has become more frequent, hacking the computers of powerful institutions more common. Open-source programming encourages wider uses and participation in generating data. Candidates for political office experience increasing difficulty to make uncomfortable data disappear.

While entropy, the systematic decay of organized matter, acts indiscriminately on the media in which data are preserved and communicated, electronic data owe their durability to the redundancy of their dissemination to more and more people able to use institutionalized technologies of transmission and computation.

### 1.2.2 Computability, Analyzability and Distinguishability

Just as texts need to be readable by someone familiar with the language in which they were written, so must scientific data be of a form amenable to formal analysis. By computability I mean the ability to apply any well-defined and unambiguously executable transformation to data of this kind, ranging from unitizing and annotating written texts and counting words to applying complex text analysis software.

It is common to distinguish between unstructured and structured phenomena. This distinction has little to do with the presence or absence of structure but with whether structures can be recognized by available analytical or computational methods. For example, handwritten notes, paintings, speeches, and theatrical performances, while linguistically and visually structured and meaningful to readers or viewers familiar with their materiality, syntax, and semantics, easily escape formal analysis for lack of computational methods that respond to their meanings. This difficulty is the sole reason for content analysts to employ human coders to transform culturally structured phenomena into analyzable data. For physicists it is easier to generate data by relying on measuring instruments that are designed to respond to theorized physical dimensions of material phenomena. Mechanical measuring instruments are causally affected by what they measure and yield numerical data, diagrams, and electronic records that are amenable to mathematical modeling or simulations. The interests of social scientists can rarely rely on or be reduced to the physical manifestations of social phenomena.

All analytical techniques, especially software, call for data that conform to specialized data languages. The most common features of such languages are:

- *Units of analysis* which must be logically or physically distinct and (within the context a particular research) indivisible to afford comparisons and enumerations
- *Variables* that account for diverse conceptual accounts of units. They are the dimensions along which units of analysis vary
- *Values*, sets of which define the variables of a data language. Values may be unordered (nominal categories) or ordered, and when ordered possess either ordinal, interval, ratio, or other metric properties
- *Links* that relate values across different variables and support the computation of correlations, logical hierarchies, and semantically meaningful networks
- *Syntactical rules* which determine the well-formedness of structural accounts of units of analysis, including its *constants* relative to the method employed
- *A logic* which governs the entailments, orderings, and equivalences among structural accounts

For example, data in the form of "A says B to C" contain three variables. A lists speakers, B messages, and C addressees. "Says" and "to" are relational constants. Such data lead to communication networks. Because A, B, and C are names or messages, data of this structure would render a variance analysis inapplicable. Measuring variances call for interval variables. But contingencies and $\chi^2$s can be computed on such assertions. The above-listed features of data languages are common but quite basic. Currently, data languages are advanced and systematized in computer science and implemented by information technology companies like Microsoft, Google, and Apple.

The use of so-called big data, referring to the astonishing volume of textual data available through the internet, social media, and numerous large governmental, commercial, and scholarly data bases, is subject to the same formal requirements. Their diversity undoubtedly has statistical advantages. However, large collections of data have existed for centuries in the form of libraries, archives, and institutional records. What distinguishes traditional collections from big data is that the latter are available electronically, encoded in a few computable formats, which enables information technologies to index them in various textual categories, connect bodies of data, and search, summarize, and analyze them for a variety of purposes. Pursuing citations from one academic paper to another is an ancient academic practice, but clicking on hyperlinks, automatically crawling URL references across websites, or establishing similarities among documents that share vocabularies creates novel kinds of networks among previously disconnected data. Creating links adds a level of abstraction to flat collections of data, which is why some prefer to call these data not big but "meta-data." The point is that connecting documents, indexing their textual qualities, and comparing them across domains is possible only when they are compatibly formatted. HTML is the language of websites. It enabled indexing and searching in them efficiently. Currently, the most widely used formats of meta-data are XLM-related; the remaining ones afford easy inter-translatability. However, fascination with unprecedented volumes of comparable data and the computational opportunities offered alone do not make big data more informative. Indeed, by computing similarities, associations and hyperlinks from short character strings, the analysis of big data rarely approximates the level of abstractions human analysts are interested in and accustomed to. Moreover, not knowing the source of big data, the contexts in which their exchanges were meaningful and to whom, or how they were played into the hands of analysts renders computational transformations of big data difficult to interpret in social terms.

There are scholars who interpret the emergence of big data as a Kuhnian paradigm shift in how we generate knowledge. Some believe that big data can do away with the idea of theories by simply following where the numbers lead to (Bollier, 2010; Graham, 2012). Opponents of such futuristic views have suggested that big data are incomprehensible in human terms as they are either the unintended byproducts of human online uses of personal computers, economic transactions, and shopping, or generated by nontrivial algorithms whose behaviors are increasingly unpredictable, even for their developers. The problem of not fully understanding the complexities of data-generating algorithms has not prevented governments and commercial enterprises from algorithmizing various social practices for the benefits they yield, effectively surrendering their research practices to accommodate available algorithms. Big data feed an "obsession with computability" (Boyd & Crawford, 2012: 663) and encourage a "rhetoric of objectivity" (Crawford et al.: 2014) by privileging textual character strings over what they mean to those who created, read, and acted on them – an issue discussed in Chapter 8.

On the other side of the debate, is important to keep in mind that all symbolically complex, socially meaningful, and context-sensitive phenomena that cannot be converted into computable data escape formal analysis. When computational techniques are applied to the mere physical manifestations of socially meaningful phenomena, for example character strings, words, or clicks with a mouse, the symbolic qualities of these phenomena ordinarily conveyed to readers mostly escape recognition and analysis. This is a serious problem for analyzing data from social media. It tends to be solved by supplementing such data by ethnographic inquiries of users. Indeed, analytically linkable data expand recognizable complexities into previously inaccessible domains and can give rise to new kinds of theories, for example, of citation networks, clusters of websites, and large systems that support numerous novel social practices ranging from offering customers products they did not know they wanted, to tracing the spread of diseases, and solving crimes.

In today's information society, computable texts become increasingly available and have vastly expanded the scope of research into previously unanalyzable domains. Recognizable structures, data languages, and computational methods to handle them are coevolving at amazing rates. However, creating computable data structures cannot be the primary goal of social scientists. When computability of texts is pursued at the expense of the context of using these texts, i.e., without reference to the motivations of people posting and responding to texts on the internet, their meanings, the emotional, organizational, and political consequences of communicating such volumes of texts makes for very shallow results. Computational content analysts have difficulties living up to their enthusiastic predictions. Without interpretations by literary, socially, and culturally competent human readers, meanings, intentions, relational consequences, and innovations escape computer text analyses. This will be the topic of the next section.

### 1.2.3 Ability to Serve as Surrogates for Analytical Concerns

All researchers need to be cognizant of the relationship between their data and the phenomena they are concerned with. Whether data consist of numbers, written documents, images, or material artifacts, analyzing them becomes pointless unless one has a sense of how they came to be, in which context they played which roles and what they meant and represented for whom. I have met researchers who had inherited large volumes of data but had only the vaguest ideas of what they are about. Similarly, I have worked with content analysts who changed their coding instructions midstream and ended up not knowing how to interpret their data or worse forgot which versions of coding instructions they gave to coders and ended up with computable stuff they could not link to what mattered to them. Scientific research is always about phenomena selected for attention, happenings that may not be readily analyzable for their complexity, transient events that disappear before they can be understood, or records that established connections of a social nature in particular contexts. The circumstances under which these phenomena come to be and how they were brought to the attention of their analysts are almost always biased, served purposes other than that of the analyst, are of unknown origin, and maybe even created to mislead the analyst. Such biases, intended or not, can hardly be ignored. Even operationally defined phenomena whose reality results from the very application of particular measurements are connected to the functionality of the measuring devices that constitute them. It follows that all data-based claims, propositions, theories, models, even mere summaries of observations of records left behind accidentally, rest on unexamined assumptions that data are reliable surrogates of phenomena that might not be so.

I want to identify four possible issues that should doubt the widely held assumption that data are the unquestionable ground for individual judgments, scientific theories, and purposive actions by examining data as proxies of phenomena of claimed analytical interest. These issues have different origins. In terms of simple propositions, they have to do with the disciplining of data, selecting an insufficient number of them for analysis, lacking shared understanding or replicability, and with the validity of data.

To start one ought to realize that all scientific inquiries literally discipline the phenomena they transform into data. This begins in interactions among experts who employ specialized vocabularies in terms of which they develop consensus on how to proceed. Psychologists may look for the psychopathologies of authors. Sociologists may focus on how social classes develop their identity in interactions between them. For critical discourse analysts power structures explain what was said, done and can be observed. Literary scholars may look for the voices that are ventriloquized. Political scientists see everything in terms of policy implications, and voters for whether a candidate for political office resonates with their political inclinations. For a builder, bricks are the elements for constructing a building. For the seller, they bring an income, for a scholar they might be book ends, and for a terrorist they are weapons. The brick affords all kinds of uses. I would say that all phenomena afford multiple disciplinary, philosophical, or pragmatic interpretations. The application of unlike disciplinary lenses creates unlike data on which unlike actions are justified. It is rarely recognized that different disciplines create their own discourse-specific data, and if it becomes noticed by practitioners of one scientific discourse, the data of the other is usually criticized as biased or inappropriate. Competing discourses, like Western and Chinese medicine, tend to dismiss the data of the other as questionable and inferior proxies of what is really going on. I am suggesting that whenever data are taken as indisputable facts they remain within particular researchers' discursively constructed bubbles. We construct our worlds in the language we have acquired, tend to take that world as being undeniably real, and construct data to provide us access, not to THE world but to OUR worlds.

Given that all research concerns phenomena of disciplinary interest, claims of the objectivity of data, already dismissed at the beginning of this chapter gets us to the epistemological problem of the surrogacy of data; what are the phenomena data a proxy of? Absent unfettered access to "true and undistorted" phenomena of declared interest, I would say that both, the data and the phenomena they are said to be about are discursively constructed and so is their relationship. It takes a considerable effort to be able to critically examine this relationship, appreciating its socially constructed nature from outside the discourse in which it is accepted – without claiming that my outsider's perspective is objective. It calls for a bold critical step out of one's disciplinary bubble and to question the convenient objectivity claims naively accepted inside it.

Data are always proxying phenomena of analytical interests whose surrogacy needs to be known before analyzing them. Blindly accepting data as facts keeps the results of their analysis inside the taken for granted disciplinary discourse.

Besides the disciplining of data which I claim occurs in all research efforts, the surrogacy of data faces three more challenges. The second issue concerns their statistical relationship to the larger population of possible data in place of which they are examined. Making sure that the analyzed data are generalizable to the population of possible data, is adequately addressed by familiar sampling theories. Therefore, this issue requires less attention than the other two. One may notice that I am deviating from the customary characterization of adequate sample sizes in terms of their ability to represent the diversity of a larger population of phenomena. The reason for deviating from this custom is that sampling theories do not and cannot address the phenomena of interest without having turned them into data. In effect, sampling theories evaluate the generalizability of a sample data from a population of possible data, hypothesized as the result of continuing the process of generating data with methods that produced the sample. As such, sampling theories are not concerned with what individual data are about, but what their distribution is about.

The third issue concerns the reproducibility of data. This is the main topic of this book. Testing the replicability of the data-making process for gaining the needed assurances that data are not polluted by irrelevant variations and that those involved agree on what the data are about. Although this relates more to the fourth issue, I want to start by making clear replicability is less important when mechanical measuring devices are used to generate data. The behavior of mechanical measuring devices is deterministic. Their use establishes one-to-one relationships between the measure and the measured. Surely, measuring devices may have systematic biases but repetitions always end in the same result. However, more important is that they produce only decontextualized data. A thermometer yields numerical temperatures whether it is of a room, of a patient in a hospital, or of an oven. What is true for mechanical measuring devices also applies to cameras, voice recorders and computers. Taking photographs of cars crossing at a red light, registering the clicks on computer user's keyboards, counting their misspelled words and phrases, or copying the email addresses from iPhones creates data that omit peoples' intentions, meanings, or consequent actions.

Humans by contrast see almost everything in context. Data of social phenomena can hardly ignore the contexts of what is turned into data. An individual's body temperature is a number all right, but when it is abnormally high, it is a fever with a probable cause that may require medical attention. A document may be converted into the 0s and 1s for a computer to do something with it, but these digits are void of meanings for their authors and readers and whatever reality the document might be enabling.

By contrast, humans do not respond to the physicality of things but to the meanings they attribute to them (Krippendorff, 2009: 47). For example, texts do not literally contain their meanings. Meanings arise in processes of reading and interpreting textual sequences (Krippendorff, 2019: 27ff). This is why content analysts cannot rely on measuring physical or material corollaries of communication. They need to rely on culturally sensitive and literally competent human observers, coders, raters, annotators, interpreters, or judges who can make sense of the phenomena of analytical interest and identify and describe their meanings in analyzable forms commonly referred to as codes.

It is of paramount importance for empirical researchers to know what their data are about. Because they tend to be several steps removed from the phenomena of interest and have no direct access to what the hired observers faced and coded, they have to obtain this knowledge indirectly. Notwithstanding observers' assurances that they have recorded their observations truthfully, such assurances do not tell whether observers' understanding is related to how the researcher expected it to be. Testing the replicability of the coding process offers the needed assurances. Replicability is measured in terms of the agreement among qualified observers who are instructed to follow the same coding instructions and apply them to the same set of phenomena. Clearly, the measures of inter-coder agreements refer only to those coders who participated in a replicability test. However, the claim of data to be replicable implies that other researchers are able to hire coders with similar qualifications and who are able to transform the same set of phenomena into data that correspond, ideally perfectly, to those used in a reliability test.

Replicability bypasses the need to know what observers had in mind when coding phenomena by giving all of them identical coding instruction to apply. In effect, replicability ensures that the written coding instructions are successfully communicable across independently working observers. It does not test whether the coding was correct or whether the phenomena of interest are correctly represented in the data; only that the observers examined and coded the phenomena as defined in the written coding instructions. A few problems might be worth mentioning:

Conceiving of the written coding instructions as communications to future replicators is violated by the frequent practice of researchers training their observers to become highly specialized coders, measuring high agreement among them that nobody else can beat. But without the ability of communicating their expertise to other observers, fellow researchers, critics, and stakeholders, their coding can hardly be considered replicable.

Measuring the agreement among coders should ideally be extended to the analyst who needs to be able to reverse the coding instruction and decode what the observers may have faced when entering a code. If analysts are unable to decode their data reliably, they do not know what they are analyzing.

The coding of complex phenomena can be difficult. Difficulties are evident when experiencing uncertainties as to how something should be coded. Guessing introduces irrelevant variations in the data that may well reduce the chance of deriving valid results from their analysis. While much can happen to data when they are analyzed, agreement measures applied to replications should always warn researchers when their data become less trustworthy. Because adequate agreements may not always be obtainable, some statisticians have proposed alternative agreement coefficients that are not related to the reliability of data. For example, Zhao et al. (2012) insist that agreement should be a function of coders' experiencing difficulties vs. ease. Coding difficulties challenge the designers of coding instructions but have nothing to do with what data are about.

Most journals already require potential authors to provide evidence of having tested the replicability of the data they are reporting, demanding the use of correct agreement measures. I am suggesting that the coding instructions that gave rise to the data be made freely available as well. Appropriate agreement coefficients enable researchers to estimate the degree to which data can be trusted in subsequent analyses (Incidentally without direct access to what the coded phenomena were). The coding instructions give analysts, critical scholars, and practitioners the ability to decode what the data are about, coming close to the phenomena (defined in terms of the coding instructions).

So-called big data highlight different problems. Data scientists – researchers who analyze what is generally referred to as huge volumes of "found data," for example, collected from electronically available records of transactions, social media messages, individual computer uses, sales or other physical manifestations of ongoing social practices – claim that their data are immune to reliability concerns. This may be so when data represent physical phenomena. However, in the domain of the social, my doubts lead to the fourth issue of surrogacy: the validity of data.

I contend that celebrating the unquestionable statistical virtues of analyzing large volumes of gathered data – invariably getting to statistically significant results – and naively equating their data with facts tends to render data scientists blind to the recognition that their data too are mere convenient surrogates of their claimed concerns. Data vary in degrees of their surrogacy and that degree is the extent to which data can be considered valid representations.

Big data researchers access their data electronically. Inasmuch as computers operate on decontextualized inputs, big data are void of the contexts that generated them, who generated them, for what purposes, and what did they mean to them, accomplish or construct. For a simplistic example consider counting polysemic words. "Playing" in "playing a piano," "playing soccer," "playing it cool," "children playing with each other," "poets playing with words" refer to entirely different activities that no English speaker would confuse but a computer could not distinguish. Polysemy is far from rare. Consider the multiple definitions that dictionaries of a natural language provide for almost every word. Of course, there is computer-aided content analysis software that disambiguates polysemic words by looking for clues in their linguistic environment. However, not all contexts are linguistic, consisting of the surrounding text of the words counted and analyzed.

Still in the domain of text analysis, take word clouds as another example. They exhibit key words occurring in a collection of documents, the size of their typeface being a function of their frequencies. When familiar with the documents, the word cloud from which it was extracted makes sense, but not the sense of reading the text. If all words in these documents had single meanings, like the names of individuals, cities, or currencies, a word cloud may well be a simplified snapshot of key concepts. However, this is rarely so in which case word clouds remain replicable representations of key words but nearly uninterpretable. What is missing is the context of these words.

For a more typical example, counting computer users' clicking on the URL of a candidate for political office does not indicate whether they like or dislike the candidate, whether they read what pops up on their screens, and what they would do with whatever they took away after clicking on the link. By all indications counting how often a candidate's website is visited leaves shallow data behind. It may not have anything to do with the chance that this candidate is elected. At best it may be indicative of recognizing a candidate's name.

Take another example from advertisers who may want to know whether their ads are compelling potential readers or viewers to buy the product they promote. Under ideal conditions they would display their ad, survey about how many saw it and how many didn't. Of those who did, they may want to know whether it made the intended sense to them, and if it did, whether it compelled them to act in the direction the advertiser hoped for and discouraged them from going to competitors. Answers to such questions cannot be found without intervention in peoples' lives. A big data researcher may advise the advertiser to count how many people clicked on the ad on their computer, how many bought their advertised product, and correlate these two variables. Not only would such data be unable to distinguish whether the ad was read or not, and whether purchasing the product was due to the ad or something else. Absent other costlier inquiries into the context they hope to influence, advertisers may act on the correlations as the best bet they can afford, but they have no clue of what happens between clicking on an ad and purchasing the product. We know that this relationship is not as simple as a correction can reveal.

The point is that computers are not social beings and process data given to them in ways humans would not. Decontextualized data may be correlated, associated, or incidentally connected with the social phenomena that big data analysts are interested in exploring. But in order to be able to infer anything about the social world, data analysts who have good reasons not to test the replicability of human coding of their data need to inquire into how their data connect to the social context in which they are found. Without empirical evidence or plausible claims of the surrogacy of big data, the validity of computational analyses of social phenomena is questionable and should be discouraged.

Precisely because all data are surrogates for phenomena of analytical interest to researchers, knowledge of their connection to these phenomena is crucial. Direct representations ensure valid data of insignificant parts of ignored contexts. To the extent these phenomena can be observed, interrogated and described, reliability – the degree to which all observers agree on seeing the same phenomena – does not provide indications of their validity. However, without a clue of how data connect to what a researcher seeks to understand, their validity remains an open question. This is implied in the popular expression: "garbage in garbage out."

### 1.2.4 Ability to Inform Research Questions

Data are relevant to the extent they provide the information about the distinct phenomena of analytical interest needed to answer given research questions (Krippendorff, 2011), visualized in Figure 1.1:

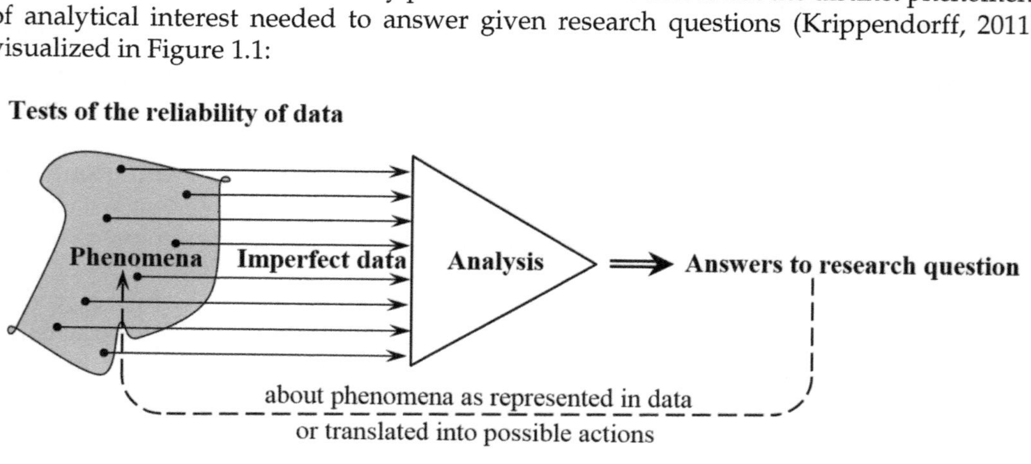


**FIGURE 1.1**
The Flow of Information From Phenomena to the Results of an Inquiry

To play this role, data have to possess sufficient diversity, variety, or variance. I already noted that grammatically, *data* is the plural of *datum*. Even when the word *data* is used as a mass noun, a single datum cannot inform anything on its own. Any one datum must be comparable with other data, minimally providing a contrast with its absence, but more generally being part of a distribution along conceptual dimensions. Omnipresent phenomena are not noticeable – as water is said to be for fish. Truly unique events are rarely intelligible. Catastrophes may be totally unexpected and frightening, but even unexpected happenings are perceived in contrast to what was expected. The difference they make amounts to at least one bit, the minimum amount of information recognizable. To make analytical use of phenomena, to explain differences between them, find patterns among them of interest to the analysts presupposes diversities in data that relate to diversities among the phenomena in question. Observers, coders, data analysts need to be assured that any analytical decisions they wish to make presuppose a diversity of phenomena they intend to study. A single decontextualized phenomenon cannot be recognized for what it is without a context of alternatives.

Section 0.2 already mentioned two extreme conditions under which data generated by observers cannot inform inquiries concerning particular phenomena. One exists when coding exhibits no variation despite obvious differences among the phenomena to be coded. The other is apparent when coding is random. To use an analogy: When a television screen is merely lit there is no connection to what a station put on the air and this is true also when the screen displays white noise. The former exhibits no variation; the latter varies without ascertainable meaning. Lack of variation is surely evident when observers fall asleep, get tired, become overconfident, generalize the nature of the phenomena and code according to their expectations rather the diversity of phenomena they are exposed to. Agreeing among several observers to code everything alike is also a devious way to achieving perfect percent agreement but not above chance. The other extreme can be observed if coders were to toss a coin to code a binary variable or throw dice to select among a larger number of categories. Of course, observers hired to record what they read or observe are not likely to engage in such practices. However, without such devices, human beings have difficulties in behaving truly randomly. It is when coders work without comprehending their assignment, are distracted by irrelevant phenomena, or want to get their job done without time-consuming attention to the needed details; the connection between the emerging data and the phenomena they are to represent is statistically severed. Chance or randomness is a convenient reference point. Under these conditions, measures designed to indicate how reliable data are, have to be zero. What observers are reporting as data has to connect to the phenomena of chosen interests.

There is much conflicting literature about the difference between data, information, and knowledge. I prefer not to review this debate except to say that for me, information is not contained in messages that could be passed on or be found in large data bases. I prefer to define information closer to Shannon and Weaver's (1949) mathematical theory of communication. Taking data as mediating between the analyzed phenomena and the research results, information about the phenomena of analytical interests is obtainable by decoding the data with the help of the coding instructions that produced them. The amount of information needed to answer given research questions about phenomena of interest has to come from the data about these phenomena.

In these terms and mathematically, the agreement measures applied to replications among independently working observers are proportions of the information in the replicated data which represent the differences among the observed phenomena. If that proportion equals 1, researchers are assured that the diversity in data matches the diversity of phenomena and is not affected by anything else. While proportions do not reveal the quantities underlying them, the actual amounts of information (above chance agreement) and noise (disagreement) in data are found in the replications required to establishing the reliability of data. It is these replications that enable the researcher to be selective of what to trust and what to avoid.

It is important not to confuse the information measured to establish the reliability of data which refers to how well the data represent the distinctions within the phenomena of interest, and the information needed to bring a research project to its valid conclusion. The latter tends to be smaller but also more abstract. In the process of moving from data to published conclusions, analysts step-by-step reduce the diversity in data to what ultimately matters to them. Analytical techniques extract or transform patterns found in data into patterns closer to what analysts want to report on, whether these patterns are manifest in frequencies, correlation, networks, or systems of hypotheses. Analytical techniques that are appropriate to the data model the properties of the phenomena represented by them, preserve relevant information in the data, and discard irrelevant quantities. Too much noise in the data may lead analysts astray and so does the use of inappropriate analytical methods. In either case, analytical methods cannot add information to what the data provide, and analysts are encouraged not to add anything either. It follows that the data always have to possess diversities much larger than the information needed to identify valid answers to a given research question.

Mathematically, the replications needed to establish the reliability of the generated data are redundant. Indeed, if agreement among them is observed to be perfect, any one replication could provide the same information for an analysis to proceed. However, if disagreements are observed among them, it is their redundancies that enable researchers to locate the noise in the data, even eliminate it selectively. The possibility of error correction is one virtue of redundant data. Redundancy is also helpful in giving analysts the option of replacing unreliable variables by reliable ones, can shift the focus of attention to something that emerged in the cause of an analysis, or can enable discovery of unanticipated pattern in data during what is called a fishing expedition for a mission to pursue in the data. In the course of any analysis of data, information is omitted, combined, re-presented and reduced to abstractions or generalizations but the results of analytical techniques are always limited by the amounts of information in data they process.

In practice, the quantitative difference of the diversity in data and the number of alternatives that researchers ultimately need to be certain of tends to be enormous. Computing an average, an index, or a correlation reduces the original diversity, hopefully without losing relevant information. Computing averages, visual displays of networks, or other abstractions are justifiable when the data provide redundancies, that is, duplications whose significance depends on their aggregated qualities. Disagreements among replications indicate the presence of noise in the relationship between data and what they are about. The presence of noise, which an analytical technique mistakes for relevant information, is what makes the analytical results questionable.

As should be clear from the above, without redundancy, noise cannot be distinguished from information within any one variable. This is the primary motivation for generating and testing replications of generating data before analyzing them. There are statistical methods that "fail safely" when noise is polluting the data. For example, in correlations noise reduces their measured strength, and when testing hypotheses noisy data reduce the otherwise reachable level of their statistical significance. However, when noise is equally distributed throughout the data, analysts may not know whether and why their findings become uncertain.

Obviously, appropriate analytical techniques preserve relevant information while inappropriate ones may not, but their appropriateness is difficult to assess without extensive knowledge of the phenomena of analytical interests and what a chosen method does with the information in data. Averages, for example, replace variances in data by a single variable. Correlations replace the diversities of two or more variables by two parameters, the correlation coefficients and their statistical significance. There are more complex methods that provide a researcher with choices. Because analysts have many choices of how to handle given data, the safest strategy to maximize the potential of achieving valid results is to establish the replicability of data at the entry point of any analysis.

By focusing attention on the flow of information from the diversity of often unstructured phenomena to discipline-specific theories, practical recommendations, or decisions on important matters, one can critically examine the transformations that data undergo in the process of communication. The reliability of each step is testable in principle and reportable. However, this is rarely done because of the belief that analyses cannot be in error. Most research follows discipline-recommended practices and yields discipline-specific results. This does not resolve them to be accountable to the stakeholders regarding their effects their published results have outside their own discipline. Tracing the data transformations through all analytical steps amounts to treating research as a channel of communication whose results need to be to recontextualize the analytical steps.

Unfortunately, the literature on empirical methods of scientific research tends to start by describing available analytical techniques and calls for their users to shape the data to their needs. So, after deciding to use an analysis of variance, data have to be gathered in the form of a finite number of interval variables. Factor analysis is built on correlations which require interval variables as well. Network analysis needs to establish the strengths of links between identified nodes. And simulations explore the consequences of explicitly stated dynamic functions. A while back, I became interested in Clyde Coombs' (1967) *A Theory of Data*. He provided pioneering insights about the handling of various data structures, now overshadowed by vastly expanded abilities to compute diverse forms of data, previously considered unanalyzable. As already noted, Coombs did not concern himself with how unreliable data affect analytical results. He pointed out, though, that all analytical techniques are designed to handle particular patterns of data and ignore patterns they cannot handle. Research that generates data suitable for a chosen analytical technique shifts the focus of analytical attention from independently conceptualized phenomena of interest to data that are tailored to demonstrate the capability a chosen method of analysis. Revealing the capabilities of particular methods of analysis has its virtues but my hope is that research turns back to studying real world phenomena.

The regrettable preoccupation of social scientists with computability at the expense of addressing issues of meaning and information in data about social phenomena continues in the literature. This is highlighted in recent excitements about the above-mentioned big data, found largely in electronic texts from the social world – a new currency in terms of which big internet corporations like Facebook and Google are thriving. They inform more about the platforms that users relate to than about the social life of the people using them. Computing data that are decontextualized from their social context tend to serve corporate interest in large numbers whether of products, consumers, mortgagees, or voters. Their social biases are well documented (O'Neil, 2017; Noble, 2018; Broussard, 2018). The bigness of such data is attractive to quantitatively oriented scientists who experience few problems in getting to statistically significant findings without the need to test their replicability, but only of shallow decontextualized data, interpreted by their analysts for naive stakeholders.

We know of some analytical consequences of using compromised information in data. However, most analytical techniques I know are unable to distinguish information and noise communicated by data. It is easy to provide evidence that the magnitude of unreliabilities in variables, the volume of noise that enters analytical processes, reduces the statistical significance of computational results, diminishing correlations, but they may also unknowingly misdirect the process of analysis toward invalid conclusions. Paul Gustafson (2004) explored by simulation the effects of measurement errors and misclassifications on research results in epidemiology. Finding unreliabilities in data signals caution. But, as already mentioned, agreement measures are proportions of information in data that say nothing about its location in noisy data. The effects of the volumes of information and noise traveling side by side through an analysis need to be explored where possible. Gustafson's explorations of these connections are exemplary but have so far been limited to very simple data. This book will visit this connection.

Employing inappropriate analytical techniques, techniques whose assumptions are incompatible with the nature of the phenomena under consideration, probably constitutes a more important threat to valid results and scientific progress than does the presence of noise in the data. Even if appropriate to the nature of phenomena under investigation, amplifying Clyde Coombs' point, analytical techniques tend to selectively amplify and systematically ignore phenomena they cannot cope with. Let me give two examples of such, I would say, mindless omissions decades apart and on vastly different levels of abstraction. The first, briefly mentioned in the previous chapter, concerns how the field of communication research limits itself by adopting an inadequate data language. The second describes how an important class of phenomena are distorted or ignored by relying on standard measures of reliability that cannot cope with them.

One of the first academic papers I wrote (Krippendorff, 1970b) was an award-winning critique of Harold Lasswell's (1948) then dominant conception of the field of communication research. He defined that field as providing answers to five separate questions: "Who? Says what? In which channel? To whom? And with what effect?" For him, the question of "who" was answered by what he called control analysis, "says what" by content analysis, "in which channel" by media studies, "to whom" by audience research, and "with what effect" by effects analysis. I argued that answering these questions separately, even differentiating academic research expertise in these terms, would generate data that could not possibly inform how interactive processes of communication dynamically shape the realities we live in. Moreover, pursuing Lasswell's research paradigm would prevent communication scholars from coming to grips with the social consequences of their own findings. It would compartmentalize the field of communication research in ways unable to capture the processes it is dedicated to explore. I proposed a data language and a computational method to analyze data that would bring the questions that Lasswell asked into a dynamic and relational framework, capable of preserving what made communication a phenomenon of vital importance to how society organizes itself. This was of course a while ago and much has been accomplished since. The example is merely intended to warn of becoming fixated on specialized analytical techniques and generating data whose analysis systematically ignores the processes that communication theorists talk about and hoped to understand but cannot study in Laswell's terms. This reminds me of the story of five blind people touching different parts of what someone without that disability recognizes as an elephant.

The second example is my justification for developing a way of assessing the reliability of multi-valued data. Virtually all available reliability coefficients and traditional statistical techniques are limited to data that consist of samples of multi-dimensionally categorized or valued units of analysis, each dimension or variable representing these units by single values. The conception of a multi-dimensional space calls for that and calculating the reliabilities for each variable makes sense when variables are logically independent of each other. However, there are very many situations in which units are more naturally described in terms of sets of values. Literary theory has argued for some time that all texts have multiple interpretations, not one. Politicians tend to make use of this possibility when trying to inspire diverse audiences with unlike perceptions of their worlds. The complexities of fictional characters are what make novels fascinating reads. Finally, academic contributions tend to rely on numerous sources, not one, and often result from several collaborating authors. All of which would call for the analysis of multi-valued data.

Required to publish the reliability of their data and limited to applying traditional agreement measures for single-valued data, content analysts have pursued essentially two strategies to cope with multi-valued phenomena. The probably worst strategy is to instruct coders to record only one, the most prominent feature of multi-valued phenomena. Suppose someone says he feels "relieved and ashamed" for what he did. Picking the most prominent emotion of the speaker's dual characterization becomes difficult in this example, ultimately arbitrary, and in effect unreliable. An alternative instruction is for coders to record the first of several characterizations mentioned. This, being a very easily executable rule, surely increases the measured agreement but omits half of the emotions expressed. It is as ridiculous as it would be to cite only the first author of a co-authored academic paper. Rules adopted that increase measured agreement at the expense of their meanings render researchers to becoming slaves of conventionally and easily available measures.

I am suggesting that, faced with diversities of complex phenomena, researchers should not reduce their complexity to satisfy available analytical methods, and rather invent methods capable of expanding the scope of what can be investigated with rigor and accountability to those who care.

### 1.2.5 Confinements to Past Happenings

A defining characteristic of data is that they always are about what happened in the past. We observe, talk, read, and write at present, but our present is always only a moment away from being the past. We can imagine futures but not observe them from where we are in time. This points to a fundamental limitation of relying on data, big or small: All data are presently examinable, but the value of their current examination is rooted in their presently unobservable past.

Nevertheless, almost all scientists aim to predict yet unobserved phenomena. These predictions rarely are surprising because all predictions are grounded in presumed deterministic histories of data, in trends assumed to continue without human interventions, in autonomous patterns of developments assumed to proceed without human interventions, or on laws assumed to govern the transformation of presently available data. Data-based predictions into unknown futures may be complex, not easily imagined, but once the rules according to which they are assumed to proceed are chosen there can be little surprise. Predictions cannot embrace human imagination, creative interpretations of past developments, technological innovations, cultural evolutions, and political revolutions. Predictions of technological developments are notorious failures. It is an illusion to think that data could solve the problems humans create in the process of living.

Data reliability is almost always focused on what happened in the past, what was observed, created, described, and widely interpreted alike. This is what replicability implies. However, analyzing data allows for novel interpretations and innovations to enter processes of discovering new ways of understanding, being and acting.

## 1.3 Diverse Disciplinary Approaches to Reliability

Literally, reliability means the ability to rely on something as intended. It expresses a sense of instrumental trust, whether in driving a car that does not break down, eating food that does not cause sickness, following advice that is not misleading, or using data that do indeed represent what one aims to analyze. There are several operationalizations of the concept of reliability, all of which acknowledge that perfect data are difficult to generate and provide measures that assess the risk of failures, mostly to taking precautionary actions.

### In engineering,

reliability is defined as the ability of a technological system to function under stated conditions and for a specified period in time. Expressed in terms of the probability of not failing, it is estimated from the known failure rate of its networked components. In the design of larger technological systems, reliability engineers tend to pursue one or more of four strategies to reduce the occurrence of technological unreliabilities: One is to strengthen vulnerable components so that the inevitable probability of a breakdown of any one component is about equal to all of them and tolerable within the lifespan of a whole system. The experience that more and more parts of a car start to fail after a period of carefree use exemplifies the effect of this principle. A second is to include an easily replaceable part that is designed to break down before the larger system fails, exemplified by fuses in electrical circuits. A third is to design systems that fail safely, affecting their users only minimally. For example, computer interfaces can lead to all kinds of breakdown; most of them are reversible at little cost. The final principle is to increase redundancy. If one component fails, others take over. The reliability of the internet is a good example of this principle.

Scientific research projects are much like technological systems designed to capture information about selected phenomena in data that would ultimately select among the answers to given research questions, support theories, or provide insights to talk of. The process of converting the phenomena of interest into analyzable data is the potentially most unreliable component of social research. Generally, the durability and language of the data are less problematic. However, if data cannot be regarded as reliable surrogates of the phenomena of interest and do not provide sufficient information for a research project to succeed, the whole research process might fail without a clear understanding of why this was so. Unlike in engineering where failures often amount to either/or propositions, in the conduct of social research failures tend to surface in degrees and become noticeable when it is too late to retract them. The following sketches a few ways reliability has been operationalized in several (non-engineering) discourses and ends with my preference for one.

**In psychometry,**

reliability tends to be equated with correlations among several test results obtained under consistent conditions. In psychometric research, "high interrater reliability means that the relation of one rated object to other rated objects is the same across judges, even though the absolute numbers used to express this relationship may differ from judge to judge." According to Tinsley and Weiss 2000: 98), "interrater reliability is sensitive only to the relative ordering of the rated objects." According to Nunnally and Bernstein (1994, p. 213), Cronbach's (1951) alpha (not to be confused with the αs that will be developed in the following chapters of this book) is the most prominent measure of reliability in psychometric research. It takes correlations among the responses by individual subjects to a test as evidence that the text consistently taps into a psychological construct underlying their varied responses. Psychometricians have good reasons to come to such conclusions as observed correlations cannot be an accident. However, correlations do not equal agreements.

Psychometric theory has given rise to the idea of measuring inter-rater, inter-coder, or inter-observer correlations which ignore the proclivities of subjects for particular responses but are negatively affected by random variations. Observed agreement among subjects is taken to be the result of a psychological theory underlying subjects' responses. However, when the reliability of data is the target of an investigation, when the question is whether data can be regarded as reliable surrogates of phenomena of analytical interest, observers are an indispensable means of recoding, but not the target of psychometric analyses. All observed disagreements among them about how given phenomena must be regarded as sources of unreliability. This suggests that Cronbach's alpha and the psychometric equation of the internal consistency of test results measure something quite unlike the reliability of data.

**Sampling theory**

offers a conception of reliability as a measure of the homogeneity of large data sets. It is established by measuring the correlation among distributions of data. Its most prominent method is called split-half reliability. A sample of data is split into two. The same test is applied to both. And the degree to which the two samples correlate is taken as a measure of their reliability. The common purpose of measuring split-half reliabilities is to establish whether the two samples have the necessary homogeneity to proceed with their analysis. If the correlation of two distributions is statistically significant, it would be sufficient to analyze any one of them, better both, but there would be no benefit in further enlarging the size of the sample. Split-half reliability may be computed, for example, from the margins of contingency tables. Cronbach's alpha has been interpreted as a generalization of the split-half reliability. It equals the mean of all split-half coefficients possible among test results. Split-half reliability suffers not only from relying on correlations rather than agreements; it also says little about how individual observations agree.

**Measurement theorists**

have adopted the metaphor of mechanical measuring instruments – clocks, yardsticks, thermometers. They offer one of the more rigorous conceptions of reliability. In Kaplan and Goldsen's (1965: 83–84) words: "The importance of reliability rests on the assurance it provides that data are obtained independent of the measuring event, instrument or person.

Reliable data, by this definition, are data that remain constant throughout variations in the measuring process." To test reliability following this conception requires repeating the measuring process under various and potentially adverse conditions, in effect testing for whether irrelevant circumstances of the measuring process affect its results. Measurement theory distinguishes between so-called true measures, which are shared by the measuring devices used, and errors, which cause the measurement to vary under conditions that are extraneous to the measuring process or random. I say "so-called," because I consider this vocabulary epistemologically misleading. For instance, consider measuring the weight of objects. Weights do not exist in nature but result from using a scale that produces numerical accounts in terms of conventional units, kilograms, pounds, US or British tons and the like. We experience different weights by making comparisons. The reliability of a weight scale can be established by weighing the same object repeatedly, by different scales, or by comparing the result of one with a trusted standard. Standards are established by agreement or conventions. They are not "true" in any objective sense. The same can be said for temperatures, lengths, times, volumes, radiation, chemical compositions, and levels of pollution. The natural sciences tend to define the qualities of interest operationally, that is, in terms of what a designed measuring instrument indicates. To know what an instrument measures, its mechanism needs to be understood, at least by its designer, and the units of measurements it produces have to be conventionally agreed upon. To me, the issue is not the extent to which measures are true or in error, but whether researchers can trust them to proceed.

As already mentioned, in the social sciences, physical measurements are rarely informative of social phenomena, mainly because humans do not respond to the physicality of things but to what they mean to them and how they interact with them (Krippendorff, 2009: 47ff). Social scientists are more likely interested in phenomena that are encoded in documents, social happenings, or interviews with people about their world. By asking observers to unitize, compare, categorize, evaluate, interpret, and describe such phenomena, the metaphor of measuring instruments is only partially applicable. For once, we may know how measuring instruments are designed to work, but not how observers understand and apply the coding instructions given to them and accordingly see, read, evaluate, judge, and describe the world. Indeed, the instructions to observers on how to transform phenomena into data plus their assurances that they followed them literately are the only clues that researchers have available to infer the phenomena from the data they produce. However, while humans are able to interpret far more complex phenomena and record them on different levels of abstractions than mechanical instruments are capable of, they are also far less stable in applying the instructions given to them. Verbal assurances aside, people rarely know how their perceptions differ from one another. Coders may have prejudices or respond to what a researcher wants to prove. They can hardly prevent learning in the process of interfacing with the phenomena they are asked to record and hence change their perceptions in the course of generating data. In addition, they may become tired or careless. Such uncertainties introduce unwanted variations in the data they generate, quite unlike how mechanical measurements behave.

#### Interpretive or qualitative scholars

tend to provide no space for measuring reliability. Interpretation is taken to be the primary path to understanding text, talk, and interactions in the context of particular social or cultural situations. Interpretations are practiced by various scholars, for example, historians, ethnographers, discourse analysts, critical social scientists, and social linguists, to name but a few. One outstanding feature shared among qualitative scholars is that they generate the very data they are analyzing. Conceiving themselves as experts in the domain of their scholarship and as their own measurement instruments does not invite quantification, least of all generating replications to ascertain their replicability.

This is not to say that qualitative scholars do not rely on formalized analytical methods and uphold discipline-specific standards for assuring the quality and communicability of their scholarly contributions, at least within their own community. For example, qualitative scholars search through archives with great care, record field notes in considerable detail, and transcribe interviews and happenings using widely shared conventions and repeatedly checking their accuracy. They assemble texts and physical traces systematically, authenticate their sources, triangulate what they say, try to read documents in their original context as well as in scholarly terms, and repeatedly examine everything they have done to be sure nothing is overlooked. The anthropologist Clifford Geertz (1973) distinguished between thin and thick descriptions; the latter concerns interpretations of the myriads of textual and observational meanings. Qualitative scholars are committed to publish key quotations of texts from exemplary sources, making sure that readers get a sense of what they are talking of. Quotations are meant to exemplify and be representative of someone's speeches, writings, or conceptualizations. They refer to literature in support of their claims, demonstrate that researchers have consulted relevant texts, and document the analytical steps that led them to their conclusions. All of these methods can be understood as efforts to safeguard against unfounded interpretations and compel readers, critics, and beneficiaries of the reported scholarship to trust the interpretations offered. Qualitative scholars may not talk of reliability; I contend, however, that their efforts are qualitative analogues to what reliability tests seek to accomplish quantitatively.

Two features that makes interpretive and qualitative scholarship unique. On the negative side is the limitation imposed by the conception of qualitative researchers as experts and their own instruments. This conception limits the size of their projects to what individual scholars can handle, especially the volume of phenomena they are able to observe, transcribe, assemble, read, and write about in their own time. It follows that their projects, even when aided by qualitative data analysis software for searching for relevant character strings in larger textual data bases cannot exceed the interpretive capacity of individual scholars. Their inquiries tend to be limited to small sets of observations or documents or take up extended periods in time to complete. This limits the generalizability of their conclusions. On the positive side, interpretations easily exceed the levels of abstractions that computational or statistical methods can achieve. Qualitative scholars tend to offer generalizations that quantitative scholars rarely dare to make, at least not when generating data from textual sources. They may conform to structured views of larger social phenomena in the form of –isms. They may mobilize social movements and they may also inspire quantitative studies. Their findings tend to be framed by highly specialized discourses and circulate largely within their discipline.

One empirical study tested the role of reliability in qualitative research. David Armstrong et al. (1997) acknowledged the conflicting attitudes of qualitative scholars toward reliability but invited six scholars, experienced in identifying genetic screening, to examine one transcript of a focus group consisting of patients with cystic fibrosis. Genetic screening was developed to predict such disabilities. Each expert was asked to identify and rank where possible up to five main themes that emerged in this discussion. The experts turned out to be in broad agreement on these themes. They differed merely in how they framed them, ranging from common prejudices and medical labels to treatment options. Armstrong et al. were concerned with the public costs of medical misdiagnoses, not with reliability; however, their findings suggested that qualitative scholars do have similar concerns.

Qualitative scholars often claim opposition to quantitative abstractions, often misleadingly calling scholars that measure reliability "positivist." Log-linear equations, networks of contingencies, probability distributions, and factor loadings do not tend to occur in their discourse. This does not mean qualitative researchers are averse to counting. Despite their dislike of talking about reliability, I would argue that the research methods of qualitative scholarship are designed to follow scholarly standards that need to withstand potential public scrutiny. Their aim is not incompatible with my synthesis.

## 1.4 My Synthesis

Let me discuss my view on reliability by highlighting four conditions I claim arise in most social research projects involving texts, talk, and observations of social or cultural interactions. The first is shared with interpretive scholarship.

*The diversity of possible interpretations of phenomena needs to be tamed or disciplined to become a shared concern.*

The reality of social phenomena is lived and enacted by many. It is experienced, interpreted in conversations, often institutionalized, and it directs everyday life, both materially and conceptually. Think of the role of an economy, elections, public events, twitter networks, celebrities, ethnic conflicts, and oppressive regimes. Such phenomena tend to be relationally complex, multi-dimensional, dynamically changing, and afford definitions from numerous even opposing perspectives. For example, the exertion of power presupposes submission. Leaders require followers. Communication amounts to co-operation among people (Goodwin, 2018). Adopting any one perspective attracts particular communities of interested actors that have a stake in the implications of how such phenomena are attended to.

By definition, stakeholders have an interest in particular phenomena, are able to articulate their interest, and can mobilize resources to act accordingly. The stakeholders of scientific research are concerned with how the phenomena of scholarly interest are defined, studied, and might affect the lives of others. They may consist of members of the researchers' community of peers, critics of their discipline, funding agencies, publishers, and those affected, whether benefiting from or being impaired by the publication of research findings.

It is a scholarly practice to develop and operationalize the objects of research from available literature. However, social scientists, being part of the very society they study, can hardly avoid paying attention to how the phenomena under consideration are talked of in ordinary language and are constituted in the interactions among their stakeholders. The obligation to publish results implies being publicly accountable for any intended and unattended consequences of their conclusions. For example, adopting metaphors of power from physics in explanations of why some actors are successful, if plausibly communicated to those observed, has the perhaps unintended consequence of entrapping some actors in oppressive relationships, limiting the possibilities of their emancipation (Krippendorff, 2009: 131–155). Similarly, naively adopting racial categories whether in public opinion surveys, genetic research or economic inquiries brings forth findings that reinforce racism in society.

All phenomena studied scientifically are inescapably framed, often from the perspective of researchers' disciplines. Definitions always tame the observed complexities, especially of the social world. However, even the pursuit of esoteric scientific topics needs to attract publishers and readers who take their time to attend to what is reported; try their best to understand what was studied, and experience excitement or express concerns. To make a difference in the social world, to be taken seriously by the stakeholders in the phenomena studied, the definitions of these phenomena, the perspective applied, and conclusions drawn from them, have to be ecologically valid, that is, understandable, actionable, or livable by potential stakeholders.

Ecological validity is the aim of ethnographic fieldwork, of inquiries into the discursive practices among the constituents of the phenomena of interest, and a necessary condition of active participation in social forms, ranging from functioning families, educational institutions and corporations, and governmental organizations. Ecological validity is manifest in the coordination of interpretations and actions within discourse communities.

Ecological validity should not be confused with external validity. External validity has to do with the truth of propositions, a subject to be discussed in the last section of this chapter. To me, ecological validity is a qualitative predecessor of reliability. It is evident in a shared understanding that enables networks of stakeholders in the phenomena in question to inquire and converse with one another about them, whether such inquiries are virtuous or objectionable, of practical benefit or too costly.

### *The process of generating analyzable data needs to be demonstrably communicable, that is, replicable elsewhere*

When the volumes of phenomena to be analyzed exceed the ability of individual scholars to interpret them, the task of generating data needs to be delegated to several observers, readers, coders, annotators, or judges who transform them into analyzable data. It calls for communicable accounts of what is being delegated; specifically how the phenomena of interest are to be framed, interpreted, and recorded.

When the concepts involved in generating data are widely shared and uniformly understood by all stakeholders in the phenomena in question, delegating the task of observing, interpreting, and coding them seems unproblematic. This, however, is rarely ever the case. Claiming to understand a phenomenon of common interest and demonstrating its inter-subjectivity are two different things. The former can be established in conversations, the latter calls for concurrent actions. Inter-subjectivity surely becomes more problematic the more the phenomena of common interest are tamed, disciplined, and carefully defined to feed into chosen analytical techniques and answer theoretically motivated research questions. Stakeholders in phenomena of common concerns who are not privy to the planning and execution of a research project – funding agencies, publishers, readers, beneficiaries, and those affected by its results – may have difficulties recapitulating the research process unless they are given access to what actually led to published claims. This is why the instructions that govern the data-making process have to be written and unambiguously communicable not just among the observers, coders, and analysts of phenomena, but to all stakeholders with legitimate interest in its consequences. Merely publishing a list of the categories and variables of the data language adopted for a study would assume their obviousness and leave out how an analysis regards or shapes the phenomena of declared analytical concerns. To ensure at least a minimum of common understanding calls for testing the replicability of all accounts of the data-making process. Such accounts may have to include:

- The criteria used to select observers, coders, and judges from a larger population of equally qualified individuals. These criteria typically include the literary competencies and background knowledge they are expected to bring to the task of converting observations or interpretations into data, as well as their ability to attend to details, and be dependable. Individuals with these qualifications need to be available elsewhere so that other researchers could replicate the process with other equally qualified individuals. Employing unique experts would make replicability unlikely.
- Detailed accounts of the training that employed observers received, the literature they were asked to familiarize themselves with, and the tests they had to pass to qualify as competent coders. This may involve making the training protocol, and all conventions that emerged during the training process, publicly available.
- The coding instructions that specify what observers were asked to look for, how the phenomena of interest are defined, framed, unitized, contextualized, and interpreted, including the specifications of all measuring devices available to aid the process of transcribing and recording.
- The data language in terms of which the interpretations of the phenomena of interest are recorded, spreadsheets, computer-aided coding software, etc.
- The specifications of recording devices used to generate the data as well as analytical methods used to transform the data into answers to research questions.

Evidence that these accounts are clearly communicable across researchers resides in demonstrations that applying them to the same or similar phenomena results in substantial agreement among qualified and suitably trained observers. Given enough variance among the phenomena in question, agreement observed among independent replications indicates the extent to which the data carry information about observed phenomena. Disagreement observed among replications indicates the extent to which data are polluted by noise or external variations unrelated to the phenomena of analytical interest. Without substantial agreement among independent replications under diverse conditions, the reliability of data is uncertain and the results of analyzing them can become questionable.

While qualitative researchers tend to refuse quantifying the reliability of their data and how their results relate to the phenomena of interest to them and their stakeholders, I claim that they do seek to demonstrate by examples, quotations, and illustrations what their data are about and do their best in making their interpretations understandable to those who share their concerns, and remain open to being held accountable for their conclusions.

*The makeup of the analytical techniques needs to be compatible with (model) the nature of the phenomena as encoded in the data.*

Once data are generated, the phenomena that gave rise to them are rarely consulted, often no longer available, even lost, yet need to be considered during all subsequent steps.

Reliability data are based on replications, as described above. Agreements observed among them are not indicative of what the replicated data are about but of the extent to which one can trust that all replications preserved the same distinctions among the phenomena of analytical interest. Preserving distinctions fits Shannon's definition of the transmission of information. Accordingly, perfect agreement indicates that the data generated by each replication are informed by the phenomena in question, and only these, and can be relied upon without question. However, what perfectly reliable data are about, what they mean, is not contained in the generated data. Rather, their meanings will have to be recovered by applying the written coding instructions in reverse to decode the generated data and decode what the observers who created them must have seen, read, or interpreted and recorded. Imperfect agreement amounts to an erosion of the certainty with which the data were encoded and can be decoded by the analysts.

What enables analysts to consider given data as surrogates of the phenomena they intend to study is the reversibility of the coding instructions that generated them. As the agreement among replications diminishes so does the analyst's confidence in knowing what they are analyzing, and their ability to convince their colleagues, critics, and stakeholders of the soundness of their data. Not achieving acceptable levels of agreement among replications may force researchers to ignore variables that cannot be trusted, limit their scholarly ambitions to what makes sense, or reconceptualize their research project and start anew.

Without the ability to apply the reverse of the coding instructions to given data, high agreement cannot say what the data represent, only that they represent something real. There are plenty of stories of analysts who assumed data to mean one thing while the coding instructions said otherwise. I have seen content analyses in which coding instructions were changed while data were generated, which rendered the resulting frequencies of recorded categories no longer uniformly interpretable and compromised the whole research project. I have been in court concerning a contentious content analysis in which the principal investigator claimed his data to be reliable by calculation, but facing examples of what was coded, his guesses were not much better than chance. A colleague of mine had preserved boxes and boxes of data that were made useless because he could not find the coding instructions responsible for them. In debates on the virtues of big data, proponents point to the unprecedented volumes easily available while critics point out that character strings retrieved from large electronic data bases without much knowledge of the context that generated them are poor data. Without knowledge of the codes used by observers who generated the data, and codes that take account of the contexts from which the coded phenomena emerged, the analysis of such data can become quite meaningless. A classic example is the interpretation of writing by extinct cultures. Without Rosetta stone-like correspondences to a known language, or adequate contextual correlates, anthropologists have the greatest difficulties in making sense of such texts.

In other words, without adequate knowledge of what given data are about, their analysis may reveal syntactical pattern in their physical manifestations but very little else. A list of numbers, strings of alphabetical characters, or arrangement of pixels are not data of anything until one can relate them to what caused them, what they meant to their creators, and what institutions they constituted. Reliability merely establishes the extent to which data represent something real. What it is that one can rely on requires some sense of what these data refer to, what they are about, which in turn can aid decisions on the analytical paths that could lead to something worth knowing.

*The public nature of scientific research calls for accounts that its diverse stakeholders can demonstrably understand, utilize, replicate and question, at least in principle*

The purpose of publishing research results is to provide access to those interested in them. Publishing the reliabilities of the analyzed data tends to provide assurances to those who care that they can trust how the phenomena they may well be familiar with were turned into analyzable data, leaving open, however, whether their analysis was appropriate and answered the research questions that motivated the published research. While inter-coder agreement among coders hired to generate is easily obtainable, to me, the idea of reliability cannot be limited to a minimum of two highly trained but unaccountable observers/coders. If reliability is a measure of shared understanding of the given coding instructions, published claims that the data are reliable surrogates of phenomena have to make sense beyond researchers' analytical focus to the diverse stakeholders in these phenomena.

To be clear, all data reside on the lowest level of abstraction; one could say one step removed from the phenomena observed, experienced, lived with and shared. They may enter a research project on a level that their stakeholders may be able to understand, make sense of, and talk about in their own terms. Assuring the ecological validity of how researchers define the phenomena they study provides the possibility that their stakeholders can communicate about them, approvingly or in opposition. However, social researchers often pursue abstract research questions of interest to large institutions, advocates of particular theories, or political actors with global missions. Concepts like volumes of communication, network densities, factor loadings, or the virality of memes may easily escape comprehension by those who live the phenomena.

Moreover, measuring the agreement between stakeholders' understanding and researchers' understanding is rarely feasible and often meaningless in view of the likelihood that stakeholders hold perspectives unlike academic researchers. The traditional criteria for publishing scientific research are favorable reviews by academic peers and widespread scholarly interest. Reviewers tend to care less about whether journal readers can follow the researchers' discourse and how publications on particular social phenomena affect the future of these phenomena. Reviewers and publishers may be the implicit gatekeepers of stakeholder networks which include competing researchers, practitioners of the studied phenomena, and people that benefit or are harmed from what is said about them. I am suggesting that all social research faces diverse stakeholder communities who have their own reasons to understand and act differently on what they read. Under these circumstances the ideal of uniform understanding of what a research establishes is simply unachievable, perhaps even undesirable. Generalizing the notion of reliability has to take a different form.

For research to be socially relevant, for it to make a difference in the networks of diverse stakeholders of which social researchers are a part as much as their colleagues, critics, beneficiaries and opponents, a single uniform understanding of the phenomena that researchers address is unlikely. The publication of research results has to take seriously the multiplicity of stakeholder communities interested in them. This entails a conception of reliability that accepts disagreements across different communities of shared understanding. Yet, even critics and opponents of reported results need to understand how data were generated and be able to judge whether they support or oppose published conclusions.

The inability to generalize the process of establishing reliability internal to the research process to diverse ways of understanding and using of scientific research by unlike stakeholder communities calls for a gestalt switch in theories of communicating scientific facts. Communication can no longer expect to yield an understanding that is shared by everyone, but it may aim at withstanding criticisms from diverse stakeholder communities without being deniable, harmful, or destructive of their worlds. I believe that awareness of the social consequences of social research, efforts to address potential critics, and concern of the possibility of being misunderstood also underlie much of qualitative scholarship.

My view on reliability could be seen as synthesizing the demand of measurement theory for the replicability of the process of generating data with what interpretative scholarship seeks to accomplish in the social world: a diversity of locally shared understandings of what research enables. It unites two methodologies that differ in rigor, scope, and terminology, and calls for reflexivity on the role that researchers play in articulating the phenomena lived by those whose lives they investigate.

## 1.5 Five Reliabilities

All claims for something to be reliable are made in language. Whereas such claims are often articulated by authorities – ranging from acknowledged experts, salespersons, to prophets – scientific research has advanced by replacing such authorities with plausible arguments and empirical demonstrations or tests. The previous sections elaborated conditions that constitute data and reviewed diverse discipline-specific conceptions of reliability. I argued that some notion of replicability was underlying most of them. The following discusses five conceptions of reliability within a coherent framework. The first three – stability, replicability, and accuracy – constitute a scale of increasing strength, incrementally including three sources of unreliability or noise. The other two – surrogacy and decisiveness – split the strongest one into two very useful indices. I start with the most intuitive coefficient: stability.

| **Reliabilities** | **Their definitions** The extent to which | **Measured agreements** are reduced by | **Strengths and limitations** |
|---|---|---|---|
| **Stability** ${}_c\alpha^1$ | Data are generated consistently and predictably by one instrument or coder | Intra-coder inconsistencies, fatigue, loss of attention, using shortcuts, and external distractions | Weakest Easiest to establish but not generalizable, not recommendable |
| **Replicability** ${}_c\alpha$ | The process of generating data is independently reproducible elsewhere | All of the above + Inter-coder disagreements on coding phenomena into analyzable data | Intermediate Easy to establish Requires at least two replications |
| **Accuracy** ${}_c\alpha^s$ | Data conform to an arguably trusted and published standard (to independent factual truths) | All of the above + Deviations from the trusted standard (Invalidities = deviations from truths) | Strongest but the availability of acceptable standards (or truths) is limited |
| **Surrogacy** ${}_c\alpha^{sm}$ | A proposed alternative can be a reliable substitute for trusted coders' collective judgements | Mismatches of the proposed alternative and trusted collective judgements by human coders | Variable as its use as a decision criterion is limited by coder's decisiveness |
| **Decisiveness** ${}_c\alpha^m$ | Independent human coders collectively agree on their interpretations of phenomena | Unaccountable diversity among independent human coders | Intermediate Easy to establish Requires at least three replications |

**TABLE 1.1**
Relations Between the Three Classical Reliabilities Plus Two

### Stability

is evident in observed agreement among repetitions of the same data generating process, measuring something again, or categorizing text a second or more times. This test-retest situation is common when establishing the reliability of mechanical measuring devices. Their makeup is or can be assumed to be unchanging, which makes observed differences explainable as extraneous influences of their intended use. For example, a nurse who finds the temperature of a patient at odds with her expectations is more likely to repeat its measurement before concluding that the thermometer is broken. Before she repeats taking the temperature of a patient, she might look for whether she had reset the device to begin with, could have influenced the measurement by her hand, or mistakenly measured the temperature at a point where it is at variance with what is to be measured. Such misapplications ruled out, apparent variations among repeated measurements that should be the same would indicate unexplained and undesirable variations interfering with the measured temperatures.

While all data-generating processes should be stable, that is, unvaryingly repeatable, human observers exhibit vulnerabilities that are rarely present in mechanical devices. Observers' attention to detail may vary with circumstances that have nothing to do with the phenomena they are asked to attend to. They may get tired from repetitions, become bored or careless, letting previous observations influence current ones, or acting out expectations that have nothing to do with the phenomena of interest.

Most qualitative scholars tend to work alone. Without hiring other observers to generate their data, the only reliability assurances they can report are whether they have been consistent in repeatedly applying their own coding instructions to the same phenomena and yielding the same data. This is common sense. Already in elementary school we are told to recalculate a sum when we are uncertain about being correct until we are satisfied. In mathematics problems have correct answers. However, when interpreting texts or judging complex observations this rarely is the case. Having nobody else to help out, generating one's own data by applying one's own coding instructions to the same data after some time has elapsed is the only way to see whether one has been consistent in generating one's data. In content analyses, repeating a coding process and measuring its stability over time is the default strategy for underfunded research projects. Economic constraints and good intentions notwithstanding, where humans serve as their own measuring instruments, I would argue that the stability of human observers does not lead to adequate assurances that the data they generate can replicated by others.

For once, stability cannot reveal repeated mistakes, habitual distortions, and systematic biases, which are typically human. For another, the measure of someone's stability cannot reveal anything about how someone else would understand the data thus generated. Observers with stable predispositions, discursive commitments, unarticulated ethnic, gender, or political prejudices, or an agenda that favors one answer to a research question over another are likely to agree with themselves over time.

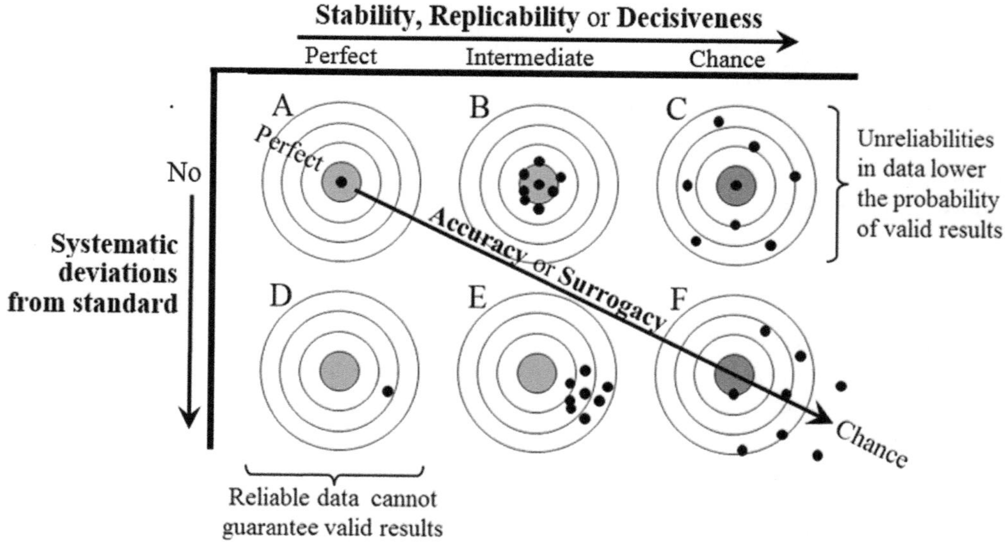


**FIGURE 1.2**
Relationships Between Stability, Replicability, Accuracy, and Surrogacy

Figure 1.2 depicts six images of eight repetitions in the form of eight dots. Ignoring the circles in which they appear for now and attending only to their distributions, in the horizontal dimension of this figure, in A, eight repetitions are indistinguishable and agree perfectly. This would indicate that the individuals who generated the data in A were perfectly consistent and in agreement over time and the data they provided were perfectly stable. Without the circles D would be as stable as A. The agreement in B is less than A and becomes even worse in C. Evidently data become increasingly polluted by random variation when moving from A to C. Just as the stability of A and D is indistinguishable so are the replicabilities of B and E and of C and E. When moving from the first to the last column the erosion of stability is the same.

Measuring the stability of human observers poses several very human problems. Unlike mechanical measuring devices, humans can hardly help but learn with repetitions and become decreasingly attentive to details. Humans try to be consistent especially when they know that their stability is a measured. When coding textual matter repeatedly, consistent repetitions can have the consequence of rendering units of analysis no longer independent of each other. Also, and especially when observing and coding larger volumes of data, individuals tend to form idiosyncratic expectations, develop unique coding habits, inventing shortcuts, and get tired, all of which may well improve the measured stability of data over time, but would not make the resulting data more trustworthy. Added to the questionable interpretation of consistency and predictable coding of data as reliability, individuals may well become distracted by external conditions, telephone calls, bad lighting conditions, or kids asking for attention. They tend to reduce the measure of stability. Measuring instruments may have different reasons to be stable over time; the wearing out of mechanical parts is like human inconsistencies over time, but external conditions, voltage fluctuations, changes in temperature, and even how an instrument is handled may make it fluctuate.

What has the potential of reducing agreement measures among repetitions of the same process of generating data is tabulated in the first row of Table 1.1. By comparison with the other reliability measures discussed below, the third column of this table suggests that stability is the weakest of the five reliabilities distinguished here. It is therefore not recommended but can be computed by the same formula as for replicability.

## Replicability

refers to the ability of several independent measuring devices, human observers, or judges who; given the same coding instructions and the same set of phenomena to evaluate provide the same coded data. The degree to which independent observers agree is measurable by their agreement. However, the replicability measured among coders following written coding instructions is meant to indicate that the process of generating data is reproducible elsewhere, by other researchers, hiring coders with the same qualifications, undergoing the same training protocol, and following the same coding instructions. The disagreements that reduce the replicability of the process respond to two sources. It adds to the intra-observer inconsistencies, accounted for by the measure of stability, the inter-observer mismatches, the sum of which amounts to a measure of replicability of the process of generating data. It is important to point out that replicability is not a measure of individual observers, nor of the possibly unlike difficulties in coding units of analysis, all of which are included but not differentiated as its focus is whether all the conditions responsible for coding the phenomena of interest can be reproduced elsewhere. High agreement among replications ensures so-called "inter-subjectively verifiable" data. Replicability can be ruined by both the instabilities of individual coders and the differences between them.

By responding to both sources of reducing the process to be replicable, replicability is a measure stronger than what stability assesses.

The idea of replicability underlies the legal practice of accepting an account as factual only when corroborated by several independent witnesses. It is evident in journalists' suspicion of single sources of information about controversial events. In the social sciences, replicability can be interpreted not only as the extent to which data are reproducible elsewhere, but also as the extent to which the phenomena of interest are recoverable from the generated data by applying the inverse of the coding instructions to the coded data.

The computation of agreements for stability and replicability is based on pair comparison and in fact employs the same formula. These two reliabilities differ in how their reliability data are obtained. In the case of stability we are concerned with one measuring device, observer, or judge whose stabilities becomes evident in measuring or coding the same set of units of analysis repeatedly, perhaps after some time has elapsed between recording their evaluations. There are $m > 1$ repetitions of valuing $N_u$ units of analysis. In assessing the replicability of the process of generating data we have $m$ independently working coders, or measuring instruments, valuing the same set on $N_u$ units of analysis. The reliability data containing repetition and replication have the same dimensions but are far from having the same meanings and the errors they respectively embrace. Regarding how irrelevant variations affect the measures of replicability and stability, the columns of the six images in Figure 1.2 are the same. The first column depicts perfect replicability which is increasingly diminished as attention shifts to the second and third column. What these losses of agreement are due to is comparatively listed in Table 1.1. Although both agreement measures serve different purposes, inasmuch as replicability is diminished not only by the intra-coder inconsistencies, which are registered in the measure of stability, but more importantly, by inter-coder disagreements, replicability is a reliability measure stronger than stability. In the above discussion of stability, I gave other reasons why the measure of stability cannot be recommended as the basis for inferring the reliability of data.

Replicability is by and large easy to demonstrate and measure. However, as mentioned in the previous section, the widespread practice of publishing merely the names of researchers' categories devised for a study and the agreement measured for each variable, typically among two well-trained human coders, may not live up to a convincing demonstration of replicability of the process by which data are generated. Let me reiterate three potential pitfalls of this practice.

First, reporting high inter-observer agreement among replications gives the impression that the data are replicable across many observers and without qualifications. However, this impression hides the likelihood that in the process of training a group of observers, the published categories have acquired group-specific interpretations that written coding instructions may not be able to convey to other coders that were not part of the training process. Training usually entails talking about questionable or ambiguously defined categories during which their original meanings are shaped until they are shared by the observers involved. Moreover, training often allows observers who encounter difficulties in understanding the instructions to be replicated to talk among each other and decide among themselves how to solve the problems they experience or receive advice from the principal investigator about the intended interpretation of the given coding instructions. The conventions that emerge in such communications undoubtedly increase the measured agreement but remain in the confines of coders, trainers, and researchers. Unless these conventions become part of the written coding instructions, the published inter-observer agreements can no longer serve as measures of the replicability of the instructions elsewhere and involving other observers. This is why talk of inter-coder agreement might well be misleading unless it is qualified by the assurance that coders worked only from the written coding instructions. Reproducibility of a process of generating data entirely depends on the ability to communicate how coders were instructed to observe, identify, and record the phenomena of concern. A nonreplicable consensus developed among the coders employed in a reliability test is a significant source of biased agreement measures and no longer interpretable as an indication of the ability to reproduce the process of generating data elsewhere.

Second, all observers employed in generating data tend to be selected for possessing literary competencies and background knowledge of the subject matter of the research project. The ability to speak the language of analyzed texts and to read the coding instructions are two obvious conditions of their employment. Cultural competencies of uniformly interpreting social phenomena in their context may be less easily testable but are equally important qualifications. The absence of shared competencies tends to show up in significant disagreements among otherwise sufficiently competent observers. The practice of selecting observers who happen to agree more readily among each other yields higher agreement coefficients but does not necessarily guarantee reproducibility as it hides the source of such agreement. Unless all qualifications for selectively employing observers are explicitly communicable, perhaps in the form of reliably tests of these qualifications, reports of high agreement are not predictive of replicability of given coding instructions by other researchers, who undoubtedly have to employ different observers. For agreement measures to be interpretable as indicators of replicability, one needs to ensure that the qualifications of the observers employed are clearly communicable; this implies that the population of equally qualified observers is large enough for other researchers to draw from.

Third, even when the process of selecting, training, and testing the competencies of observers to follow the written coding instructions is clearly communicable, using only the minimum of two observers is questionable. Training and testing observers' ability to follow given instructions before entrusting them with the task of generating the needed data always builds on unwritten and difficult to identify cultural backgrounds, and personal experiences that written instructions may only partly override. Researchers can check on the linguistic competencies of potential coders, experiences relating to the subject of research, and ideological commitments if the subject is politically sensitive. However, a sample to two coders out of many possible candidates for replicating the process or, equivalently, understanding the data by being able to decode the coded data is hardly defensible. At the same time, generating data should not be confused with psychological experiments. Replicability focuses on the process of valuing data by ideally standardized coders. Psychological experiments target the variations of individual differences. Yet, it would be difficult to argue that two coders could represent all potential coders' judgments and researchers' ability interpret the data in agreement with how they were generated by two trained coders.

There is another reason to employ more than two coders. Replicability requires demonstrations that independent replications agree well above chance. It is measured on the agreement among coders valuing a set of units of interest. One of the replications may be taken as the data eventually analyzed or becomes a part of an extended set of coded units ultimately relied upon in answering the prevailing research questions. If there are only two coders, the choice among them is arbitrary. If there are three one can identify the one who shows more disagreements than the other pair; if there are more than three one may be able to identify one who is closest to the majority and use that coder's data and not the others'. The benefits of using, say more than three coders is not only a question of whether the coders represent other potential coders, it also offers the possibility of deciding which of the coded replications are the most trustworthy, including selecting the best coders to continue generating data beyond the reliability test.

In Chapter 0, I presented good reasons for measuring agreements not in terms of entropies, correlations, or percentages, but in terms of pairwise differences within data, much like how variances are defined. To illustrate the pair comparisons that distinguish the five reliability measures, Figure 1.3 depicts what goes into each, exemplified by eight values assigned to one phenomenon. The first image in Figure 1.3 depicts what goes into the calculations of stability or replicability: For eight independent observers or eight duplications by one there are 8(8 – 1) = 56 pairs to be evaluated for their differences. Table 1.1 spells out these differences. For stability they are due to intra-coder inconsistencies and distractions experienced while coding. Replicability additionally responds to inter-coder disagreements. As already stated but now visualized in Figure 1.3, stability and replicability are paired alike, and depicted jointly in its first image. This figure also depicts the pair comparisons that constitute the three other reliability measures beyond the classical replicability measure. They will be discussed next.

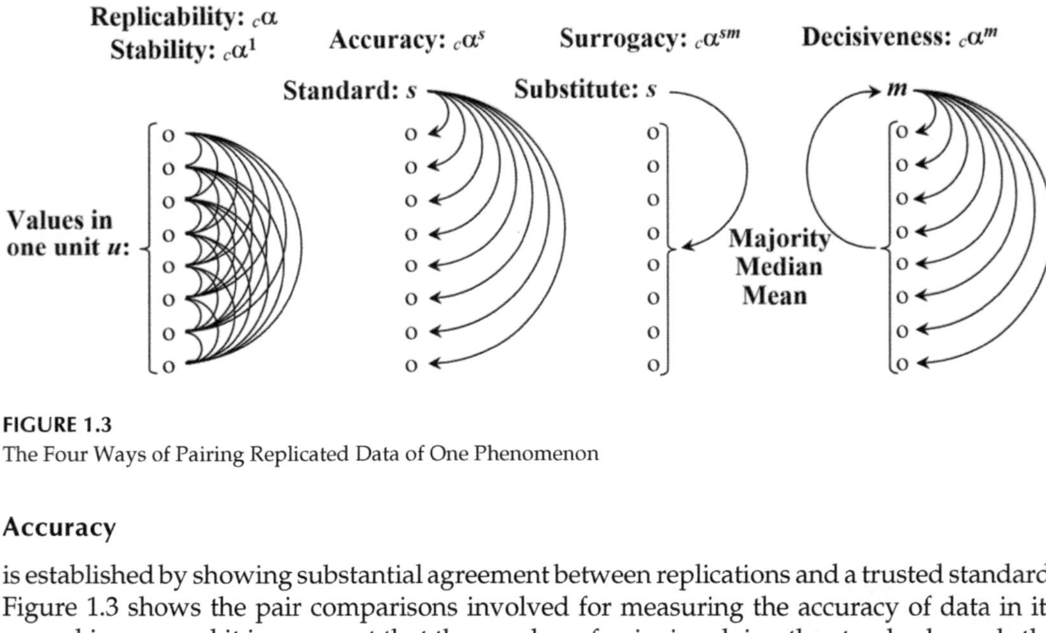


**FIGURE 1.3**
The Four Ways of Pairing Replicated Data of One Phenomenon

## Accuracy

is established by showing substantial agreement between replications and a trusted standard. Figure 1.3 shows the pair comparisons involved for measuring the accuracy of data in its second image, and it is apparent that the number of pairs involving the standard equals the number of observers who coded a phenomenon and are pairable with it. The figure also makes evident that the number of pairs that go into the accuracy measure is significantly smaller than the number of pairs that go into the measures of stability and replicability.

In content analysis, a common use of accuracy measures is during the training of observers. The standard might be established by consensus of a panel of judges who code the materials used for training of coders. Standard may also come from the literature of what should be, from dictionary definitions of words or phrases, or from a previous study that proved to be successful and is to be approximated.

Outside content analysis, accuracy measures underlie many forms of aptitude tests, used whether when applying for a job, for a driver permit, various competencies tested in the process of education, or judging whether the programming of a radio station conforms to a federal standard.

Measures of accuracy are never better than the trust in the chosen standard. When the standard is of a factual nature or equal to undisputable evidence, accuracy becomes a measure of the validity of data. In Table 1.1, the latter is articulated in parentheses found in the row for accuracy.

In Figure 1.2, the standard is visualized by the image of a target with a center that coding needs to aim at to be accurate. In its first column, A visualizes perfect stability, replicability and accuracy with all eight data points being "on target." Moving from A to D accuracy declines and so does moving from A to C. This fact suggests and Figure 1.2 depicts accuracy to decline along two dimensions: The loss of stability and replicability is depicted in moving from the first column along the horizontal dimension to the right and the increasing systematic disagreement in moving along the vertical dimension from the top row down to the row below. The increase of systematic disagreements from the chosen standard is shown moving down on the vertical dimension. In the first column of this figure, both images, A and D, would indicate perfect replicability. The accuracy of A would be perfect as well, but the accuracy of D would drop significantly. The same can be said about the other columns. The losses of replicability in the top row, moving from A to B and to C amount to a loss of accuracy as well. However, moving downwards from B to E and from C to F leaves their replicabilities intact but accuracy will suffer losses. In effect, accuracy diminishes along the diagonal dimension, being affected by both inter-coder differences and deviations from the standard. Accuracy is the strongest among the three kinds of reliability discussed so far. It is surpassed only when the standard stands for an objective truth.

Mathematics thrives on accuracy. While mechanical instruments are deterministic by design, and exhibit no randomness, they are prone to systematic deviations from a standard. A mechanical clock may be too slow or too fast relative to standard time, but this deviation is systematic, not random. In the social world there are numerous institutional norms that serve as standards against which the accuracy of human behaviors is evaluated: income tax returns, legal and contractual obligations, or vote counts. Human observers tend to exhibit systematic inaccuracies as well, for example in the form of shared prejudices, preferential uses of categories or values, and various ethnic, gender, sexual, and political biases. If such biases are shared among observers, replicability will not recognize them.

While accuracy would therefore be a measure preferable to replicability, when coding social phenomena, which are often unstructured, complex, and imbued with multiple meanings, single standards are not too easily available. Moreover, at least in the domain of content analysis, the application of standards against which generated data can be checked, poses the epistemological dilemma of having to know how a phenomenon is to be coded before coding it. It follows that standards are either redundant or their application makes sense only on subsets of data from which the accuracy of larger numbers of data might be inferred.

For content analysts, this dilemma is not threatening the logic of accuracy measurements when such measures are applied only on subsamples of the data whose accuracy is an issue. Let me describe four practical uses. First, when training observers for a larger project it becomes important for them to understand the written coding instructions – not in their own terms – but in terms of the principal researchers' interpretations of their instructions. Understanding is not observable as such, much less measurable. It becomes manifest only by how coders enact the coding instructions given to them. In the introduction, I mentioned that we gave candidates for the job of coding violence on television training material and coding instructions for which a panel of experts had decided how they should be coded. By giving them immediate feedback about their accuracy after coding a violent episode, coders learned what was expected of them until they were able to code without supervision.

Second, with or without training observers, in studying violence on television, described in Chapter 0, we made the hiring of observers dependent on the measured accuracy in coding violent episodes against our independently established standard. This was like an aptitude test for coders, which was communicable to anyone interested in replicating our coding. (At that time we had no accuracy measure and merely employed our own judgments.)

Third, on a long project, coders tend to get tired, careless, and let their expectations override the details to be attended to. This reduces the reliability of the data. If these erosions are shared, for example, when observers are allowed to talk among each other about their tasks, replicability measures would not reveal them; only a measure of accuracy can.

Fourth, in efforts to obtain data for content analyses by crowd coding, coders sign up for the job leaving researchers without an inkling of who they are, what competencies they possess, and the dedication they bring to the job they are asked to do. To test whether they are up to the task, it is common practice to include a few tasks for which the researchers know the correct response and accept the data from these workers only if they have answered these test questions accurately.

### Surrogacy

is a measure of the degree to which one coder, index, model, or algorithm can serve as a substitute or surrogate for diversely coded phenomena or vice versa, whether and inexpensively obtained crowd coded average can take the place of more costly coders.

A frequent but minor problem that surrogacy measures can address is identifying the most reliable observer among a group of observers coding the same phenomena. Absent surrogacy measures, it is a common practice to apply accuracy measures to such evaluations, place any one observer in the position of the standard and match that observer's coding with the coding of each of the other observers. This practice is flawed for two reasons: First is the direction of the evaluation – in Figure 1.3 depicted by arrows. Whereas accuracy measures proceed from a trusted standard to the uncertain quality of the replications, surrogacy proceeds from the trusted replications to the uncertain quality of a proposed surrogate. Second, and more serious, while accuracy is affected by all disagreements among data, surrogacy is concerned only with what a substitute could possibly represent. This is what all coders have in common, the judgments by majorities, medians, or means of the coded data. Thus, any possible substitute for human coding cannot be better than the collective judgments of coders regarding the phenomena they are facing.

It follows that the best possible substitute for multiple coders cannot embrace the unexplained variance among them. Figure 1.3 illustrates that difference, including direction of the arrows from what is trusted to what is being evaluated in its terms. The image for accuracy shows as many pair comparisons as replications are available whereas the image for surrogacy shows only one pair comparison between the substitute and the majority, median, or mean standing for all possibly deviant values.

Although the use of competent human observers is indispensable for generating reliable data of complex social phenomena, especially where language is involved and meanings play significant roles, the use of human beings is also expensive and time consuming. Therefore, researchers' search for more efficient methods that could replace human coders is widespread. Researchers have operationalized human coding in the form of mechanical measuring instruments, computable indices, and theoretical models of the phenomena under consideration, or algorithms that are presumed to substitute for human involvements. If the data that a mechanical device produces match the consensus among a group of observers, then its surrogacy is established. In practice, the surrogacy of mechanical devices, which work unlike how humans interpret social realities, is achievable only for routine human practices.

Advocating the use of accuracy measures for evaluating the substitutability of a method of analysis for human coders is faulty as it mistakenly responds to the individual variances among coders which have little to do with whether a surrogate represents what the data share. Figure 1.3 visualizes the pair comparisons and Table 1.1 names the disagreements that accuracy includes but surrogacy omits. The use of replicability measures is even further removed from what surrogacy has to selectively respond to.

For a general example, mentioned earlier, computational content analysts claim to be able to bypass reliability issues altogether by proposing algorithms applicable to written documents. This claim tends to be justified by pointing to the deterministic nature of computation; their unfailing stability in repeated applications. Arguing from these features alone is flawed when the issue is one of generating data that represent phenomena of analytical interest. Computational content analysts claim their methods to be superior to the common experience that the human coding of textual phenomena is rarely perfectly reliable and more importantly costly and time consuming. However, what such arguments overlook is the surrogacy of computational methods to literal human judgments. For example, the computation of word clouds, constructed from the frequency of words occurring in a lengthy document, might well be interesting for those who have a sense of what the document means to them, who the author is and what the analyzed document meant in the absent social realities. But word clouds are far from being a substitute for reading the document.

Finding computational substitutes for what coders are able do is the aim of so-called machine learning algorithms. It proceeds in two phases. The first involves exposing that algorithm to a large number of textual units with independently known characteristics, identified by human coders. During this learning phase the algorithm searches for connections between textual qualities that could distinguish among the known characteristics until a researcher is confident that it has done its best. The second phase is to apply this algorithm to textual units without knowledge of these characteristics and infer them from the textual qualities previously found to be connected to them. In practice, the characteristics to be inferred tend to be rather simple; for example, whether a particular product is mentioned, perhaps by different names, whether a politician is mentioned favorable or not, or whether suicide notes are real or used for other purposes (Ogilvie, et al., 2009). The transition from the first to the second phase requires a measure of the extent to which what the algorithm does fits what human coders identified.

For all efforts to establish the extent to which algorithms, measuring devices, formulas, even theories are about phenomena their proponents claim to study, measures of surrogacy need to be employed. As Figure 1.3 convincingly illustrates, surrogacy has little to do with replicability, or accuracy, to which one may add correlations. Ignoring measures of surrogacy when claiming that an AI substitute for human coding is superior haunts their advocates in the long run.

Three properties of surrogacy measures need to be mentioned. First, whereas accuracy can be obtained with one coder and a standard, surrogacy requires at least two replications just as do stability and replicability. Majorities, medians, and means become all the more meaningful when based on many replications.

Second, majorities, medians, and means are single values that represent what a collection of categories, ranks or values share. For example, a majority ignores the disagreements with and among the remaining minorities, just as an average represents the mean of the numbers averaged; it ignores the variance among these values. Although we make provisions for tied majorities, for a surrogate to represent the coding of many replications, it cannot be better than the majorities, medians, and means among them. The diversity of data cannot be represented by an ideally single-valued substitute.

Third is the inability of majorities, medians, and means to say anything about their strength. This can be seen most clearly in the binary case. A majority of 99% may well be unchallengeable whereas a majority of 51% could easily flip to the other side with small random variations. Also, any average does not say anything about the underlying variance, about its strength. This mathematical fact serves to make clear that surrogacy is not a measure of the reliability of data; rather of the proposed substitute.

**Decisiveness**

refers to the degree in which the individual variance in data agrees with their collective characterization in the form of majorities, medians, or means. It accounts for what an ideal surrogate cannot represent.

Figure 1.3 should make clear that computationally, decisiveness resembles the measure of accuracy insofar as it evaluates available replications to a designated one. However, whereas accuracy evaluates these replicated data by reference to a trusted standard that is external and independent of the replications being evaluated, decisiveness evaluates these replications by means of their own collective judgments, internal to the coding process.

Table 1.1 reveals another interpretation of decisiveness. In the column which lists the disagreements that reduce the applicable agreement measures, decisiveness can be seen as addressing what surrogacy omits. It follows that disagreements between a standard and the diversity of replicated data can be seen as composed of two kinds of disagreements: The disagreements between a proposed surrogate and the collective qualities of replications, and the disagreement between these collective qualities and the individual diversity of data.

The measure of decisiveness assesses exactly what its name customarily means. Dictionaries define decisiveness as definitive, conclusive, unquestionable, unmistakable, decidable, unwavering, and certain. Most examples of its conventional use relate decisiveness to making unambiguously clear decisions. In social research, decisiveness measures the variances that no surrogate, proxy, or model can represent, delegate, or model.

If replications are reasonably large, I suggest that decisiveness would be a better estimate of the reliability of data than the measure of replicability.

## 1.6 The Form of Reliability Data

Recall from Section 1.1.4 the distinction between data whose reliability is in question and data that enable their reliability to be measured. I call the latter reliability data. Reliability data must allow comparisons, and hence are tied to replications or comparisons with a standard. And, inasmuch as a reliability test may be undertaken on a sample of the data whose reliability needs to be known, reliability data need to be a representative sample of the data generated for a whole project. When there is too much uncertainty about the distribution of data to come, one ought to select examples that cover all possible cases that a given coding instruction specifies.

The canonical or most basic form of reliability data capable of calculation of the five kinds of reliabilities distinguished in the previous section is depicted in Table 1.2.

| | ← **Distinct or contiguous phenomena** → | |
|---|---|---|
| A **standard** or **surrogate** | . . . . . . *S's* Data . . . . . . . . . . . | if available |
| 1 | : | |
| 2 | : | |
| : | : | |
| **Replications:** *i* | . . . . . . *i's* Data . . . . . . . . . . . | |
| *j* | . . . . . . *j's* Data . . . . . . . . . . . | |
| : | : | |
| *m* | : | |

**TABLE 1.2**
The Canonical Form of Reliability Data, Generally for Any One Variable

In this table, 1, 2, …, $i$, $j$, …, $m$ designate the independent replications by observers, coders, raters or mechanical measuring instruments. The data tabulated in the main matrix of Table 1.2 record a diversity of phenomena, observed, read, judged, coded or measured. These phenomena, which define the columns of this matrix, may consist of sets of predefined units or segments of continua variously distinguished in the process of generating data, a distinction that will become elaborated in due course. It is not necessary that all columns are coded by $m$ replications as long as they are pairable within columns. Typically, but not exclusively, the reliability data in Table 1.2 represent the data of one variable of a larger system of variables, each of which will have its own canonical form. For calculating

- Stability and replicability require $m > 1$ replications in each column
- Accuracy requires $m > 0$ replication in each column for which a standard listed in the row above this matrix
- Surrogacy requires $m > 1$ rows whose majorities, medians, or means are paired with the values of a proposed substitute, listed in the row above this matrix
- Decisiveness calls for $m > 1$ rows whose majorities, medians, or means can be conceived of as occupying the row above this matrix, normally reserved for the standard or substitute with which the replications are paired

As detailed in Table 1.1, all reliability measures answer slightly different questions about whether the individual data of these replications, or proposed surrogates can be trusted in subsequent analyses, including whether observers need more training, coding instructions require revisions, replications need to be more numerous, and where sources of unreliability reside.

It is important to emphasize that the various observed agreements computed from Table 1.2 do not establish what the generated data are about, only the extent to which differences among the phenomena attended to are reproduced in the data. Provided data are judged sufficiently reliable, the nature of the phenomena that researchers seek to analyze with the proxies of data should be decodable by reversely applying the coding instructions responsible for generating data or by knowing the design of the instruments or algorithms involved. When agreement among replications is without exception, replicability is perfect, and the variance in the data is fully explained by the distinctions that observers or coders were instructed to make when coding. However, perfect replicability does not guarantee perfect accuracy, nor that the proposed surrogate can be considered as a surrogate for the generated data.

Nor does perfect replicability or accuracy guarantee that the amount of information in data about the phenomena of analytical interest is sufficient to answer the guiding research questions concerning these phenomena. The research questions may be stated beyond what the data can deliver; the analytical methods chosen may omit or distort the information the data provide. It follows that perfectly reliable data cannot guarantee leading to valid conclusions, but they do indicate that possibility. The disagreements that deflate the reliability measures indicate noise in the data that, if carried into the research results, reduce the possibility of valid conclusions. In the extreme case of the absence of a statistical relationship between the distinctions among phenomena and the differences that the data exhibit, which essentially zeroes all reliability measures, the conclusions derived from the analysis of such data are totally arbitrary.

All reliability data are data, of course. As such, they too have to possess the four qualities developed in Section 1.1: retrievability and durability, computability, ability to serve as surrogates of the phenomena of interest and provide relevant information. However, regarding the last two criteria, reliability data differ from the data whose reliability is being evaluated by being replications of the data-making process. As above stated, replications are redundant data. Redundancy sounds wasteful. However, the importance of redundancy for estimating the reliability of data is borne out in one theorem of Shannon's information theory (Shannon and Weaver, 1949) which states that the amount of redundancy in data limits the ability to correct coding errors. This opens the possibility of designing analytical techniques capable of correcting coding errors identifiable in observed disagreements among replications and thereby increasing the likelihood of valid conclusions even from unreliable data. However, we are not there yet. Most analytical techniques assume data to be perfectly reliable from which it follows that unreliable data introduce unpredictable variations in their results.

## 1.7 Conditions for Data to be Reliability Data Proper

Observed agreements are often naively equated with reliability and indiscriminately used to accept or reject the generated data. However, the mere observations of agreements alone are insufficient indicators of reliability. The following spells out six preparatory conditions that need to be satisfied before measures of agreements within reliability data can serve as indicators of the reliability of data for subsequent analytical methods to have a chance to yield valid results.

- Reliability data need to be statistically representative of the data whose reliability is in question. Pretesting a coding instrument before generating the needed data for a research project is common. However, representativeness must be guaranteed for all data and throughout an extended period during which data are generated. Representativeness applies to all relevant properties of the data whose reliability is in question.

- To be replicable elsewhere the process of generating data must be informed by unambiguously communicable (published, understandable, and hence reproducible) and independently executable coding instructions, individual training protocols, separate tests of coder competencies, and different measuring devices. Unlike other methods of generating data, such as by interviews and psychological tests which focus on the nature of human subjects, in assessing the reliability of data the social backgrounds and psychological makeups of individual coders including the interests of researchers should not enter the data.
- Reliability data in which measured agreement is to be interpretable as the reliability of data need to be generated by independently working observers, coders, or judges applying the same coding instructions to the same set of phenomena. When observers share their understandings of the given instructions or advise each other on how to describe or code a given phenomenon, then the observed agreement is polluted, generally inflated by unaccountable social processes.
- Assurances that data are replicable elsewhere and under diverse conditions imply not only that each replication is obtained independent of each other and informed by shared coding instructions and applied to the same phenomena. The kind of observers hired to code the given phenomena must be widely available for subsequent replications including by other researchers, critics or users of research findings. Hiring research assistants, well-meaning buddies, or a unique expert on the subject does not ensure that their observers with the same qualifications are found elsewhere. The same applies to training coders to become reliable experts without the ability to communicate what happened during that process. Such practices invalidate the relationship between measured agreements and the ability to trust the data to serve as proxies for the phenomena of interest.
- Agreement measures need to be sensitive to all possible disagreements among relevant features of the phenomena to be analyzed. For example, if units of analysis are not predefined and coders are asked to identify the locations, lengths, and kind of relevant textual continua, agreement measures that omit disagreements on the unitization of coded continua cannot possibly represent what they omit. It is not sufficient to report agreement measures that are easily measurable when other features of data are being analyzed.
- All data need to provide sufficient amounts of information about the phenomena of analytical interest in order for their analysts to select among the answers to the guiding research questions. This requirement relates to both the diversity of chosen phenomena and the coding options provided by given instructions. Data that do not possess adequate amounts of variance cannot be processed by relevant analytical techniques, and if that variance is caused by conditions other than the diversity of the phenomena of analytical interest, it most likely leads to questionable conclusions about the phenomena thought to be attended to. For data to provide information about the phenomena of declared interest, given that imperfect data may contain noise as well, data must have sufficient variance. The agreement measure chosen to indicate the reliability of data must be sensitive to the lack of information in the generated data.

These six preconditions need to be satisfied for the reliability data in Table 1.2 to give rise to agreement measures from which the reliability of data may be inferred. For example, if observers can talk with each other about how they approach their coding task, share their experiences while doing their work, or consult each other on what the written coding instructions mean, subsequent measures of inter-coder agreement most likely are improving. However, this improved agreement is deceptive as it is a byproduct of the particular social conventions that invariably emerged among observers. Noncommunicable consensus of how a coding task was interpreted makes it unlikely that it can be reproduced elsewhere and be correctly interpreted by researchers who intend to critically examine published findings or build on them.

It is important to realize that the canonical form of reliability data in Table 1.2 and the measures of agreement defined on these data cannot reveal whether these preconditions are satisfied. They have to be separately ensured. Drawing on earlier work (Krippendorff, 2011) and Chapter 0, I suggested the following conceptual formula and its visualization in the form of Figure 1.4:

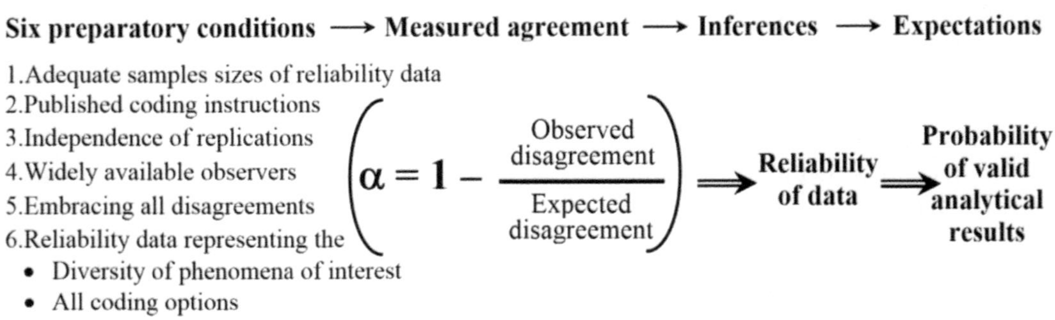


**FIGURE 1.4**
Preparatory Conditions for Agreement Coefficients to Indicate the Reliability of Data

## 1.8 Relations Between Reliability and Validity

Empirical research touches the world outside it at two points: where data are generated and where the conclusions extracted from them enter society in the form of publications and as consequences of acting on them. The quality of interest at the first point of contact is the reliability of the generated data. Reliability is an inference from measured agreements to what observers saw, read, described and recorded. Inasmuch as reliability measures the proportion of information in data about the phenomena of analytical interest, it safeguards against the use of untrustworthy noise-polluted and biased data.

The quality of the second point of contact with the world is the validity of research findings. Their validity is established by corresponding with independently available factual evidence. Validity may be of two kinds; ecological validity and what is called external validity. Evidence for ecological validity may be found in the consensual coordination among the stakeholders regarding the investigated phenomena. External validity may be direct or indirect. It is direct when research results entail predictions that are eventually corroborated. It is indirect when research results are demonstrably correlated with variables whose truth is already established or when research result are enacted and inform practices that succeed (become real) to a degree better than chance. Because data rarely answer yes or no research questions, external validity most likely takes the form of probable truths of propositions in the context of conceivable alternatives.

Reliability and validity are related. When evidence for the answers to research questions predates the conclusion of a research project, their validity is ensured, and the effort to generate data and analyze them is redundant except when alternative analytical methods are tested for reaching known results.

It has already been mentioned that direct validation can be measured in terms of accuracy, as defined in Table 1.1 and visualized in Figures 1.2 and 1.3, the validating evidence taking the place of the standard. This accuracy measure is of limited use, however, because it concerns only the reliability of data and not the potential distortions that analytical techniques could introduce.

In the reality of most empirical research, external evidence in support of validity claims is rarely immediately available. Predictions take time to prove themselves true or false. Generalizations may be validated by costly increases in sample sizes. Academic research results tend to be on a level of abstraction for which direct validating observations may not exist. Indirect validation of such abstractions may be found in subsequent correlations with other research or in whether they inform successful decisions.

Before validating evidence can be found or generated, the only hope of valid research findings rests on the assurances that data in which they were grounded are reliable surrogates of the phenomena in question and that their analysis did preserve the information relevant to the conclusions drawn from them. When validating evidence is missing, even if only temporarily, reliability remains the only assurance of the degree to which research was grounded in the real world outside.

This is why the assessment of data reliability is of epistemological importance. However, the relationship between the measured reliability and the potential that an analysis of reliable data is valid is not entirely straightforward. I offer three propositions on how reliability relates to validity. If the analytical techniques used to transform data are appropriate and without doubt, the first two propositions may be illustrated by reference to Figure 1.2.

*Unreliability in the data lowers the probability of leading to valid results.*

In a court of law, disagreements among eyewitness accounts make it difficult for a judge or jury to infer what actually happened, even whether witnesses claim to describe the same event. In criminology, it is well known that valid accounts need to be sufficiently specific and agree well above what is common and expected. When several coders of a given text offer diverging accounts of what it means, the analyst may have no basis for choosing one over the other. In content analysis, the probability of valid conclusions cannot exceed the human coders' unreliable accounts of what they observed or read.

In Figure 1.2, this proposition is illustrated in the first row of images. If the targets visualize the valid accounts to be achieved, when moving from A via B to C the loss of reliability amounts to an increase in the dispersion of the eight dots. The increasing noise does not preclude that one of the replications is perfectly valid; however, as the figure illustrates, this probability declines as reliability vanishes. The possibility that there is a valid account among the replications is of no benefit to researchers as there is no way of knowing which one is on target.

*Reliable data cannot guarantee valid results.*

This proposition applies to two situations. First, observers who share the same prejudices, ideological commitments, or interest in the research results may well agree perfectly on what they perceive and record but still be wrong by any available standard. Shared biases necessarily escape agreements measured among replications. In Figure 1.2, A and D are perfectly reliable, A is on target, but D differs systematically.

Second, researchers may make analytical mistakes even when processing reliable data. Also, given the nature of the phenomena that given data are representing, not all analytical methods are equally appropriate and may introduce biases. Whether the conclusions that researchers have drawn from given data are justifiable may be established by matching the properties of the phenomena encoded in given data with the logic of the analytical techniques applied to them, but reliability has little to say about how data are subsequently analyzed, appropriate to the nature of the phenomena attended to or not.

Both situations relate the reliability of data to probabilities of achieving valid results. It often happens that the data language that available methods of analysis require is at odds with the nature of the phenomena to be investigated. This condition leads to a third proposition of the relationship between reliability and validity. At least in the social sciences:

*In pursuit of high reliability, validity tends to suffer.*

This proposition is not demonstrable with Figure 1.2. It was anticipated in Sections 1.2.4 and 1.4. It applies to researchers' coping strategy when encountering difficulties of achieving the required levels of reportable reliabilities. The most common origin of such difficulties resides in the conflict between the need to publish high data reliabilities, the difficulties experienced in achieving acceptable levels, and the limited analytical techniques available to represent the complexities of the phenomena of scholarly interest. To cope with these frequently encountered conditions, textbooks and common sense suggest simplifying the coding instructions by replacing the use of elaborate judgments, the usual cause of unreliability, with more easily and ideally objectively identifiable qualities. The direction that such simplifications take tends to favor easily executable reliability tests and well-known analytical techniques, both of which reduce the complexity of the phenomena of interest below the initial scholarly interest.

For an obvious example, texts consist of sequences of characters. However, what authors try to say may well be spread over larger textual units of unequal importance, using parallel accounts of phenomena, embedding side stories in larger narratives, and providing contexts before moving back and forth in time. Tabulating words is highly reliable. But lists of them omit the particularities of their syntactical and semantic relationships, their contexts of use, and cannot represent the poetry that inspires reading and creates the emotions that compel actions. To make this point, I created a word cloud of the first chapter my content analysis book (Krippendorff, 2019: 198). After reading this chapter, one may well recognize why some words are more prominent in the word cloud than others. This makes sense. However, before reading this chapter one cannot possibly imagine anything about the history of content analysis that this chapter articulates. It is empirically demonstrable that the human ability to interpret written matter becomes progressively unreliable as the size and complexity of the textual units increase. Consequently, the demand for reliable data encourages researchers to limit their research questions and conception of the phenomena they wish to study to easily describable and typically small and less meaningful units of analysis at the expense of the qualities that larger and perhaps more appropriate units could capture.

Another example comes from a content analysis by Turow, Cappella, and Jamieson (1996) of a political call-in radio talk show. Such radio programs consist of interactions between their hosts and listeners. Hosts tend to raise popular controversies and select among callers who volunteer their opinion, and add to or oppose the host's accounts. Conversation analysts might distinguish turns of talk of unequal lengths and analyze who controls the interactions and how. Turow et al. realized that the units of talk in such shows are fluid, overlapping, and difficult to code. In the absence of reliability coefficients that could respond to units of variable lengths, which will be introduced in a later chapter, they settled on unitizing the shows into 30-second segments, asked coders to describe them individually, calculated the reliability of their categories, and reported on the frequencies found. 30-second segments are perfectly reliably distinguished by a mechanical clock. However, clocks are not responsive to what happens in such interactions. They are oblivious to how hosts, callers, listeners, even conversation analysts might make sense of what happened. They may accidentally capture one or more turns of talk or represent part of a longer argument. Analyzing the frequencies of such coded segments imposes units on these conversations that are alien to how listeners might understand what they hear. This is not to criticize Capella et al. for following the recommended way of coping with phenomena that are difficult to code, count, or measure. It is merely to observe that their research effort was shaped less by the nature of the phenomena studied than by the availability of reliability coefficients and analytical techniques. Inasmuch as the socio-political implications of this talk show motivated its study but could also be the primary source of validating evidence for its findings, the overemphasis on solving problems of measurement reduced the scope of the contributions it could make.

This last proposition and the two examples lead to two suggestions. One is to shy away from unduly reducing the definition of the phenomena of interest just to comply with available analytical techniques. This might be accomplished by developing more complex data languages that would allow not only measuring inter-observer agreements to establish the replicability of the process of generating data but also preserving as much as possible the complexities of the phenomena studied. The other suggestion is to improve the conception of reliability of interpretive scholarship which could enhance the inter-subjective trust and social significance of its findings.

# 2

# *Simplest Cases: The Replicability of Categorizing Predefined Units*

Recall from the preceding chapter: Reliable data must be about phenomena of analytical interests, computable by appropriately chosen analytical techniques, and bring sufficient amounts of information to bear on the possibility of answering the guiding research questions about these phenomena. The preceding chapter also distinguished between reliability data which serve as a path to estimating the reliability of the data used for drawing conclusions about phenomena in the world of interest to scholars. Section 1.4 also suggested that the most widely used form of reliability is replicability. Its reliability data consist of replications of the process by which distinct phenomena are turned into analyzable data.

This chapter presents the simplest forms of the $\alpha$-coefficient. The reference to "coefficient" pertains to a measure that is capped to be one on one extreme and has a meaningful zero point. The arguably simplest $\alpha$-coefficients apply to categorizations of predefined units of analysis. Categories are single-valued and unordered. Their use constitutes nominal data, which are unlike numerical measurements. Another source of simplicity is their reliance on the minimum of only two replications by independent observers, coders, raters, or judges. Finally, dealing with a small number of independent phenomena that have distinct identities, naturally or by definition. This chapter starts with two observers but ends by extending it to any number, including so-called missing data where some replications do not contain categories for all phenomena.

I am starting with such simple data not only to provide novices, undergraduates and researchers who have never considered analyzing the reliability of their data an easy introduction, hoping they would not be overwhelmed by subsequent chapters which will add to this simple case, one complication at a time. I have three additional reasons for limiting this chapter to the simplest kinds of $\alpha$: First, measures of agreement for unordered data enable considerable algebraic simplifications. This offers alternative ways of conceptualizing $\alpha$. Second, in the absence of other complications, several quite general properties of alpha become more readily apparent, which are the topic of the following chapter. And third, this simple version of $\alpha$ is easily comparable with the most common agreement coefficients, undertaken in Chapter 4, which are roughly on the same simple level as what this chapter seeks to formalize.

Readers who know the basics of testing the reliability of generating data may skip it and go to chapters that could inform their projects. While this book pursues paths of extending $\alpha$ in numerous directions, chapters may be read separately. Where they utilized previously introduced concepts, they are referred to, here starting with Chapter 1.

## 2.1 Reliability Data Generated by Two Observers

The simplest reliability data are generated by two observers $i$ and $j$ who assign one out of $v$ possible categories $c$ to each of $N_u$ units of analysis. The canonical or most basic form of these reliability data is depicted in Table 2.1:

| | | | | | | | |
|---|---|---|---|---|---|---|---|
| **Units:** | 1 | 2 | ... | $u$ | ... | $N_u$ |
| **Observer $i$** | $c_{i1}$ | $c_{i2}$ | ... | $c_{iu}$ | ... | $c_{iN_u}$ |
| **Observer $j$** | $c_{j1}$ | $c_{j2}$ | ... | $c_{ju}$ | ... | $c_{jN_u}$ |

**TABLE 2.1**
The Canonical Form of the Simplest Reliability Data

Table 2.1 represents reliability data in their canonical or most basic form, here as a 2-by-$N_u$ matrix comprised of:

- Predefined units of phenomena are labeled 1, 2, ..., $u$, ..., $N_u$, to be turned into coded data
- $m = 2$ replications by observers, coders, measuring devices, or tests, labeled $i$ and $j$
- Its entries are single categories $c_{iu}$ that observer $i$ assigned to unit $u$, and $c_{ju}$ by analogy
- It represents a total of $n.. = mN_u = 2N_u$ pairable categories

### 2.1.2 Observed Disagreements

One can define the reliability coefficient ${}_c\alpha$ in terms of the above canonical form of the reliability data. However, there is a conceptually more instructive form of the reliability data in Table 2.1 that omits all features of their canonical form that are irrelevant to measures of the reliability of data, namely the identities of the $m = 2$ observers and of the $N_u$ units of analysis they code. By summing all pairs of the two observers' categories over all $N_u$ units in the canonical form of reliability data in Table 2.1, we obtain what are called the observed coincidences $n_{ck}$:

$$n_{ck} = \sum_{u=1}^{N_u} 1 \text{ for each } \begin{cases} < c_{iu}, k_{ju} > \\ < c_{ju}, k_{iu} > \end{cases} \tag{2.1}$$

In words, $n_{ck}$ is the sum of all observed $c$-$k$ pairs of categories in $N_u$ units $u$ of Table 2.1, regardless of who contributed them. According to (2.1) any pair of categories $c_{iu}$–$k_{ju}$ or $c_{ju}$–$k_{iu}$ in unit $u$ are counted. With $N_u$ units $u$, there are $n.. = 2N_u$ pairs of categories in the $m = 2$-by-$N_u$ reliability data matrix as in Table 2.1.

With $v$ categories available in the given coding instructions, there are $v^2$ possible pairs of categories. We tabulate the frequencies $n_{ck}$, according to (2.1), in the $v$-by-$v$ matrix of observed coincidences of Table 2.2:

| **Values:** | 1 | . | . | $k$ | . | $v$ | |
|---|---|---|---|---|---|---|---|
| 1 | $n_{11}$ | . | . | $n_{1k}$ | . | $n_{1v}$ | $n_1$ |
| . | . | . | . | . | . | . | . |
| $c$ | $n_{c1}$ | . | . | $n_{ck}$ | . | $n_{cv}$ | $n_c = \sum_{k=1}^{v} n_{ck}$ |
| . | . | . | . | . | . | . | . |
| . | . | . | . | . | . | . | . |
| $v$ | $n_{v1}$ | . | . | $n_{vk}$ | . | $n_{vv}$ | $n_v$ |
| | $n_1$ | . | . | $n_k$ | . | $n_v$ | $n_{..} = \sum_{c=1}^{v} \sum_{k=1}^{v} n_{ck}$ |

**TABLE 2.2**
Matrix of Observed Coincidences

Note that coincidence matrices:

- Have $v^2$ $c$-$k$ cells containing the frequencies of coincidences $n_{ck}$
- Coincidences are symmetrically distributed around their diagonal: $n_{ck} = n_{kc}$
- The distributions of their marginal row and column sums $n_c$ are equal
- Coincidences with matching categories $c = k$ occupy their diagonal cells: $n_{cc}$, $n_{kk}$
- Coincidences with mismatching categories $c \neq k$ occupy their off-diagonal cells: $n_{c \neq k}$
- Coincidences are fundamental to the most common reliability: replicability (not accuracy, surrogacy, or decisiveness)

The $v$-by-$v$ matrix of observed coincidences in Table 2.2 omits from the canonical form of reliability data in Table 2, all references that are irrelevant for visualizing and measuring disagreements among these data. (2.1), the frequency of coincidences $n_{ck}$, the result of summing all references to observers' identities, $i$ and $j$, and all units $u$ which are individually and independently judged and categorized. Coincidence matrices are convenient representations of the reliability data for calculating the replicability of the process of generating data. All mismatching pairs of categories are found in its off-diagonal cells; all matching ones are found in the diagonal cells. Because categories are unordered or nominal data, observed disagreements are the simple sums of their off-diagonal cells entries. Because the frequencies $n_{ck}$ in coincidence matrices are symmetrically distributed around their diagonal, their row and column marginal sums, needed to obtain the expected disagreements, are equal.

I should mention that many researchers are easily misguided when conceptualizing reliability as inter-observer, inter-coder, inter-rater agreement. The logical consequence of this misconception is to conceptualize agreement and disagreements in the relationship among observers. This misconception leads researchers to represent reliability data in the more common contingency matrices, familiar when calculating associations, $\chi^2$, and the likes. Contingency matrices tabulate $N_u$ units within the two dimensions $i$ and $j$, whereas coincidence matrices tabulate all pairs of categories that can be formed within units. Their difference seems trivial, but it has profound consequences when defining agreement measures in either terms. Agreement measures applied to contingencies ignore the two observers' unequal proclivities for categories, whereas agreement measures applied to coincidences register observers' unequal coding habits as disagreements. For example, the oddities of Cohen's (1960) frequently used Kappa-coefficient for two coders is the result of this misconception. It will be discussed in Chapter 4.

As stated above, coincidence matrices tabulate the frequencies of all pairable categories or values within units. Observers and units are summed, no longer seen in coincidence matrix representations of the reliability date, because their identities play no role in assessing the reliability of data. Observed disagreements $D_o$ sum the coincidences $n_{c\neq k}$ in off-diagonal cells of coincidence matrices, whereas Observed agreements $A_o$ sum the matching coincidences $n_{c=k}$ in diagonal cells. This is entirely general, including for ordered values. But for unordered categories or nominal data, and only for these, the sums of $D_o$, $A_o$, and the total sum $n..$ in Table 2.2 are obviously related. This leads to the following definition:

$$D_o = \sum_{c=1}^{v} \sum_{k\neq c}^{v} n_{ck} = n.. - \sum_{c=1}^{v} n_{cc} = n.. - A_o \qquad (2.2)$$

### 2.1.3 Expected Disagreements

Expected coincidences are hypothetical coincidences that would be observed if the data generated by observers were statistically unrelated to the phenomena they are intended to be about. We do not presume that observers could actually behave randomly. The hypothesized expectations merely serve to define the zero point of α's reliability scale at which point data are totally independent of the phenomena of analytical interest, making it impossible that their analysis could yield anything about the phenomena that researchers hoped to understand. Whereas the observed coincidences in (2.1) are the sums of all pairs of observers' categorizations of $N_u$ units $u$, regardless of who contributed these categories, the expected coincidences result from pairing all categories that occur anywhere in the reliability data in Table 2.1 except with themselves and assign them to the $v^2$ cells of possible pairs of categories proportionally so as to be comparable with the observed coincidences.

The purpose of considering the hypothetical expected coincidences is to get a sense of the kind of data that would have absolutely no connection to the real world of phenomena.

One conceptualization of data that have no relationship to the phenomena they are meant to represent is to imagine them to be the result of repeatedly drawing pairs of categories from an urn that contain all $n.. = \sum_{c=1}^{v} \sum_{k=1}^{v} n_{ck}$ pairs of categories found in the reliability data – repeatedly and without replacing the first draw. Drawing pairs at random defies causal explanations of their pairing. It cannot have anything to do with a known reality except for the frequencies $n_c$ of categories in the urn. When drawing pairs of categories from that urn, the probability of drawing any category depends on how many categories of each kind are contained in the urn. When drawing one category and then a second, the probability of one category being drawn at the first draw depends on the relative frequency of drawn kind. So, the probability of drawing a $c$ at the first draw is $n_c/n..$ . That first draw reduces number $n_c$ of categories $c$ as well as the total number $n..$ of categories in the urn by 1. After drawing a $c$, the probability of drawing a second $c$ would be $(n_c-1)/(n..-1)$. However, the probability of drawing a mismatching category $k$ on the second draw would be $n_k/(n..-1)$.

A computable conception of the expected coincidences involves averaging the differences among all pairable categories except those with themselves in Table 2.1. For $n..$ categories, there would be $n..(n..-1)$ possible pairs, which, when made numerically comparable with the observed disagreement $D_o$, would require dividing the sum by $(n..-1)$.

Relying on the marginal sums, shortcuts the tabulation of $n..(n..-1)$ possible pairs, satisfying both conceptions. Accordingly, the definition of any $c$-$k$ pair of expected coincidences $e_{ck}$ is:

$$e_{ck} = \frac{n_c}{n_{..}-1}\begin{cases} n_k & \text{iff } c \neq k \\ (n_k - 1) & \text{iff } c = k \end{cases} \tag{2.3}$$

We tabulate these expected coincidences $e_{ck}$ in a $v$-by-$v$ matrix in Table 2.3, analogue to the matrix of observed coincidences in Table 2.3.

| **Values:** | 1 | . | . | $k$ | . | $v$ | |
|---|---|---|---|---|---|---|---|
| 1 | $e_{11}$ | . | . | $e_{1k}$ | . | $e_{1v}$ | $n_1$ |
| . | . | . | . | . | . | . | . |
| $c$ | $e_{c1}$ | . | . | $e_{ck}$ | . | $e_{cv}$ | $n_c = \sum_{k=1}^{v} e_{ck}$ |
| . | . | . | . | . | . | . | . |
| . | . | . | . | . | . | . | . |
| $v$ | $e_{v1}$ | . | . | $e_{vk}$ | . | $e_{vv}$ | $n_v$ |
| | $n_1$ | . | . | $n_k$ | . | $n_v$ | $n_{..} = \sum_{c=1}^{v}\sum_{k=1}^{v} e_{ck} = \sum_{c=1}^{v}\sum_{k=1}^{v} n_{ck}$ |

**TABLE 2.3**
Matrix of Expected Coincidences

Being a coincidence matrix as well, Table 2.3 exhibits the very same properties stated in conjunction with Table 2.2, including having the same marginal sums $n_c$ and $n_k$.

For our simple nominal data, the expression of the expected disagreement $D_e$ follows the definition (2.2) of $D_o$ but applies it to the expected coincidences $e_{ck}$, expressed in two ways, with and without reference to the expected coincidences in (2.3):

$$D_e = \sum_{c=1}^{v}\sum_{k \neq c}^{v} e_{ck} = n_{..} - \sum_{c=1}^{v} e_{cc} = n_{..} - \sum_{c=1}^{v} \frac{n_c(n_c - 1)}{n_{..} - 1} \tag{2.4}$$

### 2.1.4 The Alpha Coefficient Generally and Applied to Nominal Data

The general form of all agreement coefficients, already mentioned in Section 0.1, is:

$$\alpha = 1 - \frac{D_o}{D_e} \tag{2.5}$$

The origin of (2.5) was narrated in Chapter 0 and depicted in Figure 1.4. It spelled out the preparatory conditions, discussed in Section 1.7 which need to be met for $\alpha$ to be interpretable as a valid measure of agreement. The following assumes these conditions to be satisfied.

The algebraic properties of (2.5) reveal the two conditions that enable $\alpha$ to be interpretable as a reliability coefficient. When all observed coincidences match perfectly, the observed disagreement $D_o = 0$ and $\alpha = 1$. When the observed disagreements equal the expected disagreements, data are unrelated to the diversity of phenomena, $D_o = D_e$ and $\alpha = 0$. $\alpha$ may become negative when $D_o > D_e$. Given that $D_e$ equals the disagreement that would be observed in the total absence of any statistical relationship between the diversity of phenomena and of their categorizations, $\alpha$ is negative only when the observed disagreement is systematic. This cannot happen by chance – for example by drawing pairs of values from an urn. Either observers were given unequal coding instructions, or they talk and agreed to disagree, which violates one of the preparatory conditions that replications must be independent of each other, discussed in Section 1.7. In properly generated reliability data, systematic disagreements are not expected to occur. For data to be arguably reliable, $\alpha$-coefficients need to deviate only very little from perfect agreement. $\alpha$'s zero value serves merely as the reference point of the absence of any information in data about the phenomena coded. (2.5) constitutes the following scale between 0 and 1 within which $\alpha$-agreements can be interpreted as reliability measures:

$$-1 \le \left.\begin{array}{l} -\text{Systematic disagreement} \\ \pm\text{The granularity of data} \end{array}\right\} 0 \le \alpha = 1 - \frac{D_o}{D_e} \le 1 \qquad (2.6)$$

Whereas systematic disagreements always reduce $\alpha$-values, the granularity of the data is a measure of imprecision and affects $\alpha$-values bidirectionally. It is a mathematical artifact which is a function of the sample size $n..$ . It is due to the fact that fractions of integers as in (2.3) rarely yield integers and whenever they appear in the denominator of (2.5) an exact $\alpha = 0.00000$ is not too common. This imprecision applies to all $\alpha$-values except $\alpha = 1$. It is extremely small and disappears when the sample sizes become large.

For nominal or categorial data, the $\alpha$-agreement (2.5) can take several equivalent forms:

$$\begin{aligned} \alpha_{\text{nominal}} &= 1 - \frac{D_o}{D_e} = 1 - \frac{\sum_{c=1}^{v}\sum_{k\neq c}^{v} n_{ck}}{\sum_{c=1}^{v}\sum_{k\neq c}^{v} e_{ck}} = 1 - \frac{n.. - \sum_{c=1}^{v} n_{cc}}{n.. - \sum_{c=1}^{v} e_{cc}} \\ &= 1 - \frac{n.. - \sum_{c=1}^{v} n_{cc}}{n.. - \sum_{c=1}^{v} \frac{n_c(n_c - 1)}{n.. - 1}} = \frac{\sum_{c=1}^{v}(n_{cc} - e_{cc})}{\sum_{c=1}^{v}(n_c - e_{cc})} \end{aligned} \qquad (2.7)$$

The first expression of (2.7) is the generic form of (2.5). Its second is a function of the proportion of the sums of the off-diagonal cells of the two coincidences (2.2) and (2.4) in Tables 2.2 and 2.3. The third expression simplifies this proportion by summing only its two diagonals, and the fourth replaces references to the expected coincidences $e_{ck}$ by its definition in (2.4).

For nominal data, and only for these, the $\alpha_{\text{nominal}}$ coefficient in (2.7) can also be expressed in terms of what has become known as "chance-corrected agreement coefficients." They are defined and represented in (2.8) in terms of proportions of observed matches $P_o$, expected matches $P_e$, and maximally possible matches $P_{max}$. (2.8)'s second expression replaces these proportions by the number of observed, expected and maximally possible coincidences, already part of (2.7).

To prevent future confusions it is important to be aware that "chance" in so-called "chance-corrected agreement coefficients" may not be how expectations are defined in $\alpha$, to be discussed in Section 4.2.2. Nevertheless, for nominal data and only for these, $\alpha$ can be expressed as a chance-corrected agreement coefficient even for more than two coders' categorizations.

Recall from (2.2) that $D_o = n.. - A_o = n.. - \sum_{c-1}^{v} n_{cc}$

which implies $D_e = n.. - A_e = n.. - \sum_{c-1}^{v} e_{cc}$,

from which one can easily see that:

$$\alpha_{\text{nominal}} = 1 - \frac{D_o}{D_e} = \frac{D_e - D_o}{D_e} = \frac{A_o - A_e}{n.. - A_e} = \frac{\sum_{c=1}^{v} n_{cc} - \sum_{c-1}^{v} e_{cc}}{n.. - \sum_{c-1}^{v} e_{cc}} \quad (2.8)$$

### 2.1.5 A Numerical Example

To illustrate the computation of $\alpha_{\text{nominal}}$ consider a simple numerical example of two observers coding 10 predefined units of analysis by assigning one of four categories *a*, *b*, *c*, or *d* to each of them. The reliability data in their canonical form are:

| **Units:** | 1 | 2 | 3 | 4 | 5 | 6 | 7 | 8 | 9 | 10 |
|---|---|---|---|---|---|---|---|---|---|---|
| **Observers:** Inge | *b* | *d* | *d* | *a* | *d* | *b* | *b* | *d* | *b* | *a* |
| Joe | *b* | *d* | *a* | *a* | *c* | *b* | *b* | *d* | *c* | *a* |

Applying (2.1) to these data yields the cell entries of a 4-by-4 matrix of observed coincidences as in Table 2.2. Two observers who categorize 10 units jointly use a total of $n.. =$ 20 pairable categories, which appear as the sum of this coincidence matrix:

| **Categories:** | *a* | *b* | *c* | *d* | |
|---|---|---|---|---|---|
| *a* | 4 | | | 1 | 5 |
| *b* | | 6 | 1 | | 7 |
| *c* | | 1 | | 1 | 2 |
| *d* | 1 | | 1 | 4 | 6 |
| | 5 | 7 | 2 | 6 | 20 |

One may easily verify that the two marginal frequencies of this matrix are the same, and each represents the number of categories found among the 20 pairable categories. Moreover, the coincidences in Table 2.2 are symmetrically distributed around the diagonal. This accounts for the third expression in (2.7), summing not the disagreements in the off-diagonal entries but the total number of coincidences minus the diagonal entries of perfect agreements. Evidently, coincidences account for the co-occurrences of these categories in the data, not for observers' predilections. Note that the two observers' use of the four categories is rather different. For example, Joe uses two *c*s, Inge doesn't even use one. Inasmuch as Joe's *c*s do not coincide with any other *c*s, it follows that its $n_{cc}$ cell is empty.

Applying the fourth expression of (2.7), which sums the agreements in the diagonal cells of the 4-by-4 matrix of observed coincidences and expresses α without reference to the expected coincidences yields:

$$\alpha_{\text{nominal}} = 1 - \frac{n.. - \sum_{c=1}^{v} n_{cc}}{n.. - \frac{1}{n.. - 1}\sum_{c=1}^{v} n_c(n_c - 1)}$$

$$= 1 - \frac{20 - (4 + 6 + 4)}{20 - \frac{1}{20 - 1}(5(5-1) + 7(7-1) + 2(2-1) + 6(6-1))} = 0.601$$

The expected coincidences in Table 2.3, computed by (2.3) from the margins of the matrix of observed coincidences in Table 2.2, are tabulated in the following 4-by-4 matrix:

| **Categories:** | $a$ | $b$ | $c$ | $d$ | |
|---|---|---|---|---|---|
| $a$ | 1.05 | 1.84 | 0.53 | 1.58 | 5 |
| $b$ | 1.84 | 2.21 | 0.74 | 2.21 | 7 |
| $c$ | 0.53 | 0.74 | 0.11 | 0.63 | 2 |
| $d$ | 1.58 | 2.21 | 0.63 | 1.58 | 6 |
| | 5 | 7 | 2 | 6 | 20 |

Applying the fourth expression of the chance-corrected agreement coefficient in (2.8) which, given the separately obtained expected coincidences, offers the shortest computational path to $\alpha_{nominal}$:

$$\alpha_{nominal} = \frac{\sum_{c=1}^{v} n_{cc} - \sum_{c=1}^{v} e_{cc}}{n_{..} - \sum_{c=1}^{v} e_{cc}} = \frac{(4+6+4)-(1.05+2.21+0.11+1.58)}{20-(1.05+2.21+0.11+1.58)} = 0.601$$

The last form of (2.7) shows the proportions of the sums of the difference in the diagonal cells of observed and expected coincidences and maximum and expected coincidences:

$$\alpha_{nominal} = \frac{\sum_{c=1}^{v}(n_{cc} - e_{cc})}{\sum_{c=1}^{v}(n_{c} - e_{cc})} = \frac{(4-1.05)+(6-2.21)+(0-0.11)+(4-1.58)}{(5-1.05)+(7-2.21)+(2-0.11)+(6-1.58)} = 0.601$$

There obviously are numerous ways to come to the same results. This is the virtue of nominal data, soon lost starting with Chapter 5.

## 2.2 The Alpha Coefficient for Binary Data

Binary or dichotomous data are special cases of nominal data that feature only two values. $\alpha_{binary}$ affords further simplifications of $\alpha_{nominal}$:

$$\alpha_{binary} = 1 - \frac{D_o}{D_e} = 1 - \frac{\sum_{c}^{k}\sum_{k \neq c}^{c \neq k} n_{ck}}{\sum_{c}^{k}\sum_{k \neq c}^{c \neq k} e_{ck}} = 1 - \frac{n_{c \neq k}}{e_{c \neq k}} = 1 - (n_{..} - 1)\frac{n_{c \neq k}}{n_{c.} n_{.k}} \quad (2.9)$$

For a computational example, we lump the categories $b$, $c$, and $d$ of the previous example into one category. This yields the following coincidence matrices:

Observed coincidences

| **Categories:** | $a$ | $bcd$ | |
|---|---|---|---|
| $a$ | 4 | 1 | 5 |
| $bcd$ | 1 | 14 | 15 |
| | 5 | 15 | 20 |

Expected coincidences

| | $a$ | $bcd$ | |
|---|---|---|---|
| $a$ | 1.05 | 3.95 | 5 |
| $bcd$ | 3.95 | 11.05 | 15 |
| | 5 | 15 | 20 |

Whereupon all three computable expressions in (2.9) yield:

$$\alpha_{\text{binary}} = 1 - \frac{\sum_c^k \sum_{k \neq c}^{c \neq k} n_{ck}}{\sum_c^k \sum_{k \neq c}^{c \neq k} e_{ck}} = 1 - \frac{1+1}{3.95+3.95} = 1 - \frac{n_{c \neq k}}{e_{c \neq k}} = 1 - \frac{1}{3.95}$$

$$= 1 - (n_{..} - 1)\frac{n_{c \neq k}}{n_c n_k} = 1 - (20-1)\frac{1}{5 \cdot 15} = 0.747$$

Note that $\alpha_{\text{binary}} = 0.747 > 0.601 = \alpha_{\text{nominal}}$. This is not a difference between two coefficients but the results of the data in this example. Compare two other partitions into binary data:

| **Categories:** | $a$ | $b$ | $c$ | $d$ | |
|---|---|---|---|---|---|
| $a$ | 4 | | | 1 | 5 |
| $b$ | | 6 | 1 | | 7 |
| $c$ | | 1 | | 1 | 2 |
| $d$ | 1 | | 1 | 4 | 6 |
| | 5 | 7 | 2 | 6 | 20 |

$\alpha_{\text{binary}} = 0.604$

| | $a$ | $b$ | $c$ | $d$ | |
|---|---|---|---|---|---|
| $a$ | 4 | | | 1 | 5 |
| $b$ | | 6 | 1 | | 7 |
| $c$ | | 1 | | 1 | 2 |
| $d$ | 1 | | 1 | 4 | 6 |
| | 5 | 7 | 2 | 6 | 20 |

$\alpha_{\text{binary}} = 0.415$

## 2.3 Reliability Data with Any Number of Replications

Although binary data and two replications are the most reduced form of reliability data possible, Section 1.7 argued why a larger number of replications would be preferable. Two observers may not be representative of the larger population of observers who could be asked to replicate the data-making processes by other researchers, trying to replicate the data-making process or expanding published findings, working under different conditions.

Using just two replications is common for three fallible reasons. First, the most intuitive conception of agreement is between two and expressed in terms of percent or proportions, a topic that will be addressed in the next chapter. Chance-corrected agreement coefficients, exemplified in (2.8), are stated as a function of percent agreements. Percent agreement cannot be generalized to more than two observers. Second, perhaps related to this conception is the belief that a larger number of observers are less likely to agree. This is certainly so when agreement is defined as consensus. However, the two disagreements that go into the α-coefficient in (2.7) are defined in terms of pairs of categories or values which avoid being affected by the number of observers participating in the needed replications. They are also not affected when some observers do not code all units of analysis which will become clear in the following example. Third is the unfortunate fact that most agreement coefficients are defined for two observers, perhaps for the first two reasons, also resulting in available software, all of which severely limits the ability of researchers to report agreements among more than two replications. The α-agreement measure does not have such limitations.

The canonical form of reliability data for any number $m$ of replications, Table 2.4, is a simple extension of Table 2.1:

| **Units:** | 1 2 3 . . | $u$ | . . . . . . . $N_u$ | |
|---|---|---|---|---|
| **Replications:** 1 | $c_{11}$ . . . . | $c_{1u}$ | . . . . . . . $c_{1N_u}$ | |
| : | : | : | : | |
| $i$ | $c_{i1}$ . . . . | $c_{iu}$ | . . . . . . . $c_{iN_u}$ | |
| $j$ | $c_{j1}$ . . . . | $c_{ju}$ | . . . . . . . $c_{jN_u}$ | |
| : | : | : | : | |
| $m$ | $c_{m1}$ . . . . | $c_{mu}$ | . . . . . . . $c_{mN_u}$ | |
| **Sums of pairable values** > 1 | $m_1$ . . . . | $m_u$ | . . . . . . . $m_{N_u}$ | $n.. = \sum_{u=1}^{N_u} m_u$ |

TABLE 2.4
Reliability Data with Any Number of Replications/Observers

Table 2.4 includes Table 2.1 as a special case. Its terms refer to:

- A predefined set of phenomena, labeled 1, 2, …, $u$, …, $N_u$, the units of analysis $u$
- Up to $m$ replications labeled 1, 2, …, $i$, $j$, …, $m$, by observers, measuring devices, or tests
- All units $u$ are single-valued, i.e., are assigned at most one category $c_{iu}$ by observer $i$
- $m_u$ is the number of pairable values assigned to unit $u$. Pairability means $m \geq m_u > 1$
- As replications may not cover all units, the total number of pairable values is $n.. \leq m \cdot N_u$

All but one of the previous definitions of α-agreements apply to these reliability data. The exception being the definition of the observed coincidences $n_{ck}$. In (2.1) it is defined for two observers which is the special case of (2.10) which enumerates $m_u(m_u-1)$ pairs of categories within all of $N_u$ individual units $u$:

$$n_{ck} = \sum_{u=1}^{N_u} \frac{1}{m_u - 1} \sum_{i=1}^{m_u} \sum_{j \neq i}^{m_u} 1 \text{ iff } c_{iu} \text{ and } k_{jh} \tag{2.10}$$

These observed coincidences $n_{ck}$ can be seen as summing all $c$–$k$ pairs of values within units. As there are $m_u(m_u-1)$ possible pairs among $m_u > 1$ pairable values, excluding valued paired with themselves, to give each value the same weight of 1 in their coincidence matrix Table 2.2, the number of possible pairs of values has to be divided by $m_u-1$. So, the total sum of coincidences in Tables 2.2 and 2.3 equals the number of pairable values in reliability data Tables 2.1 and 2.4.

From the observed coincidences $n_{ck}$ in (2.10), tabulated in Table 2.2, the observed disagreement $D_o$ is obtained by (2.2). From the expected coincidences $e_{ck}$ in (2.3), tabulated in Table 2.3, the expected disagreement $D_e$ in (2.4) can be obtained either by summing the off-diagonal cells of Table 2.3, subtracting the sum of the diagonal cells from the total $n_{..}$ of that table, or by computing it from the marginal sums of Tables 2.2 or 2.3 excluding their diagonals.

With $D_o$ and $D_e$ defined above, (2.7) leads to the four alternative ways to conceptualize and calculate the $\alpha_{nominal}$ agreement coefficient for predefined units coded with single nominal values or categories by any number of observers, coders, judges, or measuring devices.

### 2.3.1 An Example

For example, let the reliability data be:

| **Units** $u$: | 1 | 2 | 3 | 4 | 5 | 6 | 7 | 8 | 9 | 10 | 11 | 12 | |
|---|---|---|---|---|---|---|---|---|---|---|---|---|---|
| **Observers**: A | 📖 | ✉ | ☎ | ☎ | ✉ | 📖 | 💻 | 📖 | ✉ | | | | |
| B | 📖 | ✉ | ☎ | ☎ | ✉ | ✉ | 💻 | 📖 | ✉ | 📂 | | | |
| C | | ☎ | ☎ | ☎ | ✉ | ☎ | 💻 | ✉ | ✉ | 📂 | 📖 | ☎ | |
| D | 📖 | ✉ | ☎ | ☎ | ✉ | 💻 | 💻 | 📖 | ✉ | 📂 | 📖 | | |
| **Pairable icons** $m_u > 1$: | 3 | 4 | 4 | 4 | 4 | 4 | 4 | 4 | 4 | 3 | 2 | 0 | $40 = n_{..}$ |

These reliability data represent four observers recording twelve occasions regarding whether the source of a belief was a book, letter, phone call, file, or email. The twelfth occasion is coded only by observer C. This makes the lone value ☎ not pairable with any other category at this occasion, hence $m_{12} = 0$ by definition of pairability. The total number of pairable categories in 12 units is $n_{..} = 40$, less than the 48 cells in this 4-by-12 table of reliability data.

The observed coincidences are obtained by (2.10) and tabulated according to Table 2.2. From its marginal sums the expected coincidences are calculated by (2.3) and tabulated as in Table 2.3.

Both coincidence matrices are shown below, side by side. Inasmuch as coincidence matrices are symmetrical, their off-diagonal cells are numerically identical and sum to the same totals, here 40 each.

**Categories:**

Observed coincidences

| | 📖 | ✉ | ☎ | 💻 | 📂 | |
|---|---|---|---|---|---|---|
| 📖 | 7 | 1.33 | 0.33 | 0.33 | . | 9 |
| ✉ | 1.33 | 10 | 1.33 | 0.33 | . | 13 |
| ☎ | 0.33 | 1.33 | 8 | 0.33 | . | 10 |
| 💻 | 0.33 | 0.33 | 0.33 | 4 | . | 5 |
| 📂 | . | . | . | . | 3 | 3 |
| | 9 | 13 | 10 | 5 | 3 | 40 |

Expected coincidences

| | 📖 | ✉ | ☎ | 💻 | 📂 | |
|---|---|---|---|---|---|---|
| 📖 | 1.85 | 3 | 3.31 | 1.15 | 0.69 | 9 |
| ✉ | 3 | 4 | 3.33 | 1.67 | 1 | 13 |
| ☎ | 2.81 | 3.33 | 2.31 | 1.28 | 0.77 | 10 |
| 💻 | 1.15 | 1.67 | 1.28 | 0.51 | 0.38 | 5 |
| 📂 | 0.69 | 1 | 0.77 | 0.38 | 0.15 | 3 |
| | 9 | 13 | 10 | 5 | 3 | 40 |

In terms of the observed coincidences $n_{ck}$, i.e., without referring to the separately calculated expected coincidences, the third computable form of $\alpha_{nominal}$ in (2.7) yields:

$$\alpha_{\text{nominal}} = 1 - \frac{D_o}{D_e} = 1 - \frac{n_{..} - \sum_{c=1}^{v} n_{cc}}{n_{..} - \frac{1}{n_{..} - 1}\sum_{c=1}^{v} n_c(n_c - 1)}$$

$$= 1 - \frac{40 - (7 + 10 + 8 + 4 + 3)}{40 - \frac{9 \cdot 8 + 13 \cdot 12 + 10 \cdot 9 + 5 \cdot 4 + 3 \cdot 2}{40 - 1}} = 1 - \frac{8}{31.18} = 0.743$$

Alternatively, with references to the independently obtained expected coincidences and as a chance-corrected agreement coefficient defined in (2.8):

$$\alpha_{\text{nominal}} = \frac{\sum_{c=1}^{v} n_{cc} - \sum_{c=1}^{v} e_{cc}}{n_{..} - \sum_{c=1}^{v} e_{cc}}$$

$$= \frac{(7 + 10 + 8 + 4 + 3) - (1.85 + 4 + 2.31 + 0.51 + 0.15)}{40 - (1.85 + 4 + 2.31 + 0.51 + 0.15)} = 0.743$$

We shall resume the development of $\alpha$ in Chapter 5 and devote the following two chapters to several side issues that will surface later as well.

# 3

## *Some Basic Properties of Alpha*

Although this chapter relies on the terminologies developed in Chapter 2 which is limited to assessing the replicability of data with unordered categories, most of the properties of $\alpha$ examined here apply to all $\alpha$-agreement measures, unless noted otherwise. These properties resurface in Chapter 4, which compares $\alpha$ with other agreement coefficients.

### 3.1 Alpha's Range

The previous chapter defined $\alpha$ in (2.5) and already stated $\alpha$'s numerical range in (2.6):

$$-1 \leq \left.\begin{matrix} -\text{Systematic disagreement} \\ \pm\text{The granularity of data} \end{matrix}\right\} 0 \leq \alpha = 1 - \frac{D_o}{D_e} \leq 1$$

When all replications match perfectly, the observed disagreement $D_o = 0$ and $\alpha = 1$. When the observed disagreements equal the expected disagreements, $D_o = D_e$, $\alpha = 0$. Between 0 and 1, the $\alpha$-agreement can be interpreted as a measure of the reliability of data. Negative values, $D_o > D_e$, cannot happen by chance and are due largely to systematic disagreement and in minor ways to the granularity of the data.

Chapter 2 also discussed conditions under which $\alpha$ could be negative. Recall the two conceptions of the expected disagreement $D_e$ as resulting either from drawing pairs of categories from an urn or pairing all categories in the data regardless of any relationships to what they could represent. Systematic disagreement is likely when observers unknowingly apply different coding instructions to the same phenomena or talk among themselves and agree to disagree. The latter is conceivable but improbable. The granularity is an algebraic artifact due to $\alpha$ being defined as a proportion which rarely produces full integers, rarely exactly $\alpha = 0.000$. This imperfection is always small and disappers as reliability data become large.

## 3.2 Alpha's Degrees of Freedom

As already discussed in conjunction with (2.6), as the sample sizes of the reliability data increase, the granularity vanishes. However, it is not just the sample size that matters but also the degree of freedom of particular coincidence matrices.

For 2-by-2 matrices with values 0 and 1 and marginal sums of $n_{0\cdot}$, $n_{1\cdot}$, $n_{\cdot 0}$, and $n_{\cdot 1}$ the smallest possible disagreement is $D_{min} = n_{\cdot\cdot} - min(n_{0\cdot}, n_{\cdot 0}) + min(n_{\cdot 1}, n_{1\cdot})$. For symmetrical coincidence matrices, $D_{min} = 0$. And $D_{max} = min(n_{0\cdot}, n_{\cdot 1}) + min(n_{\cdot 0}, n_{1\cdot})$ in either case. Whereupon their degrees of freedom are:

$$d.f. = \frac{D_{max} - D_{min}}{2} \tag{3.1}$$

This can be illustrated by three binary examples of coincidence matrices whose two marginal sums are the same, $n_{\cdot\cdot} = 20$. When the marginal sums are equal, its degree of freedom is $d.f. = 10$ and the number of possible matrices with these marginal sums equals $d.f. + 1 = 11$, which also equals the number of possible $\alpha$s:

| | | | | | | | | | | | |
|---|---|---|---|---|---|---|---|---|---|---|---|
| | 10 0 | 9 1 | 8 2 | 7 3 | 6 4 | 5 5 | 4 6 | 3 7 | 2 8 | 1 9 | 0 10 |
| | 0 10 | 1 9 | 2 8 | 3 7 | 4 6 | 5 5 | 6 4 | 7 3 | 8 2 | 9 1 | 10 0 |
| $\alpha =$ | 1.000 | 0.810 | 0.620 | 0.430 | 0.240 | 0.050 | –0.140 | –0.330 | –0.520 | –0.710 | –0.900 |

The effect of the above mentioned granularity on $\alpha$ near zero can be observed in this example of a rather small sample size.

The marginal sums of the second example are more constraining. Its degree of freedom is $d.f. = 1$ which implies two possible coincidence matrices and two $\alpha$s. In this case, $\alpha$ has lost its negative values:

| | | |
|---|---|---|
| | 19 0 | 18 1 |
| | 0 1 | 1 0 |
| $\alpha =$ | 1.000 | 0. 000 |

The third example has no degrees of freedom, $d.f. = 0$:

| | |
|---|---|
| 20 | 0 |
| 0 | 0 |

$$\alpha = 1 - \frac{0}{0} = 0 \text{ by definition}$$

For larger $v$-by-$v$ matrices of nominal variables with $v > 2$ categories containing frequencies $n_{ck}$ as in Table 2.2 with marginal sums $n_{c\cdot}$, $n_{\cdot c}$, $n_{k\cdot}$, and $n_{k\cdot}$, $D_{min} = n_{\cdot\cdot} - \sum_{c=1}^{v} min(n_{c\cdot}, n_{\cdot c})$. However, $D_{max}$ has to be obtained by an iterative algorithm detailed in Chapter 14, as (14.14) in particular. With the degrees of freedom unlike for binary data in (3.1):

$$d.f. = D_{max} - D_{min} \tag{3.2}$$

## 3.3 Alpha's Scale

The above list of all eleven possible 2-by-2 tables within their marginal sums of 10 each demonstrates an important characteristic of α. α constitutes an equal interval scale of agreements. For 2-by-2 coincidences its intervals are $2/D_e$. In the above example all intervals are demonstrably $^2/_{10.53} = 0.19$ between each neighboring α-value, all the way from α = 1.000 to α = –0.900. For all reliability data with more than two nominal values, the equal intervals are $1/D_e$. For ordered data the intervals are smaller and not as easily expressed. The point that this example demonstrates is that α does not weight agreement unevenly. It weighs deviations from perfect agreement like those from its near zero values.

## 3.4 Alpha as an Information Measure

Because disagreement is the complement of agreement, subtracting the proportion of two disagreements $D_o/D_e$ from 1 renders α a proportion as well, an agreement coefficient.

More specifically, if the observed disagreement $D_o$ is a measure of the undesirable variance within units of Table 2.1 and the expected disagreement $D_e$ is a measure of the total variance in the reliability data, then the observed disagreement $D_o$ within units amounts to the noise in the data, and the difference $D_e - D_o$ becomes a measure of the amount of information in the data about the phenomena of analytical interest. Therefore, α is to be interpreted as the proportion of information about phenomena contained in the data:

$$\begin{aligned}\alpha &= 1 - \frac{D_o}{D_e} = 1 - \frac{\text{Variance within units}}{\text{Total variance among all units}} \\ &= \frac{D_e - D_o}{D_e} = \frac{\text{Information in data about phenomena}}{\text{Total variance} = \text{information} + \text{noise}}\end{aligned} \tag{3.3}$$

Its complement $D_o/D_e$ is the proportion of noise unrelated to any of the qualities of the phenomena that observers have faced when generating data.

According to (3.3), if $D_o = 0$, the total variance is tied to what the data are intended to be about. If $D_o = D_e$, data are unrelated to the phenomena of interest and consist of random variations or noise. Therefore, α expresses the proportion of information in the data about the phenomena of analytical interest.

More specifically, if the observed disagreement $D_o$ is a measure of the undesirable variance within units of Table 2.1 and the expected disagreement $D_e$ is a measure of the total variance in the reliability data, then the observed disagreement $D_o$ within units amounts to the noise in the data and the difference $D_e–D_o$ becomes a measure of the amount of information in the data about the phenomena of analytical interest. Therefore, α is to be interpreted as the proportion of information about phenomena contained in the data.

## 3.5 Alpha's Independence from the Number of Replications, Units, and Missing Data

The probability of unanimous agreement among independently coding observers becomes increasingly unlikely the more observers are involved. $\alpha$ avoids being affected by this condition; first, by not being a function of percent agreement, as most chance-corrected agreement coefficients are, and second, by giving each observer's contributions equal weight – regardless of how many contribute to the evaluation of units as long as there are two to compare.

Table 2.4 in the previous chapter represents the reliability data in their canonical form of $m$ replications by $N_u$ units and contains the values $c_{iu}$ of the $i^{\text{th}}$ replication or observer assigned to unit $u$.

From these reliability data, (2.10) defines the coincidences that are entered in a matrix of all coincidences found in the reliability data:

$$n_{ck} = \sum_{u=1}^{N_u} \frac{1}{m_u - 1} \sum_{i=1}^{m_u} \sum_{j \neq i}^{m_u} 1 \text{ iff } c_{iu} \text{ and } k_{jh}$$

Wherein $m_u > 1$ is the number of pairable values in unit $u$. In the computational example of Section 2.3.1, there are $m = 4$ observers. Eight contribute $m_u = 4$ values to the reliability data, two of them contribute $m_u = 3$, one $m_u = 4$. Unit 12 is categorized by only one observer, making it not pairable. A unit with no or one unpairable value cannot contribute to whether data are reliable or not and cannot be counted.

Note in (2.10) above that $\sum_{i=1}^{m_u} \sum_{j \neq i}^{m_u} 1$ enumerates $m_u(m_u-1)$ $<c_{iu} - k_{jh}>$ pairs of values in any one unit $u$. To weigh each of the $m_u$ values (of a replication by observers, coders, or judges) contributing to these pair comparisons equally, the total number of pairs is divided by $1/(m_u-1)$.

The tabulation of all <*c*–*k*>-coincidences found in the reliability data depends neither on the total number $m$ of replications, nor on whether some valuations of units are missing as long as there are pairable values in $u$.

In as much as $\alpha$ is a function of the mismatching pairs of values, $\alpha$ is not affected by the $m$ replications, nor by the number $N_u$ of units being observed. Unaffected as long as the reliability data contain $m_u > 1$ pairable values in units.

Moreover $\alpha$ is immune to the proportion of $N_u/m$, of the number of units to the number of replications in the reliability data – except for its granularity.

## 3.6 Alpha's Responsiveness to Small Sample Sizes

Recall that observed coincidences result from pairing all values assigned to individual units, and expected coincidences from pairing all values found in the reliability data regardless of the units to which they are originally assigned – both exclude pairing of values with themselves. In (2.10) and (3.1), observed coincidences are a function of the number $m_u(m_u-1)$ of possible pairs of $m_u$ values in units $u$. The definition (2.3) of the expected coincidences also excludes pairing values with themselves:

$$e_{ck} = \frac{n_c}{n..-1}\begin{cases} n_k & \text{iff } c \neq k \\ (n_k - 1) & \text{iff } c = k \end{cases}$$

I explained the motivation for this definition in three ways; by reference to blindly drawing pairs of values from an urn containing all, without replacing the first draw; by pairing all values in the reliability data with each other except with themselves; and by computing the latter as a function of their frequencies. When randomly drawing two values from an urn, in order not to draw the same one the second time, one cannot put the first draw back into the urn. Any draw of a value $c$ from the urn reduces the number of values remaining in the urn from $n_c$ to $n_c-1$ and from $n..$ to $n..-1$. If the first draw was $c$, a second $c$ would have to be selected among the remaining $n_c-1$ values in the urn. However, if a value other than $c$ is selected from that urn, the number of such values available for selection is not affected by the selection of the first. This is where the $(n..-1)$ and $(n_k-1)$ in (2.3) above come from.

Expressing expectations by pairing all values including with themselves has three undesirable consequences. First, inasmuch as observed coincidences do not pair values with themselves, doing this in estimating the expected disagreements would introduce an internal inconsistency into the agreement measure. Second, this inconsistency has the effect of underestimating expected disagreements and thereby reducing the agreement measures that take this route. And third, it mistakenly assumes reliability data to be of infinite size. This assumption is built into two close relatives of $\alpha_{\text{nominal}}$: Scott's (1955) $\pi$ (pi) and Fleiss' (1971) $K$ (mistakenly equated with kappa).

While the numerical effect of this inconsistency disappears when the sample sizes become large, when sample sizes are small the inconsistency in $\pi$ and $K$ has the effect of being insensitive to small sample sizes:

| | | | | | |
|---|---|---|---|---|---|
| | 6 1<br>1 2 | 30 5<br>5 10 | 60 10<br>10 20 | 300 50<br>50 100 | 600 100<br>100 200 |
| $\alpha =$ | 0.571 | 0.533 | 0.529 | 0.525 | 0.524 |
| $\pi = K =$ | 0.524 | 0.524 | 0.524 | 0.524 | 0.524 |

Evidently, $\pi$ and $K$'s inconsistencies have the effect of deflating the measured agreement below that of $\alpha$. Many researchers, who compare agreement measures with data that are equal in size, have noted that $\alpha_{\text{nominal}}$, $\pi$, and $\kappa$, even including $K$ are numerically similar. However, even small numerical similarities do not mean that they measure the same. Researchers who measure their reliability data by different measures with similar results may be safe, but this would not suggest that their similarities can be generalized. Similarities do not reveal what their small differences are due to, what qualities of data they omit or illegitimately include. Small numerical differences may become large when the data are pushed to their extremes. This is what the above demonstration attempted: An agreement measure that is defined for infinite sample sizes cannot reveal the inconsistencies that become more prevalent when samples sizes are small.

A more drastic example is Cohen's $\kappa$. It produces the same value as $\pi$ and $K$ if and only if the contingency matrix to which it is applied is symmetrical and its row and column sums are uniformly distributed. As soon as they differ in either of these two ways, $\kappa$ deviates from $\pi$ and $K$. The culprit of $\kappa$ taking a different route lies in Cohen's conception of "chance," already alluded to in the previous chapter, to be further elaborated in Chapter 4. Here, it suffices to point out that definitions of chance introduce other undesirable properties than those revealed between π, *K*, and α which are relatively harmless, merely responding differently regarding the sample sizes.

The point is that finding differences by simulating the performance of several agreements rarely reveals what they are due to, nor do similarities suggest that they respond to the same qualities of data. π, κ, and *K* assume reliability data to be of infinite sample sizes; α is sensitive to the size of the reliability data, converging to π and *K*, when samples become large but not κ.

## 3.7 Alpha's Reliance on Estimating the Distribution of Values in the Population of Data Whose Reliability Is in Question

Consider the computational example in section 2.3.1. The reliability data consist of 12 units of analysis yielding 40 pairable values. Nothing is known about these units except a researcher's definition of them; we have no clue how well these 12 units represent the data whose reliability is the issue. We can decode the reliability data by applying the reversed coding instructions to them. However, this does not get us access to what the four observers saw and interpreted, only what the coding instructions said they should have looked at. The epistemological fact is we never know the phenomena the observers coded, only what the coding instructions told them to identify.

From the measured agreement we want to infer the reliability of the data. However, whether we are assured that the reliability data are representative of the population of data whose reliability is in question or whether we have replicated their coding for all, all we have is what the observers coded. There is no better estimate as to what the data are than what the collectivity of all coders say they have seen and made into the data we have in front of us.

For these reasons α takes the distribution of these values as the best and only estimate one has of the distribution of the values in the population of data of ultimate concerns. Section 2.3.1 reproduces the matrix of observed and expected coincidences side by side. Their marginal sums are identical and their cell entries, a function of these sums, enter the calculated α = 0.743. While π and *K* are a function of these population estimates as well in the previous section I addressed why they differed from α. More important differences, which will be discussed in the next chapter, are agreement coefficients such as Bennet et al.'s (1954) *S* which makes no reference to such estimates and cannot serve as indicators of the reliability of data; and Cohen's κ which relies on the frequencies of two observers and cannot say anything beyond them.

## 3.8 Relations between Contingencies and Coincidences

Chapter 2 defined α in terms of coincidence matrices. Coincidence matrices are unlike contingency matrices familiar in statistical representations of correlations or associations between a pair of variables whose statistical relationship is in question. Their cell entries are asymmetrical, their marginal sums are specific to the two variables, and their total sums equal the number of units categorized or valued by two methods.

Coincidence matrices, by contrast, tabulate pairs of values regardless of who contributed them, and which units were coded. Their cell entries are symmetrical with respect to their diagonal, $n_{ck} = n_{kc}$, which contains perfectly matching pairs $n_{cc}$ and $n_{kk}$. The distribution of their row and column sums is equal and sums to $n..$. This is so for observed coincidences as in Table 2.2 and for expected coincidences as in Table 2.3.

Chapter 1 distinguished five reliabilities. Stability and replicability are obtained from coincidence matrices; accuracy, surrogacy, and decisiveness are obtained from contingency matrices. Both result from different pair comparisons. Coincidences and contingencies are often confused, which deserves some clarification.

There is a simple relationship between coincidences and contingencies. In (2.10), coincidences $n_{ck}$ enumerate all $c$–$k$ pairs of values found within units. Coincidences eliminate references to observers. By contrast contingencies $x_{c_ik_j}$ are defined for pairs of observers $i$ and $j$. Coincidences sum to the total number $n..$ of pairable values. Contingencies sum to the total number $x._i._j$ of units valued by the two observers $i$ and $j$.

$$n_{ck} = \frac{1}{m-1}\sum\nolimits_i^m \sum\nolimits_{j \neq i}^m (x_{c_ik_j} + x_{c_jk_i}) \tag{3.4}$$

In words coincidences are the sum of all pairs of observers' contingencies $x_{c_ik_j}$ plus their transpose $x_{k_ic_j}$ divided by $(m-1)$ to give each observer the same weight.

From the computational example in Section 2.1.5, the observed coincidences among just two observers $(x_{c_ik_j} + x_{c_jk_i})$ is demonstrated by their 4-by-4 contingency matrix between Inge and Joe, plus their transpose:

Inge-Joe Contingencies

| **Categories:** | Joe: *a* | *b* | *c* | *d* | |
|---|---|---|---|---|---|
| Inge: *a* | 2 | | | | 2 |
| *b* | | 3 | 1 | | 4 |
| *c* | | | | | 0 |
| *d* | 1 | | 1 | 2 | 4 |
| | 3 | 3 | 2 | 2 | 10 |

+

Joe-Inge Contingencies

| | Inge: *a* | *b* | *c* | *d* | |
|---|---|---|---|---|---|
| Joe: *a* | 2 | | | 1 | 3 |
| *b* | | 3 | | | 3 |
| *c* | | 1 | | 1 | 2 |
| *d* | | | | 2 | 2 |
| | 2 | 4 | 0 | 4 | 10 |

=

Coincidences

| | *a* | *b* | *c* | *d* | |
|---|---|---|---|---|---|
| *a* | 4 | | | 1 | 5 |
| *b* | | 6 | 1 | | 7 |
| *c* | | 1 | | 1 | 2 |
| *d* | 1 | | 1 | 4 | 6 |
| | 5 | 7 | 2 | 6 | 20 |

Note that (3.4) does not apply to the expected contingencies. The latter need to be computed from the marginal sums of values in the coincidence or contingency matrices. The distinction between coincidences and contingencies is important not only because they will play different roles in obtaining the reliabilities mentioned in Chapter 1; confusing them has severe consequences. For example, calculating the expected disagreements from contingency matrices excludes unequal marginal distributions as disagreements whereas calculating them from coincidences includes them. There are good reasons for both.

## 3.9 Predicting Observed Coincidences From Alpha

One can interpret the $\alpha_{nominal}$ coefficient in (2.7) as expressing the extent to which the observed coincidences fail to occupy the diagonal cells of coincidence matrices as opposed to being distributed randomly throughout all cells of these matrices. It suggests decomposing the observed coincidences into two parts, one that is a function of the expected coincidences and one due to the coincidences of perfect agreement. There is a caveat. $\alpha$ averages agreements and disagreements. It follows that $\alpha$ cannot predict particular coincidences, but coincidences that sum to the observed. For 2-by-2 coincidence matrices the observed coincidences are predicted exactly; and this relationship hold for all $\alpha$s regardless of their metric:

$$\begin{array}{|cc|}\hline n_{cc} & n_{ck} \\ n_{kc} & n_{kk} \\ \hline\end{array} = (1-\alpha)\begin{array}{|cc|}\hline e_{cc} & e_{ck} \\ e_{kc} & e_{kk} \\ \hline\end{array} + \alpha \begin{array}{|cc|}\hline n_c & \\ & n_k \\ \hline\end{array}$$

However, for larger coincidence matrices, (3.5) states these predictions but limited to $\alpha_{\text{nominal}}$:

$$\bar{n}_{ck} = (1-\alpha_{\text{nominal}})\, e_{ck} + \begin{cases} \alpha_{\text{nominal}}\; n_c & \text{iff } c = k \\ 0 & \text{iff } c \neq k \end{cases} \tag{3.5}$$

Wherein $\bar{n}_{ck}$ satisfies:

$$\alpha_{\text{nominal}} = 1 - \frac{D_o}{D_e} = 1 - \frac{\sum_{c=1}^{v}\sum_{k\neq c}^{v} n_{ck}}{D_e} = 1 - \frac{\sum_{c=1}^{v}\sum_{k\neq c}^{v} \bar{n}_{ck}}{D_e}$$

In the numerical example in Section 2.1.5 which illustrated the computation of $\alpha_{\text{binary}} = 0.747$, (3.3) predicts $n_{ck}$ without exception:

$$\begin{array}{|cc|}\hline 4 & 1 \\ 1 & 14 \\ \hline\end{array} = (1-0.747)\begin{array}{|cc|}\hline 1.05 & 3.95 \\ 3.95 & 11.05 \\ \hline\end{array} + 0.747 \begin{array}{|cc|}\hline 5 & \\ & 15 \\ \hline\end{array}$$

Accordingly:

| | |
|---|---|
| The observed coincidence | $n_{cc} = (1-0.747)\ 1.05 + 0.747 \cdot 5 = 4$ |
| The observed coincidence | $n_{kk} = (1-0.747)\ 11.05 + 0.747 \cdot 15 = 14$ |
| And the observed coincidences | $n_{ck} = n_{kc} = (1-0.747)\ 3.95 = 1$ |

## 3.10 The Necessity of Information in the Variance of Reliability Data

For scholarly inquiry to rely on empirical data, data must provide enough information about the phenomena it explores, analyzes, theorizes, or is otherwise concerned with. Any acceptable reliability coefficient should and $\alpha$ does respond to the presence (or absence) of information that data represent about the phenomena studied.

Recall from Section 1.1.2 that all data need to be computable, from Section 1.1.3 that reliable data need to represent relevant differences among the phenomena of analytical interest, and from Section 1.1.4 that information about these phenomena need to be able to ultimately inform the research questions that guide an inquiry into the phenomena selected for attention. Data provide the channel of communication between the targeted phenomena and the answers to research questions they need to answer, and when there is an intention to act on analytical findings, to adequately address the phenomena studied – recall the arrow from analytical results to what they claim to be about in Figure 1.1. Section 1.5 gave reasons for why reliable data cannot guarantee valid results, but that unreliable data may reduce the probability of reaching valid conclusions even with perfectly appropriate analytical techniques. Just as the word "data" is the plural of "datum," all computational techniques, even qualitative research methods, presuppose at least the ability to draw distinctions within data and compare what is distinguished. This possibility presupposes that data are as diverse as the phenomena to be studied and needed to select among the possible answers to the given research questions.

It follows that the essence of informative data resides in their diversity, variety, or variance. Without variation, qualitative comparisons, triangulations, and quantitative techniques such as correlations, factor analyses, and other statistical methods are not available. After reviewing several approaches to measuring reliability in Section 1.2, Section 1.4 suggested that measures of agreement among replications can serve as indicators of the reliability of data if and only when replications satisfy several preparatory conditions, depicted in Figure 1.4. Most of these preconditions ensure that there is enough diversity in the phenomena coded, coding instructions rich enough to capture them, sufficient numbers of observers available to ensure their replicability, etc.

This section began by pointing out the importance of distinguishing the information in the variance of data about the phenomena of interest, and unrelated variance or noise that pollutes the channel of communication that an analysis of data seeks to establish between the phenomena of analytical interest and what an inquiry claims to reveal about them. Unfortunately, reliability tests cannot distinguish the wheat from the chaff, only inform researchers about their proportions. In all of these efforts, variance needs to be acknowledged in opposition to the following:

Feinstein and Cicchetti (1990) criticized Cohen's (1960) κ (kappa) for embodying two undesirable paradoxes. According to these authors, the first paradox occurs under a condition of observing near 100% agreement while κ measures are near zero. Lombard et al. (2002), Gwet (2002), Zhao et al. (2012), and other commentators have echoed Feinstein and Cicchetti's puzzlements and noted that Feinstein and Cicchetti's paradox: "high agreement and low kappa," also applies to Scott's (1955) π, and I should mention to Krippendorff's α as well. These authors take the mathematical fact that percent agreement and chance-corrected agreement coefficients point in opposite directions as paradoxical and as a good reason for disqualifying these coefficients as unsuitable for assessing reliability when data exhibit low variances. However, on closer examination, I am suggesting that these authors are drawn into this paradox by their own use of language in interpreting this mathematical fact. I consider it:

- Regressive to disregard the history of reliability measurement. It progressed from using raw percent agreements among two coders to agreement coefficients that used a conception of chance to correct many of its reliability-unrelated properties. I suspect, holding the percent agreement against agreement coefficients designed to correct its problems results from conceiving so-called "chance-corrected agreement coefficients" as fundamentally percent agreement measures with only a small "correction," not as of entirely different kinds.
- Mistaken to equate the observation of agreement with reliability. Obviously, agreement among replications is observable and can be measured. However, the ability to rely on data for subsequent analysis can only be inferred from that agreement, provided that (i) the preparatory conditions, discussed in Section 1.7, are met; and (ii) the chosen coefficient does not lead to invalid representation of phenomena of interest in the data.

- Misguided to follow an intuition of what good reliability measures should look like overruling its purpose of establishing whether data can serve as reliable proxies of the phenomena of scholarly interest. Serving this purpose can be demonstrated only if observers' distinctions among phenomena turn up as analyzable differences in the data slated for analysis. Phenomena that do not possess enough diversity or measuring instruments that fail to distinguish among them cannot possibly provide evidence of what data are about. To trust processes of generating data, the variance in data has to be demonstrably related to what is to be studied, not to variables extraneous to the research questions.
- Misleading to introduce concepts into agreement measures that generate larger numerical results but ignore the methodological role of data in scientific inquiries. For one example, Zhao et al. (2012) and Feng (2014, 2015) want to equate reliability with the difficulties that coders experience when applying coding instructions given to them. It is well established that experiences of coding difficulties result in unreliable data. However, such difficulties present a challenge to the designers of coding instructions to state them less ambiguously or add procedures for the correct uses of codes. Reliability measures ought to indicate the extent to which data represent something real not about who wrote the coding instructions and who experienced which difficulties when applying them. Gwet proposed an agreement coefficient heralded for producing higher agreement measures but without a clear mathematical justification rooted in whether data represent distinct phenomena. Others argue for equating reliability with the existence of majority judgments among coders. However, majorities can be overwhelming or slim (see Sections 6.3.3 and 6.3.4). Equating the existence of a majority with the reliability of data yields agreement measures that ignore the very minorities that introduce unreliabilities into data.
- Omitting the vital role of the variance in data. A large proportion of information in data is what enhances the probability that their analysis leads to valid research conclusions. Without variations in data, one cannot draw distinctions, entertain comparisons, ensure whether one's measuring instruments work, correlate variables, and move an analysis to higher levels of abstraction. Research results derived from variety-impoverished data become questionable.

To illustrate the above authors' contentions, Zhao et al. (2012) published a "dramatic example." I am relying on the frequencies of their example but take the liberty of placing them in a more convincing context. How the setting is narrated should not influence the quantitative conclusions derived from observed disagreements.

Suppose two medical experts seek to establish the reliability of a new device for detecting a rare disease. Since the disease is rare the sample of subjects needed to detect it has to be large. Suppose both experts test the same 1,000 subjects. Suppose each expert finds 999 subjects negative and one positive. The tests agree regarding 998 subjects but do not agree regarding the remaining two. In terms of contingencies, these reliability data are as follows:

| | – | + | |
|---|---|---|---|
| – | 998 | 1 | 999 |
| + | 1 | 0 | 1 |
| | 999 | 1 | 1000 subjects tested by two doctors |

In this example, $\kappa$, $\pi$, and $\alpha$ measure near zero, –0.001, –0.001, and –0.0005 to be precise. Zhao et al. find the concurrence of observing 99.8% agreement and chance-corrected agreement coefficients measuring near zero incomprehensible and a sufficient reason to retire the latter altogether.

I am suggesting that Zhao et al.'s conclusions compare "apples with mice." By echoing Feinstein and Cicchetti's paradox of "high % agreement and low κ" (π and α included but not mentioned), their argument privileges percent agreement $P_o$ over agreement coefficients that are defined to overcome the manifest inadequacies of percent agreement. It is important to realize that in these hypothetical reliability data, there is not even one agreement on the positive identification of the disease. In my reading, these reliability data cannot possibly support the conclusion that the medical test reliably distinguishes between the presence or absence of the disease. The near zero value of κ, π, and α would concur with my conclusion of these findings. I cannot imagine anyone arguing that the test certifies the reliability of the medical test. To be sure, the same conclusion would have to be reached if the test were to have found 998 subjects to have a common disease but experts could not agree on its absence regarding the two remaining subjects. Nor would it matter whether the compliance with a law was at stake, which is Zhao et al.'s original example of the use of warning labels on a consumer goods. Claiming that a reliability test found it to be 99.8% reliable has an objective ring and might well sound convincing to those unable or unwilling to examine the nature of the data. However, it would surely be unconscionable to sell a test as reliable based on such data. The above example is another demonstration that percentage agreement is not a suitable reliability measure as it does not respond to the needed variance in the data. In Zhao et al.'s example, the near zero values of κ, π, and α should lead to the perfectly justifiable conclusion that such data cannot be trusted to provide information about the phenomena of interest.

Feinstein and Cicchetti's paradox of "high agreement and low κ" needs to be reframed. A simple conceptual gestalt switch makes the condition that puzzled them completely sensible:

*As information in data (about phenomena of interest) declines so does their reliability.* (3.6)

Information is co-variation between two variables, for Shannon (Shannon & Weaver, 1049) between a sender and a receiver, here between a diversity of phenomena of analytical interest and the data generated to analyze them. Zhao et al.'s example demonstrates that in the absence of variance, high percent agreement provides no information to a researcher. Perfect percent agreement would indicate perfect reliability only when data do have enough variance. Evidently, percent agreement alone says nothing about the reliability of data.

The following examples are intended to show how α declines and the amount of information disappears as the variance in data drops to zero, while the percent agreement remains largely unrelated to the loss of reliability:

| | A | B | C | D | E | F |
|---|---|---|---|---|---|---|
| | 100 0 | 99 1 | 90 10 | 170 10 | 198 1 | 200 0 |
| | 0 100 | 1 99 | 10 90 | 10 10 | 1 0 | 0 0 |
| α = | 1.000 | 0.995 | 0.608 | 0.447 | 0.000 | $1-{}^{0}/_{0}=0$ by definition |
| Information[1] = | 1 bit | 0.919 bits | 0.521 bits | 0.090 bits | 0.0001 bits | N/A |
| % agreement = | 100% | 98% | 98% | 99% | 99% | 100% |

[1] Information = $H$(Rows) + $H$(Columns) – $H$(Rows × Columns). (Shannon & Weaver, 1949: 21; Ashby, 1965). See also (0.1).

Thus, it is only the variance in A that allows %-agreement to be interpretable as a measure of the reliability of data – see also Table 4.1. B exhibits comparatively minuscule disagreements. Its α is barely noticeably below 1.000. B and C show the same disagreements in their off-diagonal cells, but they differ in the amount of information they could respectively convey, largely because of C's unequal marginal distribution. This unequal distribution becomes more pronounced when moving from D, to E and finally to F. From the perspective of informing the analytical process, data in D, whose α is below what some consider worthy of tentative explorations, provide nearly no information. Data in E support not a single decision, exhibit no variance whatsoever, and appropriately result in $\alpha = 0$ by definition. In contrast, %-agreement is not even correlated with the declining reliability in these examples.

In conclusion, although α does not literally distinguish information-bearing parts of data from polluted ones, it represents the proportion of the variance in data that relate to real phenomena as opposed to irrelevant variations.

As data lose their variance, the distinction between the information about the phenomena of concern and noise becomes more problematic and α invariably follows. Incidentally, percent agreement bears no relationship to the variance in the data and is therefore not interpretable as an indicator of the reliability of data. Zhao et al.'s example above approaches near-perfect percent agreement without exhibiting a single agreed-upon distinction. In the absence of any variance, agreement would always be 100% at which point no reliability test would be sensitive to what it is meant to differentiate. This includes the case when the measuring instrument is broken.

# 4

# *Alpha Compared With Primarily Nominal Agreement Measures*

Historically, current agreement measures evolved in two lineages. One developed from percent agreement $P_o$, and morphed into chance-corrected measures, in which $P_e$ is the agreement expected when chance governs the data-making process, both limited by $P_{max}$. The other originated with Karl Pearson's product-moment correlation coefficient $r_{xy}$, became the intraclass coefficient $r_{xx}$ and then part of ${}_c\alpha$ as well. α's subscript "*c*" is meant to indicate the coding of predefined units, which is common to the measures discussed here. The subscript "metric" indicates its applicability to forms of data other than unordered categories shared in the first and interval data which characterizes the second lineage. ${}_c\alpha_{\text{metric}}$ embraces both and is the subject of further extensions in the following chapters.

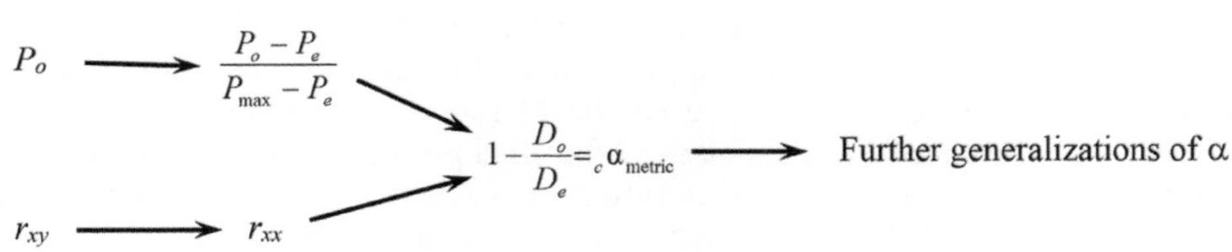


**FIGURE 4.1**
Two Lineages in the History Leading to α

Although at least one chance-corrected agreement coefficient, Fleiss' (1971) *K*, can handle more than two observers, most other agreement measures are limited to two. This brief sketch is not meant to be exhaustive. There are many other coefficients measuring different kinds of reliabilities. The following explores the suitability of those in line with Figure 4.1 in view of their ability to infer the reliability of data, i.e., their relationship to the phenomena for which they are intended to be proxies, by description, measurement, or coding, as defined in (2.7) to start.

## 4.1 Percent Agreement

The percent of agreement $P_o$ between two observers, coders, raters, or judges defines a scale between 0 and 1 or 0% and 100%. Its proponents appeal to its simplicity and intuitiveness.

However, on closer examination this is rarely so. Percent agreement can serve as an indicator of reliability only under two conditions. The first pertains to observing agreements without exception, $P_o = 1$, and the second pertains to data exhibiting sufficient variance. Under these two conditions, percent agreement corresponds to all common agreement coefficients I know of, but this marks only one point on its scale between 0 and 1. For once, $P_o$ is independent of the variance in data which, according to (3.3), is a prerequisite for data to inform subsequent analyses. Recall Zhao et al.'s (2012) example discussed in Section 3.9. He wanted to convince his readers of the importance of percent agreement by constructing a case in which 1,000 units were tested regarding the presence or absence of a trait. 998 units tested one way but on the other two, the tests were inconclusive. To me, this is evidence of the test's inability to reliably distinguish between the two traits and sufficient justification for a valid reliability measure to measure near zero. Going further than Zhao's extreme, if all 1,000 units had tested alike, 100% agreement could not possibly convince anybody of the test's reliability. The instrument could have been broken, the observers been asleep, or knowing that their performance would be measured in percent of agreement between them, deviously agreed to code everything the same. The point made in Section 3.9; data need to inform their analysts about the phenomena of analytical interest. According to (3.3), variance matters.

Another problem is the range of $P_o$. Its smallest value, $P_o = 0$ is unlikely to occur by chance and manifests extreme systematic disagreements, already suggested in the previous two chapters. Systematic disagreements can be observed under essentially three conditions. One is when observers unknowingly apply incompatible coding instructions to the same set of phenomena; for example, when one instructs coders to rate phenomena on a scale from 1 to 10 whereas the other calls for rating the same from 10 to 1. A second can arise when agreements measures are applied to unlike sets of phenomena. Such administrative mistakes can happen. The third may happen when two observers communicate with one another and agree to disagree on coding each unit. None of these conditions has anything to do with the reliability of data, which makes all the other points of $P_o$'s scale uninterpretable.

For example, if replicability is defined as the extent to which one can rely on data, the range of a useful measure should be between when one cannot rely on the data and when relying on them is unquestionable. It turns out that $P_o$'s range of meaningful reliability interpretations varies among several conditions by the number of categories available for coding. When observers have two categories to choose from, by chance, $P_o$ can range between 50% and almost 100% (remember Zhao's example of 99.8% is an incidence of chance). When they have three categories at their disposal, by chance $P_o$ can range between 33% and nearly 100%, etc. Any observed $P_o < 1$ leaves wide open how reliable the coding of data is.

Also, the proportion $P_o$ is the sum of all matching pairs of values assigned to a set of units. It limits reliability considerations to two replications, observers, coders, or judges. Most software featuring percent agreement measures is so limited. When more than two observers are involved, unanimity becomes less likely, $P_o = 1$ becomes quickly improbable and percentages cannot cope with a mixture of matching and mismatching observers' category assignments. To avoid the latter complications, most researchers employ just two observers and software that provides percent agreement measures is limited to calculating agreements between two coders.

There have been proposals to give percent agreements a different meaning than the one adopted above. As just one example, Aickin (1990) argues, much like Zhao et al. did 22 years later, that agreement should take the difficulties of coding into consideration. Operationally, Aickin partitions a set of units into two subsets: Units on which raters agree without exception are considered easy to code. Units on which raters show some disagreement are considered difficult to code. His coefficient is the proportion of units that are easy to code to the total number of units. This percent agreement is more nuanced than the simple $P_o$ but is marred by the same indeterminacies mentioned above.

Incidentally (13.28) generalizes percentage agreement to any number of coders and different metrics – but this does not change the above assessment.

## 4.2 Various Chance-Corrections of Percent Agreements

The indeterminacies of $P_o$ as an indicator of reliability motivated the emergence of so-called chance-corrected agreement coefficients depicted in Figure 4.1 – "so-called" because the word chance easily enters ordinary arguments without clarification of what it refers to. Its denotation is far from being conceived consensually. Benini's (1901) β (beta) seems to have been the first coefficient of this kind. Goodman and Kruskal (1954, pp. 756–758), building on Guttman's (1946) ρ (rho), relied on this form when proposing their $\lambda_r$ (lambda) to compare pairs test results or individuals judging subjects. It is a coefficient that can reach the value 1 under conditions of unequal use of categories by the two tests or judges as intended by Benini but they correct $P_o$ not by a measure of chance agreement but by their modal probabilities.

Much of content analysis literature attributes the emergence of chance-corrected agreement coefficients to Bennett, Alpert, and Goldsteins's (1954) *S*. Their coefficient was subsequently reinvented with only minor variations as Holley & Guilford's (1964) *G*, Maxwell's (1977) *R.E.*, Janson & Vegelius' (1979) *C*, Brennan & Prediger's (1981) $K_n$ , and Perreault & Leigh's (1989) $I_r$ , and most recently built into software called Lotus by Fretwurst (2015). All of these authors intended to overcome one of $P_o$'s undesirable qualities, its dependence on the number of categories available in a coding instrument. Neither version pays attention to the actual use of categories in the population of data. *S* was succeeded by Scott's (1955) π (pi). Scott noted the difficulties with *S*'s conception and replaced it by reference to the frequencies of the joint use of categories by both observers. He was correct in making this modification as these frequencies are the best estimates of the distribution of categories in the population of data whose reliability was in question. Thereafter, Cohen (1960) modified Scott's π and proposed his popular κ (kappa)-coefficient which replaced this estimate by the coders' individual proclivities for the categories. Most recently Gwet's (2008) $AC_1$ modified Scott's π to avoid Feinstein and Cicchetti's (1990) paradoxes.

All of these coefficients modify the observed agreement $P_o$ and are as such confined to $P_o$'s limitations to nominal data and two observers. The reason for ${}_c\alpha$ to follow a different form, as seen in (2.5) and Figure 4.1 lies in the need to overcome these two and other limitations. However, for nominal data, but only for these, (2.8) has shown that ${}_c\alpha$ can be stated as a chance-corrected coefficient as well.

Unfortunately, the common form of chance-corrected agreement coefficients $(P_o–P_e)/(P_{max}–P_e)$ and the use of the single word "chance" hides fundamental conceptual differences among diverse conceptions of how observed agreement is corrected. They can be distinguished by how $P_{max}$ and $P_e$ are defined. To reveal the conceptual and quantitative implications of diverse conceptions of chance and to evaluate whether and how they relate to the reliability of data, let me use a very simple example of reliability data generated by two observers *i* and *j* who assigned one of just two values *c* or *k* to 10 predefined units of analysis. Using binary data for this comparison has the advantage of not being confounded by factors that arise with more complex data. The following represents the same reliability data in two ways:

**Contingencies**
of 10 units between two observers

| Observer *i*'s values: | $c$ | $k$ | |
|---|---|---|---|
| Observer *j*'s values: $c$ | 2 | 0 | 2 |
| $k$ | 4 | 4 | 8 |
| | 6 | 4 | 10 |

$P_o = {}^{(2+4)}/_{10} = 0.6$

**Coincidences**
among 20 pairs of values

| Values: | $c$ | $k$ | of both observers |
|---|---|---|---|
| $c$ | 4 | 4 | 8 |
| $k$ | 4 | 8 | 12 |
| | 8 | 12 | 20 |

$P_o = {}^{(4+8)}/_{20} = 0.6$

Recall from Section 3.3 how these two matrices differ. Contingency matrices are asymmetrical and tabulate units, here 10, coded by two observers whose proclivities for particular values can be recognized in their unequal marginal sums and asymmetrical distribution of cell entries.

(3.4) suggested to construct a coincidence matrix by summing the entries of a contingency matrix and of its transpose. Coincidence matrices are symmetrical and tabulate pairs of values, here 20, that occur within the 10 units of reliability data. Their marginal sums no longer refer to observers' unequal proclivities for particular categories. The observed agreement $P_o$ is the same for both representations and its form is common to all chance-corrected agreement coefficients for two observers.

I shall limit the following elaborations to a few prototypical conceptions.

Regarding the definition of $P_{max}$, Benini's (1901) β is an exception from the more common chance-corrected coefficients compared below. Benini defined $P_{max}$ as the maximum agreement possible within the constraints of the marginal frequencies of a matrix of *c-k* contingencies $x_{ck}$. In the above example, β's $P_{max} = \frac{1}{x_{..}}\sum_{c=1}^{v} min(x_{c\cdot}, x_{\cdot c}) = \frac{1}{10}(2+4) = 0.6$.

In this case, $P_{max} = 0.6 = P_o$ which indicates that the observed agreement $P_o$ is the largest agreement possible under conditions of the marginal constraints of this contingency matrix. It follows that in our example, β = 1. The definition of β ensures that β could always measure 1 when applied to contingencies with unequal margins. The range of β is unlike Cohen's κ which does not take the degrees of freedom within the marginal constraints of the observed contingencies into considerations and therefore cannot reach 1 whenever observers have unequal proclivities for particular values. Benini was a demographer and not concerned with reliability issues. However, his idea of a coefficient that can reach 1 is important in the following comparison of the five prototypical chance-corrected agreement coefficients.

Recently, von Eye (2006) proposed a coefficient $\kappa_s$, suggested an alternative to Cohen's κ, but adopted several features of Benini's β. $\kappa_s = 1$ indicates that the observed agreement is the largest one possible under conditions of unequal marginal frequencies. By defining chance as the statistical independence of observers, $\kappa_s$ is not sensitive to disagreements due to observers' unequal uses of categories. While $\kappa_s$ is conceptualized for two or more observers, by counting disagreement as the lack of unanimous consent among observers, $\kappa_s$ gets smaller with an increase in the number of participating observers, rendering its results not comparable across different kinds of reliability data with different numbers of replications. Its contingency matrix should not be confused with the coincidences among all pairs of observers' judgments.

In all chance-corrected agreement coefficients the expected agreement $P_e$ defines their zero point and locates the proportion of observed agreement $P_o$ on their particular scale.

Except for β's $P_{max}$, all of the five agreement coefficients compared below define $P_{max} = 1$. Because $P_{max}$ and the observed agreement $P_o = 0.6$ of our example are common to all five coefficients to be compared, I focus on how their $P_e$ is defined and what difference it makes for indicating the reliability of data.

To facilitate this comparison, the expected agreements $P_e$ of each of these coefficients is visualized graphically by the shaded areas within the square, $(P_{max})^2$. The proportions of the marginal frequencies of the contingency matrix are recognizable in the axes of the square for Cohen's κ, and the proportions of the marginal frequencies of the coincidence matrix are recognizable in the three remaining squares for Gwet's $AC_1$, Scott's π, and ${}_c\alpha$.

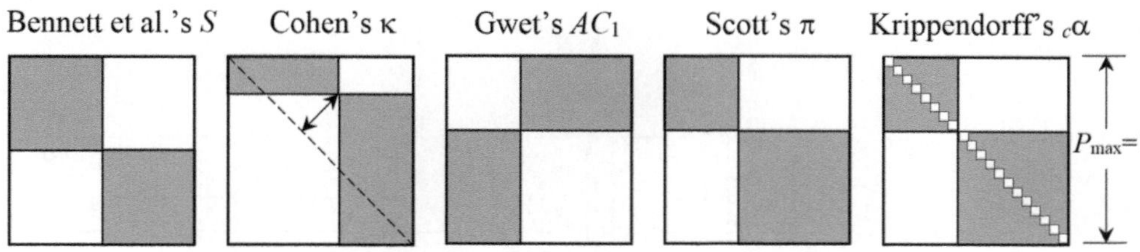


$P_e = \quad 2\left(\frac{1}{2}\right)^2 = 0.5 \qquad \frac{2\cdot 6+8\cdot 4}{10^2} = 0.44 \qquad 2\frac{2+6}{2\cdot 10}\left(1-\frac{2+6}{2\cdot 10}\right) = 0.48 \qquad \left(\frac{2+6}{2\cdot 10}\right)^2+\left(\frac{8+4}{2\cdot 10}\right)^2 = 0.52 \qquad \frac{8(8-1)+12(12-1)}{20(20-1)} = 0.495$

$S = 0.200 \qquad \kappa = 0.286 \qquad AC_1 = 0.231 \qquad \pi = 0.167 \qquad {}_c\alpha = 0.208$

**FIGURE 4.2**
Comparing the Expected Agreements Across Four Prototypical Agreement Coefficients

### 4.2.1 Bennett et al.'s S

The shaded area for Bennett et al.'s $P_e$ exhibits no relationship to the marginal frequencies of contingencies or coincidences. $S$'s $P_e$ is a function of the number $v$ of categories available in the coding instrument by which the data are generated. With $v = 2$ values, in this example, $P_e = {}^1/_v = {}^1/_2$. $S$ accomplishes its authors' intention to define an agreement coefficient that is independent of the number and frequency of the categories used to generate $P_o$. However, by correcting both the observed agreement $P_o$ and $P_{max} = 1$, by a proportion that is rooted in the categorical logic of researchers and manifest in their coding instrument, not in the nature of the data coded, $S$ cannot be interpreted as a coefficient of the reliability of data. This point becomes more obvious when a researcher applies $S$ to data with more than two categories. The number $v$ of categories chosen by the analyst affects the coefficient $S$ regardless of whether they actually occur in the coded data. Unless there are rules to select $v$, this enables analysts to manipulate the coefficient, rendering it less interpretable as indication of the reliability of data.

### 4.2.2 Cohen's κ (kappa)

The expected agreement $P_e$ in Cohen's κ equals the agreement that could be observed when the two observers' categorizations are statistically independent of each other. As previously mentioned, Cohen's definition of chance is related neither to the number of matching categories possible among several observers' collective use of them as in (2.7), (2.8), and (2.9), nor to the expected agreements in the definition of $S$. Rather, it is defined as the absence of associations between them, familiar in conceptions of the absence of correlations between two variables along a regression line. In Figure 4.1, the arrow in the diagram for κ suggests by how much κ's regression would have to move if its $P_e$ could be used in a replicability-related agreement measure. I already mentioned that κ cannot reach 1 when marginal frequencies are uneven. In our numerical example, the degree of freedom for κ and β is *d.f.* = 2. By comparison the degree of freedom for α for the very same example is *d.f.* = 8. The three possible contingency matrices and the corresponding values of κ and β are shown together with the nine possible coincidence matrices and corresponding values of $_c\alpha$ and Scott's π to be discussed below in Section 4.2.4:

| | $\kappa_{observed} = \kappa_{largest}$ | | $\kappa_{smallest}$ |
|---|---|---|---|
| Matrix | 2 0 / 4 4 | 1 1 / 5 3 | 0 2 / 6 2 |
| κ = | 0.286 | −0.071 | −0.429 |
| β = | 1.000 | 0.250 | −0.500 |

| | $_c\alpha_{largest}$ | | | | $_c\alpha_{observed}$ | | | | $_c\alpha_{smallest}$ |
|---|---|---|---|---|---|---|---|---|---|
| Matrix | 8 0 / 0 12 | 7 1 / 1 11 | 6 2 / 2 10 | 5 3 / 3 9 | 4 4 / 4 8 | 3 5 / 5 7 | 2 6 / 6 6 | 1 7 / 7 5 | 0 8 / 8 4 |
| $_c\alpha$ = | 1.000 | 0.802 | 0.604 | 0.406 | 0.208 | 0.010 | −0.188 | −0.385 | −0.583 |
| π = | 1.000 | 0.792 | 0.583 | 0.375 | 0.167 | −0.041 | −0.250 | −0.458 | −0.667 |

Here, I am less concerned with the numerical differences these two coefficients yield but with their scales. While all four of them define linear scales, however, the numerical ranges of κ and β evidently are much smaller than for $_c\alpha$ and π. While β was conceived to measure 1 when the distribution of contingencies is strongest, κ does not recognize this condition. At this point κ responds to some disagreements but not all.

What speaks against using κ to infer the replicability of data is that it fails to register observers' disagreements, which is manifest in their marginal distributions of categories (Krippendorff, 1978). When observers disagree on their predilections for categories, κ yields values higher than when they do agree. My observation of κ's counter indication echoes what has been said for some time. Brennan and Prediger (1981) concluded that κ penalizes observers for agreeing on their use of categories. This bias is due to reliability data being represented as contingencies which tabulate units of analysis, not pairs of values from which follows that the number of units with mismatching categories *c-k* are unlike those categorized *k-c*. For reliability it should not matter how off-diagonal contingencies are distributed. However, when they are unevenly distributed, κ is larger than when they are evenly distributed. Zwick (1988), citing others who mention this flaw as well, suggested testing the marginal homogeneity of contingency matrices as a precondition for κ's applicability – an unrealistic burden for advocates of Cohen's κ. As soon as the marginal frequencies of contingency matrices differ, κ becomes inflated. Most of κ's oddities boil down to its definition of $P_e$ in terms of the statistical independence among observers, not as the disconnect between the distinctions among the phenomena of interest and how they appear in the data. It follows that κ's zero point at $P_o = P_e$ has no meaningful relationship to the reliability of data. Surely, Cohen's κ has a place in other statistical applications but it simply cannot serve as an unbiased indicator of the reliability of data. There are numerous publications concurring with this conclusion, for example, Gwet (2002), albeit for different reasons.

A brief historical note: Scott published π in 1955. He was a content analyst and public opinion researcher with a keen understanding of the effect of unreliable data on predictions. His π served the practical purpose of measuring agreement. By contrast, Cohen was an influential statistician who criticized Scott's $\pi$, wrongly, I would say, for not taking the statistical independence of two observers as the basis for evaluating agreement. I suspect his reasoning was biased in favor of "inter-coder agreement," which led him not to recognize Scott's focus on the reliability of data, not on how coders agree. Cohen mistakenly argued that that Scott's $\pi$ omits "one source of disagreement between a pair of raters (which) is precisely their proclivity to distribute their judgments different over the categories" (Cohen, 1960: 41). He then replaced the definition of $P_e$ in Scott's $\pi$ by one that is consistent with the more common measures of association, correlation, and multiple regression, which informed much of his statistical work. However, equating chance with the statistical independence of the two observers' unequal judgments had the opposite effect of inflating the $\kappa$-coefficient by eliminating what he claimed to have included. The evidence for this failure could have been anticipated by the difference between $\kappa = 0.286$ and $\pi = 0.208$ in the above binary examples but is clearly demonstrable in the following 3-by-3 contingency matrices with identical agreements in their diagonals but unlike marginal sums:

| Categories: | | Observer A *a* | *b* | *c* | |
|---|---|---|---|---|---|
| | *a* | 12 | 9 | 9 | 30 |
| Observer B | *b* | 9 | 14 | 9 | 32 |
| | *c* | 9 | 9 | 20 | 38 |
| | | 30 | 32 | 38 | 100 |

$P_o = 46\%$
$\boldsymbol{\kappa = 0.186}$
$\pi = 0.186$
${}_c\alpha = 0.190$

| | | Observer A *a* | *b* | *c* | |
|---|---|---|---|---|---|
| | *a* | 12 | 18 | 18 | 48 |
| Observer B | *b* | 0 | 14 | 18 | 32 |
| | *c* | 0 | 0 | 20 | 20 |
| | | 12 | 32 | 56 | 100 |

$P_o = 46\%$
$\boldsymbol{\kappa = 0.258}$
$\pi = 0.186$
${}_c\alpha = 0.190$

Evidently, Cohen's modification of Scott's $\pi$, intended to include these disagreements, does just the opposite. $\kappa$ removes them from what $\pi$ measures and inflates the $\kappa$-coefficient with them. In the absence of unlike distributions of off-diagonal contingencies, $\kappa = \pi$, both differing only minimally from ${}_c\alpha$, due to $\kappa$ and $\pi$ assuming infinite sample sizes while ${}_c\alpha$ does not. Readers of Cohen's 1960 paper cite his words while the numerical facts do not bear out his claim. The popularity of Cohen's $\kappa$ over Scott's $\pi$ has little to do with what they measure.

Artstein and Poesio (2005) concur with most of my above assessments. They advocate a measure of bias, defined as the difference between $\pi$'s $P_e$ and $\kappa$'s $P_e$. They also propose a slight modification of ${}_c\alpha$, called $\alpha'$, obtained by replacing ${}_c\alpha$'s $P_e$ by $\pi$'s $P_e$. Although references to $\pi$ and $\kappa$ are confined to percent agreements, they too observe that the difference between $\pi$ and $\alpha'$ diminishes with increasing sample sizes. They write (2005: 14) that the motivation for ${}_c\alpha$'s $D_e$ is not clear to them. My simple answer: ${}_c\alpha$ estimates the expected disagreements to be consistent with how the observed disagreements are enumerated – avoiding conceptual confusions.

### 4.2.3 Gwet's $AC_1$

Gwet's $AC_1$ is the latecomer among the chance-corrected agreement measures. It is billed as a solution of Cicchetti and Feinstein's (1990) paradox, already discussed in the previous chapter based on their correct observation that $\kappa$, ($\pi$, and ${}_c\alpha$) becomes very small as percent agreement $P_o$ increases due to the loss of variance in data. In Section 3.9, I suggested that the paradoxical nature of this observation resulted from conceptually privileging percent agreement $P_o$ over the variance required for data to supply an analysis with information about the phenomena under consideration. Information-bearing variance is an essential ingredient of reliable data – see Section 1.1.4, Figure 1.1, and Krippendorff (2011). For dichotomous or binary data, which is the most reduced form for comparing various agreement measures, Gwet's $AC_1$ replaces what serves other chance-corrected agreement coefficients as the expected agreement by its complement, by the expected disagreement $D_e = 1-P_e$. This is easily recognized in Figure 4.2 by comparing the shaded areas in the square for $AC_1$ with the one for Scott's $\pi$ and their numerical values. Subtracting $1-P_e$ from both $P_{max} = 1$ and $P_o$ leads to a measure that has lost its interpretability as a chance-corrected agreement coefficient. Compare $AC_1$ with $\kappa$ in the following inequality. In this binary case, $P_e$ and $D_e$ in $\kappa$ are switched in $AC_1$!

$$AC_1 = \frac{P_o-(1-P_e)}{1-(1-P_e)} = \frac{P_o-D_e}{P_e} \quad \neq \quad \frac{P_o-P_e}{D_e} = \frac{P_o-P_e}{1-P_e} = \kappa \tag{4.1}$$

I am not the only one for whom the switching of $\kappa$'s $D_e$ and $P_e$ in $AC_1$ is uninterpretable in terms of reliability. Moreover, expressed relative to the expected agreement $P_e$ raises doubts in the ability to interpret $AC_1$ as a reliability coefficient as expressed by Ejima, Aihara and Nishiura (2013). Feng (2013) used Monte Carlo simulations to compare the numerical values of $AC_1$, with $\pi$, $\kappa$, and $\alpha$ among others and concludes to no surprise that $AC_1$ produces higher values than the other measures do. However, without offering a plausible interpretation of how $AC_1$'s mathematical structure relates the data to the phenomena of concern, switching $\kappa$'s $D_e$ and $P_e$ just to avoid Feinstein and Cicchetti's (1990) paradox and get higher values when variance becomes low can only invite deceptions.

When data are dichotomous, as in the above example, there are only two conditions under which Gwet's $AC_1$ correlates with the reliability of data: when $P_o = 1$ and data possess enough variance for $P_e<1$, and when $D_e = P_e = 0.5$. At both statistically rare occasions $AC_1 = S = \pi = \kappa = \beta$.

### 4.2.4 Scott's π (pi) and Fleiss' *K* (mistakenly equated with kappa)

Scott's $\pi$ is limited to nominal data and two observers. It was extended by Fleiss' (1971) $K$ to a larger but fixed number of observers – misleadingly billed as a generalization of $\kappa$. Figure 4.2 reveals their estimated $P_e$ to resemble that of ${}_c\alpha$ except for including values paired with themselves which ${}_c\alpha$ omits. This difference is seen in the 20 empty cells in the diagonal of ${}_c\alpha$'s $P_e$ not counted. Excluding them from ${}_c\alpha$'s $P_e$ solves three of $\pi$ and $K$'s conceptual problems. First, ${}_c\alpha$ avoids inflating its expected agreement. Second, it calculates $P_e$ in a manner consistent with how the observed coincidences $P_o$ of all chance-corrected coefficients are calculated. Nobody would consider inflating observed agreements by pairing the values of any one observer with themselves. Third, ${}_c\alpha$'s $P_e$ responds to the finite nature of actual reliability data enumerated in $P_o$. In effect, as already mentioned in Section 3.6, $\pi$, $K$, and $\kappa$ assume data to be infinite in size whereas ${}_c\alpha$ takes all sample sizes as they are. In an empirical reality, data are always finite; infinity can at best serve as an unwarranted assumption. As reliability data become large, $\pi$, $K$, and ${}_c\alpha$ become asymptotically indistinguishable. This explains the often-noted numerical similarities among the three coefficients, especially when applied on large data. Figure 4.2 illustrates what accounts for their albeit often small difference: ${}_c\alpha \geq \pi$ and $K$.

## 4.3 Pearson's Intraclass Correlation

In the history of agreement coefficients with reliability interpretations, the other lineage started with Karl Pearson's (1901) famous product-moment correlation coefficient $r_{xy}$ and his subsequent proposal for a "coefficient of racial likeness" (Tildesley, 1921, Pearson, 1926), later developed into the intraclass correlation coefficient $r_{xx}$. Correlations are computed on interval data and measure the degree to which two variables *X* and *Y*, here two observers, are linearly correlated. Linear dependencies between two observers' valuation of units do not equal agreement and can say little about the reliability of their valuation. Cronbach's (1951) alpha is defined in terms of correlations, called a reliability coefficient, but actually measures the consistency among multiple test results. Its numerous applications in psychometric research notwithstanding, it does not relate to the reliability of generated data.

As already noted, Pearson's product-moment correlation coefficient $r_{xy}$ measures the strength of linear relationships between two interval variables. It becomes weaker as deviations from a regression line become larger and more numerous. However, measures of agreement need to respond to deviations from a 45° regression line, from the diagonal of contingency or coincidence matrices. Pearson's ingenious proposal for his "coefficient of racial likeness" $r_{xx}$ – initially called racial because it was developed to measure the capacity of human skulls in Burma (Tildesley, 1921; Pearson, 1926) – amounts to applying his product-moment correlation coefficient on a cross-tabulation of interval data that consists of summing the *X*–*Y* contingencies and their transposes, the *Y*–*X* contingencies into a single table. This construction was also demonstrated in Section 3.7 but only for nominal data. This joined representation of interval data is always symmetrically distributed around a 45° regression line. Deviations from its diagonal – disagreements – are due to the deviations from the two regression lines plus the difference between one and its transpose. The resulting matrix is an analogue of our coincidence matrices, in Pearson's case of interval variables. When applied to dichotomous or binary data, $r_{xx}$ behaves just like Scott's $\pi$ and is implicitly represented in the above comparisons of five chance-corrected agreement coefficients. Pearson deserves credit for this idea. (3.4) could be interpreted as a generalization of Pearson's way of creating a coincidence matrix from a contingency matrix and its transpose.

This second trajectory in Figure 4.1 concerned the reliability of interval data. With Pearson's coefficient of racial likeness $r_{xx}$, the origin of the intraclass correlation was limited to two measurements and assumes infinite sample sizes. Unaware of this history, the interval ${}_c\alpha_{\text{interval}}$ for a fixed number of observers (Krippendorff, 1970a) generalized Pearson's coefficient of racial likeness to any number of observers. The ${}_c\alpha$-coefficient embraces several other coefficients as well, which will be discussed in Chapter 5.

## 4.4 Comparisons and Properties of Several Agreement Coefficients

The preceding sections evaluated six prototypical chance-corrected agreement coefficients regarding their ability to serve as indicators of the reliability of data and mentioned a few others in passing. Their differences were illustrated by means of binary or dichotomous reliability data, the extreme form to which all agreement coefficients apply – without demonstrating their strength in more complex situations. This served to highlight their fundamental differences and unequal sensitivities to properties of data which might otherwise be difficult to recognize in data with large numbers of categories. The literature is full of simulations that compare the numerical results of applying known coefficients to the same reliability data. Not only are such simulations limited to the nature of the data to which they are applied, they also say little about their unlike sensitivities unless they are traced to the mathematical properties of these coefficients. The latter is what this chapter attempted.

At this point, it may be helpful to offer a comparison of the above and a few other agreement measures regarding the kinds of data to which they are applicable plus the conditions under which they could possibly serve as indicator of the reliability of data. Table 4.1 lists a total of eleven coefficients. This table is not meant to be a survey of the field. Readers who discover other coefficients are invited to analyze their mathematical structures and ask and answer the questions of what they respond to, omit, and under which conditions they apply.

| Coefficient | Predefined units of analysis | Unitizing (units of various lengths) | Three or more coders | Missing data | Binary data | Nominal data | Ordinal data | Interval data | Ratio data | Relates to population estimates | Acknowledges small sample sizes | Condition under which it can serve as an indicator of the reliability of data |
|---|---|---|---|---|---|---|---|---|---|---|---|---|
| **Percent $P_o$** | ✓ | | | | ✓ | ✓ | | | | | | $P_o = 1$ in the presence of variance |
| **Bennet et al.'s $S$** | ✓ | | | | ✓ | ✓ | | | | | | All categories are used equally often |
| **Gwet's $AC_1$** | ✓ | | ✓ | | ✓ | ✓ | | | | ✓ | | Two categories are used equally often |
| **Benini's ß** | ✓ | | | | ✓ | ✓ | | | | | | Both coders agree on freq. of categories |
| **Cohen's $\kappa$** | ✓ | | | | ✓ | ✓ | | | | | | Both coders agree on freq. of categories |
| **Scott's $\pi$** | ✓ | | | | ✓ | ✓ | | | | ✓ | | Large sample sizes |
| **Fleiss' $K$** | ✓ | | ✓ | | ✓ | ✓ | | | | ✓ | | Large sample sizes |
| **Krippendorff's $_c\alpha_{metric}$** | ✓ | | ✓ | ✓ | ✓ | ✓ | ✓ | ✓ | ✓ | ✓ | ✓ | Always |
| **Krippendorff's $_u\alpha_{metric}$** | ✓ | ✓ | ✓ | ✓ | ✓ | ✓ | ✓ | ✓ | ✓ | ✓ | ✓ | Always |
| **Pearson's $r_{xy}$** | ✓ | | | | ✓ | | | ✓ | | | | Both coders agree on freq. of values |
| **Pearson's $r_{xx}$** | ✓ | | | | ✓ | | | ✓ | | ✓ | | Large sample sizes |
| **Lin's $\rho_c$** | ✓ | | | | ✓ | | | ✓ | | ✓ | | Large sample sizes |

**TABLE 4.1**
Comparison of Several Agreement Coefficients: The Reliability Data to Which They May Apply and the Conditions Under Which They Could Serve as Unbiased Indicators of the Reliability of Data

All coefficients listed in Table 4.1 are applicable to single-valued accounts of predefined units. Most of them are also limited to measuring the agreement between two observers, coders, raters, or judges, which is the minimum for any agreement measure and the special case of coefficients capable of handling more than two replications, already introduced in Chapter 2. Whether these coefficients are conceived of as agreement coefficients for nominal data or ordered data, all of them reduce to binary versions when only two values are used. This makes them more easily comparable. This chapter did not mention coefficients for evaluating multi-valued data, to be introduced in Chapter 7. As well as those applicable to unitizing and coding of textual continua, addressed in Chapters 8, 9, and 10. Lin's (1989) $\rho_c$ for interval data corresponds to Krippendorff's (1970a) $\alpha$-coefficient, which is not further elaborated here.

Regarding the above Table 4.1, I want to highlight five points made earlier, now more focused on comparisons.

The percent of observed agreement $P_o$ is virtually useless as its scale between 0 and 1 or 100% has no place for the variance in the reliability data on which reliability depends and it has no fixed point at which one can say that the data and the phenomena of analytical interest are totally unrelated, leaving all $P_o < 1$ difficult to interpret.

Inasmuch as in Bennett et al.'s $S$, $P_e = 1/v$ is a function of the fixed number $v$ of categories available in a coding instrument, it reveals progressively less about the reliability of data as human coders have no reasons to use categories equally often. $S$ is also vulnerable to researchers' decisions to add or remove unused categories from its calculations, opening the doors to inflate agreement without being related to the reliability of data. $S$ has other uses, however. For example, it can be used to establish the reliability of coding instructions (as opposed to the particular data they generate; an issue addressed in Section 14.2.3.2) provided one can find or create reliability data in which all categories occur equally often.

Cohen's $\kappa$ becomes inflated with increasing disagreement among the two coders' unequal predilection of categories. Zwick (1988) suggested to apply $\kappa$ only after the marginal homogeneity of categories tests positively. The need for this additional test renders Scott's $\pi$ superior to $\kappa$.

Gwet's $AC_1$ is interpretable as a reliability coefficient only when data are binary or dichotomous and the marginal frequencies of categories are uniformly distributed. This empirically rare condition renders $AC_1$'s application to larger data with unequal marginal frequencies no longer interpretable. $AC_1$-measures being generally larger than the other chance-corrected coefficients, it can deceive researchers into believing their data are reliable when the evidence that data represent something real remains in question.

Coefficients that calculate expectations by assuming reliability data to be infinite in size, Scott's $\pi$ and Fleiss' $K$, undervalue the reliability of small data. But as the size of the reliability data increases, $\pi$, $K$, and ${}_c\alpha$ become asymptotically equal.

The versatility of Krippendorff's ${}_c\alpha$ will be developed in the following chapters.

# 5

# *Metric Differences Between Single-Valued Units*

This chapter starts where Chapter 2 ended, with the definition of reliability data for replicability involving any number of observers assigning any number of single categories to any number of predefined units of phenomena. It generalizes the α for unordered categories, for nominal data, also discussed in the preceding Chapters 3 and 4. Its Table 4.1 compared twelve coefficients also regarding their applicability to metrics other than nominal – ordinal, interval, and ratio metrics – without defining them, which is the aim of this chapter.

While Chapter 2 provided the basic idea of the α-coefficient, Chapter 3 discussed its properties, and Chapter 4 compared $_c\alpha$ to other agreement coefficients, this chapter will not assume that the reader has worked through the three previous chapters. Definitions that are relevant for this chapter will be restated here. Also, to distinguish agreement coefficients obtained from reliability data that represent the coding of predefined units of analysis from other kinds of units, we add the prescript "*c*" to α and refer to this class of α-coefficients as $_c\alpha$-coefficients – already used in Table 4.1. The role of metrics in data is entirely general but exemplified by means of the simpler cα-coefficients.

As mentioned in Chapter 1, scholarship encourages quantification as soon as the volume of phenomena to be scrutinized exceeds the capability of individual researchers, needs to be communicated to hired coders and potentially reproduced, critically examined, or built upon by members of their communities. Statistical accounts, counting of observed phenomena, make sense primarily when phenomena appear in the form of independent units of analysis, whether they are individuals, objects, propositions, answers to interview questions, test subjects, medical diagnoses, accident reports, or votes. This chapter addresses reliability concerns that take the unitization of phenomena for granted, as natural or given, and in any case predefined. For many disciplines units of phenomena are not questioned. For example, psychologists, medical professionals, public opinion researchers, and criminal law take the distinction of individual human beings as natural, at least in the western world. Even sociologists who theorize larger units of analysis take individuals as parts of systems. The same can be said about words, sentences, paragraphs, book chapters, documents, and genres. While they are countable, their meanings depend on the larger context that determines them. Counting decontextualizes the units counted, renders their context unavailable for scholarly inquiries. However, in the social sciences, the physical forms of observable events, readable texts, video recordings, and artifacts are literally meaningless unless they are placed in the context that created them, relates the units counted to each other, or refers to what makes them meaningful in the context of their use.

For these reasons, literate human observers, coders, judges, are instrumental in creating data that conform to written coding instructions and represent what mechanical measuring instrument cannot possibly capture. To generate analyzable data, the task is to identify, unitize, and code relevant phenomena so that the units of analysis preserve the contexts of their origins and connections between them. Part of the context of decontextualized but appropriately coded units is how they relate to each other; their metric.

This chapter focuses on what I call the single-valued coding of predefined units of analysis by independently working observers. Predefined units should not be confused with the results of physical distinctions – 30 seconds of speech, pages in a book, checkmarks on multiple choice questions, individual words. Textual matter can be unitized variously, in terms of propositions, attributions of sentiments, speech acts, or scenes. Predefined units are analytically distinct but coded by the same coding instructions. While coding instructions may specify a complex web of variables, each consisting of different sets of mutually exclusive values, single-valued means that units are described in terms of one value from each variable. Subsequent chapters will expand these limitations, expanding replicability to other forms of reliability in Chapter 6, expanding single-valued to multi-valued coding in Chapter 7, and generalizing predefined units to units identified at the discretion of observers in Chapters 8 and 9.

This chapter expands the limitation of previous chapters to categorizations of predefined units, to capturing the intricacies of different data languages that define values relative to each other responding to contexts of phenomena that imply several kinds of orderings.

## 5.1 Reliability Data for the Replicability of Coding Predefined Units

Chapter 2 ended by defining the reliability data for replicability in its canonical form as in Table 2.4, renumbered here Table 5.1.

Just as in Tables 2.1 and 2.4, Table 5.1 is defined by:

| | | | | | |
|---|---|---|---|---|---|
| **Units:** | | 1 2 3 . . | $u$ . . . . . . . | $N_u$ | |
| **Replications:** | 1 | $c_{11}$ . . . . | $c_{1u}$ . . . . . . . | $c_{1N_u}$ | |
| | : | : | : | : | |
| | $i$ | $c_{i1}$ . . . . | $c_{iu}$ . . . . . . . | $c_{iN_u}$ | |
| | $j$ | $c_{j1}$ . . . . | $c_{ju}$ . . . . . . . | $c_{jN_u}$ | |
| | : | : | : | : | |
| | $m$ | $c_{m1}$ . . . . | $c_{mu}$ . . . . . . . | $c_{mN_u}$ | |
| **Sums of pairable values** > 1 | | $m_1$ . . . . | $m_u$ . . . . . . . | $m_{N_u}$ | $n_{..} = \sum_{u=1}^{N_u} m_u$ |

**TABLE 5.1**
Canonical Form of Reliability Data for Any Number of Replications

- A predefined set of distinct phenomena: 1, 2, …, $u$, …, $N_u$, the units of analysis
- Up to $m$ replications: 1, 2, …, $i$, $j$, …, $m$, observers, coders, measuring devices, or test results

- Units are single-valued. The value $c_{iu}$ is assigned by observer $i$ to unit $u$
- The values assigned to units $u$ are pairable, their number is $m \geq m_u > 1$
- Because replications may not cover all units, the total number of pairable values $n.. \leq m \times N_u$

For such data, observed coincidences are defined, just as in (2.10). There are $m_u(m_u - 1)$ pairs among $m_u$ values within units $u$. In order for each value to contribute exactly one to the observed coincidences $n_{ck}$ as in Table 2.2, (5.1) divides the number of pairs by $m_u - 1$:

$$n_{ck} = \sum_{u=1}^{N_u} \frac{1}{m_u - 1} \sum_{i=1}^{m} \sum_{j \neq i}^{m} 1 \text{ iff } c_{iu} \text{ and } k_{ju} \text{ and } m_u > 1 \tag{5.1}$$

These coincidences are tabulated in the center of Table 5.1

The expected coincidences are defined as in (2.3) in terms of the marginal sums in the matrix of observed coincidences $n_c$ and $n_k$ and their totals $n..$ by:

$$e_{ck} = \frac{n_c}{n..-1} \begin{cases} n_k & \text{iff } c \neq k \\ (n_k - 1) & \text{iff } c = k \end{cases} \tag{5.2}$$

Recall from Chapter 3 the importance of being consistent in not pairing values with themselves when calculating observed and expected coincidences. In (5.1) this is prevented by the inequality of the second sum. It excludes values $c_{iu} = k_{iu}$ from being counted in $n_{ck}$. In the definition of the expected coincidences (5.2) this exclusion is accomplished by the $n_k - 1$ in its numerator and $n..-1$ in its denominator.

These coincidences gave rise to two $v$-by-$v$ matrices in Tables 2.2 and 2.4 to which Table 5.2 adds on its left a third coincidence matrix containing only perfect matches of all coincidences.

| **Values:** | 1 | . | c | k | . | v |
|---|---|---|---|---|---|---|
| 1 | $n_1$ | . | . | . | . | . |
| . | . | . | . | . | . | . |
| c | . | . | $n_c$ | . | . | . |
| k | . | . | . | $n_k$ | . | . |
| . | . | . | . | . | . | . |
| v | . | . | . | . | . | $n_v$ |
| | $n_1$ | . | $n_c$ | $n_k$ | . | $n_v$ |

$_c\alpha = 1.000$

| | 1 | . | c | k | . | v |
|---|---|---|---|---|---|---|
| 1 | $n_{11}$ | . | $n_{1c}$ | $n_{1k}$ | . | $n_{1v}$ |
| . | . | . | . | . | . | . |
| c | $n_{c1}$ | . | $n_{cc}$ | $n_{ck}$ | . | $n_{cv}$ |
| k | $n_{k1}$ | . | $n_{kc}$ | $n_{kk}$ | . | $n_{kv}$ |
| . | . | . | . | . | . | . |
| v | $n_{v1}$ | . | $n_{vc}$ | $n_{vk}$ | . | $n_{vv}$ |
| | $n_1$ | . | $n_c$ | $n_k$ | . | $n_v$ |

$1 \geq {}_c\alpha_{\text{metric}} \geq -1$

| | 1 | . | c | k | . | v | |
|---|---|---|---|---|---|---|---|
| 1 | $e_{11}$ | . | $e_{1c}$ | $e_{1k}$ | . | $e_{1v}$ | $n_1$ |
| . | . | . | . | . | . | . | . |
| c | $e_{c1}$ | . | $e_{cc}$ | $e_{ck}$ | . | $e_{cv}$ | $n_c$ |
| k | $e_{k1}$ | . | $e_{kc}$ | $e_{kk}$ | . | $e_{kv}$ | $n_k$ |
| . | . | . | . | . | . | . | . |
| v | $e_{v1}$ | . | $e_{vc}$ | $e_{vk}$ | . | $e_{vv}$ | $n_v$ |
| | $n_1$ | . | $n_c$ | $n_k$ | . | $n_v$ | $n..$ |

$_c\alpha = 0 \pm$ Granularity

**TABLE 5.2**
Matrices of Perfect, Observed and Expected Coincidences

The entries of all coincidence matrices are symmetrical around their diagonal, $n_{ck} = n_{kc}$, and rows and columns sums are the same for each value:

$$n_c = \sum_{k=1}^{v} n_{ck} = \sum_{k=1}^{v} e_{ck}$$

$$n_k = \sum_{c=1}^{v} n_{ck} = \sum_{c=1}^{v} e_{ck}$$

$$n.. = \sum_{c=1}^{v}\sum_{k=1}^{v} n_{ck} = \sum_{c=1}^{v}\sum_{k=1}^{v} e_{ck}$$

This is how far Chapter 2 went. The following generalizes its nominal ${}_c\alpha$-agreement coefficient to all metrics:

## 5.2 Metric *c*-Alphas

Unordered values either match or they don't. There is no middle ground. It follows that all mismatches among unordered values are of the same magnitude. In coincidence matrices, their sums complement the sums of perfect matching values in diagonal cells. The uniform evaluation of mismatching values encouraged algebraic simplifications that the previous chapters took advantage of.

However, any ordering of values introduces unequal differences between pairs of values. They do not affect the diagonals of coincidence matrices but make a difference in their off-diagonal cells. Following the history of α, narrated in Chapter 0, this chapter recognizes that the differences between mismatching pairs of values may vary in magnitude reflecting the strength of different metrics which weigh the off-diagonal coincidences accordingly.

All metric difference functions ${}_{\text{metric}}\delta^2_{ck}$ have the following properties. They are:

Symmetrical: $${}_{\text{metric}}\delta^2_{ck} = {}_{\text{metric}}\delta^2_{kc} \quad (5.3)$$

Measure: $${}_{\text{metric}}\delta^2_{c=k} = 0 \quad (5.4)$$

And are standardized: $$0 \le {}_{\text{metric}}\delta^2_{ck} \le 1 \text{ (convenient comparatively)} \quad (5.5)$$

Standardization is not a formal requirement. However, it makes the tabular representations of metric-related differences more comprehensible.

All difference functions weigh the coincidences discussed in Chapter 2, here defined by (5.1) and (5.2) and featured in Table 5.2. The observed disagreement $D_o$, defined in (2.2), accommodates only nominal data by summing the off-diagonal coincidences, implicitly giving each the same weight. This is where different metric difference function ${}_{\text{metric}}\delta^2_{ck}$ apply, already discussed in Chapter 0:

$${}_{\text{metric}}D_o = \sum_{c=1}^{v}\sum_{k=1}^{v} n_{ck}\ {}_{\text{metric}}\delta^2_{ck} \quad (5.6)$$

The same logic applies to the first and third expression of the expected disagreements $D_e$ in (2.4) yielding:

$${}_{\text{metric}}D_e = \sum_{c=1}^{v}\sum_{k=1}^{v} e_{ck}\ {}_{\text{metric}}\delta^2_{ck} = \frac{1}{n..-1}\sum_{c=1}^{v} n_c \sum_{k=1}^{v} n_k\ {}_{\text{metric}}\delta^2_{ck} \quad (5.7)$$

In these terms the metric *c*-Alphas become:

$$
{}_c\alpha_{\text{metric}} = 1 - \frac{{}_{\text{metric}}D_o}{{}_{\text{metric}}D_e} = 1 - \frac{\sum_{c=1}^{v}\sum_{k=1}^{v} n_{ck\ \text{metric}}\delta_{ck}^2}{\sum_{c=1}^{v}\sum_{k=1}^{v} e_{ck\ \text{metric}}\delta_{ck}^2}
= 1-(n..-1)\frac{\sum_{c=1}^{v}\sum_{k=1}^{v} n_{ck\ \text{metric}}\delta_{ck}^2}{\sum_{c=1}^{v} n_c \sum_{k=1}^{v} n_{k\ \text{metric}}\delta_{ck}^2} \quad (5.8)
$$

${}_c\alpha_{\text{metric}}$ in (5.8) satisfies the numerical condition of an agreement coefficient with reliability interpretations by conforming to the required definition of its range. According to (5.4) differences between matching values are zero, ${}_{\text{metric}}\delta_{cc}^2 = 0 = {}_{\text{metric}}\delta_{kk}^2$. If all coincidences match perfectly, and occur in the diagonal of a matrix of observed coincidences, the observed disagreement is ${}_{\text{metric}}D_o = 0$, which renders ${}_c\alpha_{\text{metric}} = 1$. Regarding the zero point of ${}_c\alpha$'s scale, when data are unrelated to the phenomena coded, the observed and expected coincidences are equal, ${}_{\text{metric}}D_o = {}_{\text{metric}}D_e$, and ${}_c\alpha_{\text{metric}} = 0$. As discussed in previous chapters, ${}_c\alpha_{\text{metric}}$ may become negative when ${}_{\text{metric}}D_o > {}_{\text{metric}}D_e$, a condition that is unlike to occur by chance. It can happen mainly when the preparatory conditions for generating reliability data, discussed in Sections 1.3 and 2.1.4 are violated. For data to be arguably reliable, α-coefficients need to deviate only very little from perfect agreement. However, α's zero value serves as the point of reference of the degree to which data provide information about the phenomena coded. (2.6) defined the scale within which α can be interpreted as a reliability measure by:

$$
-1 \le \left.\begin{matrix}-\text{Systematic disagreements}\\ \pm\text{The granularity of data}\end{matrix}\right\} 0 \le {}_c\alpha_{\text{metric}} \le +1 \quad (5.9)
$$

Whereas systematic disagreements always reduce α-values, often unnoticed when α is large, the granularity of data is due to imprecision and affects α-values bidirectionally. When agreement is perfect, α = 1, granularity plays no role. When α = 0 the granularity may cause minuscule deviations from zero due to the simple mathematical fact that fractions as in (5.2) and (5.8) rarely yield integers. However, when reliability data are large, the effects of granularity vanish.

It is worth realizing that the simplifications discussed in Chapters 2 and 4 are due to taking advantage of the algebraic properties of the nominal difference function, already mentioned (0.4), now stated in the terms to be used from here on:

$$
{}_{\text{nominal}}\delta_{ck}^2 = \begin{cases} 0 & \text{iff } c = k \\ 1 & \text{iff } c \neq k \end{cases} \quad (5.10)
$$

(5.10) inserted in (5.8) yields the algebraic simplifications (2.7), discussed in Chapter 2:

$$
{}_c\alpha_{\text{nominal}} = 1 - \frac{{}_{\text{nominal}}D_o}{{}_{\text{nominal}}D_e} = 1 - \frac{\sum_{c=1}^{v}\sum_{k=1}^{v} n_{ck\ \text{nominal}}\delta_{ck}^2}{\sum_{c=1}^{v}\sum_{k=1}^{v} e_{ck\ \text{nominal}}\delta_{ck}^2}
= 1 - \frac{n.. - \sum_{c=1}^{v} n_{cc}}{n.. - \sum_{c=1}^{v} e_{cc}} = 1 - \frac{n.. - \sum_{c=1}^{v} n_{cc}}{n.. - \sum_{c=1}^{v} \frac{n_c(n_c-1)}{n..-1}}
$$

(2.8) also showed the condition under which ${}_c\alpha_{\text{nominal}}$ can formally resemble chance-corrected agreement coefficients:

$$ {}_c\alpha_{\text{nominal}} = 1 - \frac{D_o}{D_e} = \frac{\sum_{c=1}^{v} n_{cc} - \sum_{c=1}^{v} e_{cc}}{n_{..} - \sum_{c=1}^{v} e_{cc}} = \frac{P_o - P_e}{1 - P_e} \neq \kappa $$

As discussed in Chapter 4, formal resemblances can be deceptive. The difference between the measure of replicability ${}_c\alpha_{\text{nominal}}$ and Cohen's (1960) κ is due to unlike expressions for the proportion of the expected agreement $P_e$. In ${}_c\alpha$, $P_e = \left(\sum_{c=1}^{v} n_c(n_c - 1)\right) / n_{..}(n_{..} - 1)$ is the proportion of agreements among many pairs of observers, expected when no relationship between data and the coded phenomena exists. In κ, $P_e = \left(\sum_{c=1}^{v} n_{c\cdot} \sum_{k=c}^{v} n_{\cdot k}\right) / n^2_{..}$ is the proportion of agreements expected when two coders with unequal frequencies of using categories *c* and *k* are statistically unrelated. The latter has little to do with the reliability of data. Moreover, the effect of not relating values with themselves, neither on the observed disagreement nor in the expected disagreement, renders ${}_c\alpha$ responsive to small sample size, unlike π, κ, and *K* which express $D_e$ (but not $D_o$) as if data were of infinite size.

In addition to these divergent conceptions of chance, there are two more reasons for not conceiving ${}_c\alpha$ as a chance-corrected agreement coefficient. First, coefficients like Bennett et al.'s (1954) *S*, Scott's π (1955), and Cohen's (1960) κ, are a function of percent agreements among two replications. A focus on percentages makes it more difficult to generalize agreement to more than two independent observers, coders, or judges. The involvement of more than two replications has encouraged coefficients taking disagreements as the absence of unanimous consent, as proposed by Craig (1981), von Eye (2006), and others, which cannot capture nuances of the agreement/disagreement dichotomy. Second, inasmuch as the ordering of values – ranks, Likert-type scales, or proportions – involve more nuanced differences, the either-or differences of nominal data, the conceptions of majorities, and the form of chance-corrected agreement coefficients are difficult to maintain. Therefore ${}_c\alpha$ is defined in terms of the proportion of disagreements starting with (0.1), (2.5), (2.7), and (5.8).

## 5.3 Metric Differences for Single-Valued Data

(5.3), (5.4), and (5.5) introduced the concept of metric difference functions, and (5.7) defined the one for nominal data and showed why the latter results in simple forms of ${}_c\alpha$, discussed in Chapters 2, 3, and 4. (5.11) illustrates how (5.8) copes with all sorts of metrics underlying reliability data by weighing the entries in observed and expected coincidence matrices according to the properties of these metrics:

$$ {}_c\alpha_{\text{metric}} = 1 - \frac{{}_{\text{metric}}D_o}{{}_{\text{metric}}D_e} = 1 - \frac{\sum_{c=1}^{v}\sum_{k=1}^{v} n_{ck}\;{}_{\text{metric}}\delta^2_{ck}}{\sum_{c=1}^{v}\sum_{k=1}^{v} e_{ck}\;{}_{\text{metric}}\delta^2_{ck}} = 1 - \frac{\sum \boxed{n_{ck}} \times \boxed{{}_{\text{metric}}\delta^2_{ck}}}{\sum \boxed{e_{ck}} \times \boxed{{}_{\text{metric}}\delta^2_{ck}}} \qquad (5.11) $$

The following defines four common and two less common difference functions, giving their mathematical form and a numerical illustration of how this function behaves, starting with the nominal difference, now obvious.

### 5.3.1 The Nominal Metric Difference Function

The difference function for nominal data was defined in (2.3) and (5.6) and sufficiently discussed and is mentioned here only for comparison:

*Nominal* metric differences: $_{\text{nominal}}\delta^2_{ck} = \begin{cases} 0 & \text{iff } c = k \\ 1 & \text{iff } c \neq k \end{cases}$ where $c$ and $k$ are categories

Table 5.3 exemplifies these differences within 10 categories, the first five of which were used in the example discussed in Chapter 2:

| **Categories:** | 🕮 | 🖂 | ☎ | 🖳 | 🗁 | 🖫 | 🖎 | 🗏 | 🗋 | 🕭 |
|---|---|---|---|---|---|---|---|---|---|---|
| 🕮 | **0** | 1 | 1 | 1 | 1 | 1 | 1 | 1 | 1 | 1 |
| 🖂 | 1 | **0** | 1 | 1 | 1 | 1 | 1 | 1 | 1 | 1 |
| ☎ | 1 | 1 | **0** | 1 | 1 | 1 | 1 | 1 | 1 | 1 |
| 🖳 | 1 | 1 | 1 | **0** | 1 | 1 | 1 | 1 | 1 | 1 |
| 🗁 | 1 | 1 | 1 | 1 | **0** | 1 | 1 | 1 | 1 | 1 |
| 🖫 | 1 | 1 | 1 | 1 | 1 | **0** | 1 | 1 | 1 | 1 |
| 🖎 | 1 | 1 | 1 | 1 | 1 | 1 | **0** | 1 | 1 | 1 |
| 🗏 | 1 | 1 | 1 | 1 | 1 | 1 | 1 | **0** | 1 | 1 |
| 🗋 | 1 | 1 | 1 | 1 | 1 | 1 | 1 | 1 | **0** | 1 |
| 🕭 | 1 | 1 | 1 | 1 | 1 | 1 | 1 | 1 | 1 | **0** |

**TABLE 5.3**
A Matrix Exemplifying Nominal Metric Differences

### 5.3.2 The Ordinal Metric Difference Function

Ordinal data consist of ranked values. In the social sciences ordinal data are the second most common form of data after unordered categorizations. This is so especially when verbal attributes are coded. Almost all comparisons are expressed in terms of more or less, better or worse, faster or slower; for example in Likert-scale variables, such as "never," "rarely," "sometimes," "often," and "always." Some researchers assign numerical values to ranks such as 1st, 2nd, 3rd, etc. However, this can only mean that the 1st rank precedes, is preferred, or has more of any subsequent rank. Using numbers to designate ranks – 1st, 2nd, 3rd, …, last – does not justify algebraic operations like adding to or subtracting from ranks. The tallest and the shortest basketball player of a team can be measured in feet, inches, or centimeters, but sports commentators are more likely to refer to their sizes relative to the sizes of the members of their team. We talk of cars to be most, average, or least fuel efficient. Once miles per gallon are converted into ranks, we can converse about them but not calculate with them. Often quantitative relationships can be seriously distorted by their attributes. For example, Landis and Koch (1977) partitioned the scale of Cohen's kappa values into six intervals and associated each with ranked qualities:

| | | | | | |
|---|---|---|---|---|---|
| 1.00–0.81 > | 0.80–0.61 > | 0.60–0.41 > | 0.40–0.21 > | 0.20–0.00 > | negative |
| Almost Perfect > | Substantial > | Moderate > | Fair > | Slight > | Poor |

Replacing a numerical scale by rank-ordered verbalizations may make it easier to conceptualize the meaning of a scale. However, in this example ranks are quite misleading. For example, in view of the discussion of (5.9), to consider negative κ-values indicative of poor reliability is vastly misleading.

However, I am interested here in defining ordinal differences between ranks without a judgment of their appropriateness by other criteria, merely in terms of their use. I am expressing ordinal differences as a function of the number of ranks that distinguish one rank from another. This conforms to how Spearman's (1904) rank order correlation coefficient rho accounts for the difference between any two ranks.

For symmetrical coincidence matrices, with common marginal sums, and only these, for any two ranks $c$ and $k$:

$$\textit{Ordinal}\text{ metric differences: } {}_{ordinal}\delta^2_{ck} = \left( \frac{\sum_{b=c}^{b=k} n_{b\cdot} - \frac{n_{c\cdot} + n_{k\cdot}}{2}}{n_{\cdot\cdot} - \frac{n_{\text{first}\cdot} + n_{\text{last}\cdot}}{2}} \right)^2 \qquad (5.12a)$$

Table 5.4 exemplifies the ordinal differences for *coincidences* among five rankings of 10 units:

$$n_1 = 2,\ n_2 = 3,\ n_3 = 1,\ n_4 = 2, \text{ and } n_6 = 2$$

both tabularly and summarily:

| **Ranks:** | 1st | 1st | 2nd | 2nd | 2nd | 3rd | 4th | 4th | 6th | 6th |
|---|---|---|---|---|---|---|---|---|---|---|
| 1st | **0** | **0** | 0.10 | 0.10 | 0.10 | 0.32 | 0.56 | 0.56 | 1 | 1 |
| 1st | **0** | **0** | 0.10 | 0.10 | 0.10 | 0.32 | 0.56 | 0.56 | 1 | 1 |
| 2nd | 0.10 | 0.10 | **0** | **0** | **0** | 0.06 | 0.19 | 0.19 | 0.47 | 0.47 |
| 2nd | 0.10 | 0.10 | **0** | **0** | **0** | 0.06 | 0.19 | 0.19 | 0.47 | 0.47 |
| 2nd | 0.10 | 0.10 | **0** | **0** | **0** | 0.06 | 0.19 | 0.19 | 0.47 | 0.47 |
| 3rd | 0.32 | 0.32 | 0.06 | 0.06 | 0.06 | **0** | 0.04 | 0.04 | 0.19 | 0.19 |
| 4th | 0.56 | 0.56 | 0.19 | 0.19 | 0.19 | 0.04 | **0** | **0** | 0.06 | 0.06 |
| 4th | 0.56 | 0.56 | 0.19 | 0.19 | 0.19 | 0.04 | **0** | **0** | 0.06 | 0.06 |
| 6th | 1 | 1 | 0.47 | 0.47 | 0.47 | 0.19 | 0.06 | 0.06 | **0** | **0** |
| 6th | 1 | 1 | 0.47 | 0.47 | 0.47 | 0.19 | 0.06 | 0.06 | **0** | **0** |

Reduces to:

| | 1st | 2nd | 3rd | 4th | 5th | 6th |
|---|---|---|---|---|---|---|
| 1st | **0** | 0.10 | 0.32 | 0.56 | 0.88 | 1 |
| 2nd | 0.10 | **0** | 0.06 | 0.19 | 0.56 | 0.69 |
| 3rd | 0.32 | 0.06 | **0** | 0.04 | 0.31 | 0.44 |
| 4th | 0.56 | 0.19 | 0.04 | **0** | 0.13 | 0.25 |
| 5th | 0.88 | 0.56 | 0.31 | 0.13 | **0** | 0.13 |
| 6th | 1 | 0.69 | 0.44 | 0.25 | 0.13 | **0** |

$c =$

$\sum_{b=c}^{b=k} n_{b\cdot} - \frac{n_{c\cdot} + n_{k\cdot}}{2} =$

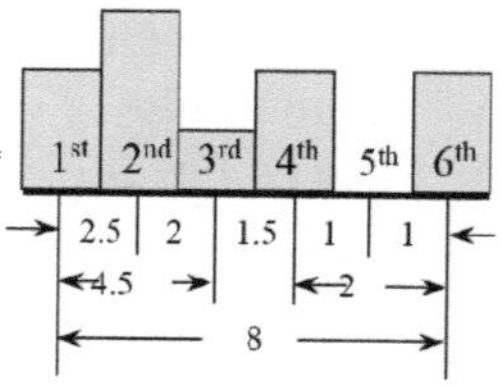


**TABLE 5.4**
A Matrix Exemplifying Ordinal Metric Differences Among Coincidences

Note that (5.12a) applies to coincidence matrices which are symmetrical, and their row and column sums have identical distributions. When reliability data are represented by asymmetrically distributed contingencies: $x_{ck} \neq x_{kc}$ and $x_{c\cdot} \neq x_{\cdot c}$, such as when accuracies are measured, ordinal metric differences must account for these asymmetries by generalizing (5.12a) to (5.12b):

$$\textit{Ordinal}\text{ metric differences: } {}_{\text{ordinal}}\delta_{ck}^2 = \left( \frac{\sum_{b=\text{first}}^{b=c} x_{b\cdot} - \frac{x_{c\cdot}}{2} - \sum_{b=\text{first}}^{b=k} x_{\cdot b} + \frac{x_{\cdot k}}{2}}{n_{\cdot\cdot} - \frac{min(x_{\text{first}\cdot}, x_{\cdot\text{first}}) + min(x_{\text{last}\cdot}, x_{\cdot\text{last}})}{2}} \right)^2 \qquad (5.12b)$$

To illustrate the results of this function, Table 5.5 tabulates the ordinal differences of a numerical example involving marginal sums of a contingency matrix of ranks. Columns sums $x_{\cdot c}$ result from pairing coders' ranks with a trusted standard and row sums $x_{c\cdot}$ represent the frequency and ranks used by coders:

By standard: $x_{\cdot 1} = 1$, $x_{\cdot 2} = 1$, $x_{\cdot 3} = 1$, $x_{\cdot 4} = 1$, $x_{\cdot 5} = 1$, $x_{\cdot 6} = 1$, $x_{\cdot 7} = 1$, $x_{\cdot 8} = 1$, and $x_{\cdot 9} = 2$
By coders: $x_{1\cdot} = 2$, $x_{2\cdot} = 3$, $x_{3\cdot} = 1$, $x_{4\cdot} = 2$, and $x_{6\cdot} = 2$

| **Standard's ranking:** / **Coders' ranking:** | 1st | 2nd | 3rd | 4th | 5th | 6th | 7th | 8th | 9th | $x_{c\cdot}=$ |
|---|---|---|---|---|---|---|---|---|---|---|
| 1st | 0.004 | 0.004 | 0.03 | 0.09 | 0.17 | 0.28 | 0.42 | 0.58 | 0.89 | 2 |
| 2nd | 0.12 | 0.06 | 0.01 | **0** | 0.01 | 0.06 | 0.12 | 0.22 | 0.42 | 3 |
| 3rd | 0.35 | 0.22 | 0.12 | 0.06 | 0.01 | **0** | 0.01 | 0.06 | 0.17 | 1 |
| 4th | 0.58 | 0.42 | 0.28 | 0.17 | 0.09 | 0.04 | 0.004 | 0.004 | 0.06 | 2 |
| 6th | 1 | 0.78 | 0.58 | 0.42 | 0.28 | 0.17 | 0.09 | 0.04 | **0** | 2 |
| $x_{\cdot k}=$ | 1 | 1 | 1 | 1 | 1 | 1 | 1 | 1 | 2 | |

**TABLE 5.5**
A Matrix Exemplifying Ordinal Metric Differences Among Contingencies

Note: As Table 5.2 demonstrates, in coincidence matrices, the row and column sums are identical, and the diagonals of the matrix of all differences are zero. When marginal sums are unequal, as in contingency matrices, zero differences appear only where two ranks assume the same position in the distribution of ranks. Table 5.5 includes three such situations. Obviously, the largest rank of the standard equals the largest rank judged by coders. Also the standard's 6th rank and the coders' 3rd rank agree regarding the number of smaller and the number of larger ranks.

### 5.3.3 The Interval Metric Difference Function

Interval data are familiar through the use of scales whose values can be subjected to algebraic additions and subtractions. There is no need to assume that values are equally spaced along a scale, only that their algebraic differences matter. The interval metric difference function is defined by:

$$\textit{Interval}\text{ metric differences: } {}_{\text{interval}}\delta_{ck}^2 = \left( \frac{c-k}{c_{\text{largest}} - c_{\text{smallest}}} \right)^2 \qquad (5.13)$$

and their differences are exemplified in Table 5.6:

| **Interval values:** | 0 | 1 | 2 | 3 | 4 | 5 | 6 | 7 | 8 | 9 |
|---|---|---|---|---|---|---|---|---|---|---|
| 0 | **0** | 0.01 | 0.05 | 0.11 | 0.20 | 0.31 | 0.44 | 0.60 | 0.79 | 1 |
| 1 | 0.01 | **0** | 0.01 | 0.05 | 0.11 | 0.20 | 0.31 | 0.44 | 0.60 | 0.79 |
| 2 | 0.05 | 0.01 | **0** | 0.01 | 0.05 | 0.11 | 0.20 | 0.31 | 0.44 | 0.60 |
| 3 | 0.11 | 0.05 | 0.01 | **0** | 0.01 | 0.05 | 0.11 | 0.20 | 0.31 | 0.44 |
| 4 | 0.20 | 0.11 | 0.05 | 0.01 | **0** | 0.01 | 0.05 | 0.11 | 0.20 | 0.31 |
| 5 | 0.31 | 0.20 | 0.11 | 0.05 | 0.01 | **0** | 0.01 | 0.05 | 0.11 | 0.20 |
| 6 | 0.44 | 0.31 | 0.20 | 0.11 | 0.05 | 0.01 | **0** | 0.01 | 0.05 | 0.11 |
| 7 | 0.60 | 0.44 | 0.31 | 0.20 | 0.11 | 0.05 | 0.01 | **0** | 0.01 | 0.05 |
| 8 | 0.79 | 0.60 | 0.44 | 0.31 | 0.20 | 0.11 | 0.05 | 0.01 | **0** | 0.01 |
| 9 | 1 | 0.79 | 0.60 | 0.44 | 0.31 | 0.20 | 0.11 | 0.05 | 0.01 | **0** |

**TABLE 5.6**
A Matrix Exemplifying Interval Metric Differences

A characteristic of interval differences is that its lines of equal differences are parallel to its diagonal of zero differences. As numerical differences $|c–k|$ among pairs of values increase, ${}_{\text{interval}}\delta^2_{ck}$ increases exponentially to its maximum of 1.000.

The interval metric difference is also implicit in the analysis of variance, already discussed with the help of (0.2), and numerous related statistics. Its use in the ${}_{c}\alpha_{\text{interval}}$ measure assesses the very unreliabilities that negatively affect virtually all interval statistics.

When the frequencies of all ranks are equal, ordinal differences equal the interval differences provided the ranks are considered consecutively numbered. The interval difference function can be applied to coincidence and contingency matrices alike, although when the latter does not have the same values in rows and columns, their matrix representation is not as neat as in Table 5.6.

### 5.3.4 The Ratio Metric Difference Function

Data that possess a ratio metric are valued by reference to a zero point: proportions, distances traveled, time elapsed, age, and temperature in degrees Kelvin have no negative values and no upper limit. Hence, the values $c$ and $k$ of ratio data cannot be negative. They must be $\geq 0$. Ratio metric differences between any two values are expressed relative to how far they are from their absolute zero. They conform to the experience that losing a dollar means much more to someone who has only a few than to someone who is rich. It accounts for the fact that guessing the age of a baby is easier than guessing the age of an adult. The ratio metric difference function is defined by:

$$\textit{Ratio} \text{ metric differences: } {}_{\text{ratio}}\delta^2_{ck} = \left(\frac{c-k}{c+k}\right)^2 \qquad (5.14)$$

Its differences are exemplified in Table 5.7 for its first ten values:

| **Ratio values:** | 0 | 1 | 2 | 3 | 4 | 5 | 6 | 7 | 8 | 9 |
|---|---|---|---|---|---|---|---|---|---|---|
| 0 | **0** | 1 | 1 | 1 | 1 | 1 | 1 | 1 | 1 | 1 |
| 1 | 1 | **0** | 0.11 | 0.25 | 0.36 | 0.44 | 0.51 | 0.56 | 0.60 | 0.64 |
| 2 | 1 | 0.11 | **0** | 0.04 | 0.11 | 0.18 | 0.25 | 0.31 | 0.36 | 0.41 |
| 3 | 1 | 0.25 | 0.04 | **0** | 0.02 | 0.06 | 0.11 | 0.16 | 0.21 | 0.25 |
| 4 | 1 | 0.36 | 0.11 | 0.02 | **0** | 0.01 | 0.04 | 0.07 | 0.11 | 0.15 |
| 5 | 1 | 0.44 | 0.18 | 0.06 | 0.01 | **0** | 0.01 | 0.03 | 0.05 | 0.08 |
| 6 | 1 | 0.51 | 0.25 | 0.11 | 0.04 | 0.01 | **0** | 0.01 | 0.02 | 0.04 |
| 7 | 1 | 0.56 | 0.31 | 0.16 | 0.07 | 0.03 | 0.01 | **0** | 0.004 | 0.02 |
| 8 | 1 | 0.60 | 0.36 | 0.21 | 0.11 | 0.05 | 0.02 | 0.004 | **0** | 0.003 |
| 9 | 1 | 0.64 | 0.41 | 0.25 | 0.15 | 0.08 | 0.04 | 0.02 | 0.003 | **0** |

$$\frac{c-k}{c+k} = \tan\theta$$

**TABLE 5.7**
A Matrix Exemplifying Ratio Metric Differences

Notice that all lines of equal differences emerge from the zero point of a ratio scale. Numerical differences as in (5.13) count more near that zero point than when they are among larger values. For example, the difference between 8 and 9 is 0.003. The difference between 1 and 2 is 0.11, and all differences with zero are at their maximum of 1.00. Mathematically the proportion $(c-k)/(c+k)$ equals the tangent, ranging from 0° to 45°.

### 5.3.5 The Polar Metric Difference Function

Polar or bi-polar data are numerical data that are confined by two extremes, called poles. Whereas interval data are open ended on both sides of a scale, polar data are defined by known extremes. For example, when nurses ask patients to rate their pain on a scale from 1 to 10. The absence of pain and its maximum are the defining values of a pain scale between which patients are asked to locate the pain they experience. The semantic differential, an investigative tool common in psycholinguistics, asks subjects to judge people, products, feelings, or ideas on a numerical scale between –3 and +3. Polar opposites labeled "good/bad," "strong/weak," "fast/slow," "attractive/repulsive," etc. define their extremes that cannot be exceeded. Subjects are expected to know the meanings of these poles and are asked to indicate their responses within this seven-point scale without further instruction. The only intuition is its midpoint, zero, which can be interpreted as neither.

In the psychometric literature there is an ongoing debate on whether such scales should be considered ordinal, interval or polar data. Ordinal metrics do not handle quantitative differences, and interval metrics assume that the scale has no endpoints. Both characteristics are questionable on conceptual grounds. The polar metric difference function recognizes that distinctions near the defined extremes are finer and their differences weigh more than distinctions in the undefined middle. It is defined by:

$$\textit{Polar}\text{ metric differences}: {}_{\text{polar}}\delta^2_{ck} = \frac{(c-k)^2}{(c+k-2c_{min})(2c_{max}-c-k)} \tag{5.15}$$

Its differences within a ten-point scale are exemplified in Table 5.8:

| **Bi-polar values:** | 0 | 1 | 2 | 3 | 4 | 5 | 6 | 7 | 8 | 9 |
|---|---|---|---|---|---|---|---|---|---|---|
| 0 | **0** | 0.06 | 0.13 | 0.20 | 0.29 | 0.38 | 0.50 | 0.64 | 0.80 | 1 |
| 1 | 0.06 | **0** | 0.02 | 0.07 | 0.14 | 0.22 | 0.32 | 0.45 | 0.60 | 0.80 |
| 2 | 0.13 | 0.02 | **0** | 0.02 | 0.06 | 0.12 | 0.20 | 0.31 | 0.45 | 0.64 |
| 3 | 0.20 | 0.07 | 0.02 | **0** | 0.01 | 0.05 | 0.11 | 0.20 | 0.32 | 0.50 |
| 4 | 0.29 | 0.14 | 0.06 | 0.01 | **0** | 0.01 | 0.05 | 0.12 | 0.22 | 0.38 |
| 5 | 0.38 | 0.22 | 0.12 | 0.05 | 0.01 | **0** | 0.01 | 0.06 | 0.14 | 0.29 |
| 6 | 0.50 | 0.32 | 0.20 | 0.11 | 0.05 | 0.01 | **0** | 0.02 | 0.07 | 0.20 |
| 7 | 0.64 | 0.45 | 0.31 | 0.20 | 0.12 | 0.06 | 0.02 | **0** | 0.02 | 0.13 |
| 8 | 0.80 | 0.60 | 0.45 | 0.32 | 0.22 | 0.14 | 0.07 | 0.02 | **0** | 0.06 |
| 9 | 1 | 0.80 | 0.64 | 0.50 | 0.38 | 0.29 | 0.20 | 0.13 | 0.06 | **0** |

**TABLE 5.8**
A Matrix Exemplifying Polar Metric Differences

When comparing Tables 5.6 and 5.8 one may notice that the interval differences correspond to the polar differences in the center of the polar scale, but only those. As mismatching values come closer to their poles they increase significantly. In fact, near both poles, the polar differences behave like the ratio metric differences, but near their middle, the polar differences behave like interval metric differences.

Whenever verbal descriptions of polar opposites leave open how coders judge what lies between them, there is much to be said in favor of using polar difference functions rather than interval difference functions. For one, in practice, most rating scales are finite. Likert-type more-or-less judgments almost inevitably imply conceptual extremes. For another reason, especially when scale-points between defined poles are left open for coders to decide, one can rarely assume them to be judged as having equal intervals. Small deviations from any one pole tend to be more serious that small deviations in the neutral middle of a polar scale. The polar metric difference function in (5.15) approximates this intuition.

### 5.3.6 The Circular Metric Difference Function

Circular data emerge in recursions. A clock defines intervals between any two points it revisits, typically a minute, an hour, and 12 hours. A city bus makes its round and comes back to where it started – starting and end points are quite arbitrary, for any rider who joins the loop may have their own. Many biological phenomena are conceptualized in cycles. Workers on assembly lines perform their task over and over again, at each moment thinking how far they have advanced and how far they need to go. Circular data can be described as interval data in a loop with a circumference $U$. By definition:

*Circular* metric differences:

$$_{\text{circular}}\delta^2_{ck} = \left(\sin\left[180\frac{c-k}{U}\right]\right)^2 \quad \text{in degrees} \tag{5.16}$$

$$_{\text{circular}}\delta^2_{ck} = \left(\sin\left[\pi\frac{c-k}{U}\right]\right)^2 \quad \text{in radians}$$

Graphically:

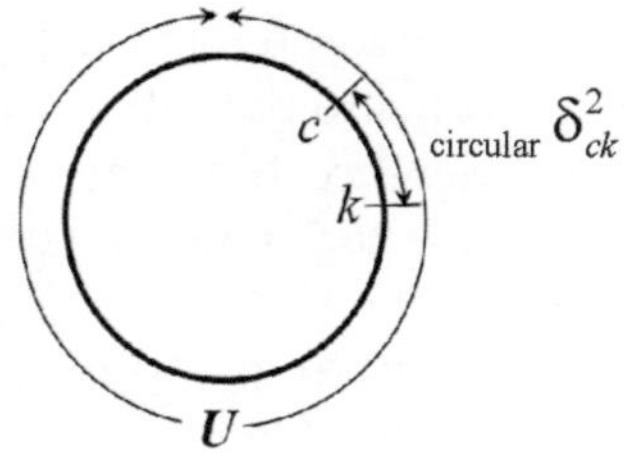


**FIGURE 5.1**
A Visualization of the Circularity of Differences

Figure 5.1 depicts circular data confined by the circumference of $U = 10$. Whereas polar data are confined by its two poles that cannot be exceeded. Ratio data have one fixed point but are open ended on the other extreme. Interval data are by definition infinitely extendable in both directions. Circular data are in principle infinitely extendable in both directions but are measured by being in a loop. Their actual values are always mapped back into the defined loop. On a clock, reading the time 24 hours later should show the same time. Circular data conform to what mathematicians call a modular algebra. Figure 5.1 suggests four properties of circular data:

For integers, its circumference: $U = 1 + v_{\text{largest}} - v_{\text{smallest}}$

Its largest circular difference is ½$U$ for values opposite the circle.

When $c-k$ equals any multiple of $U$, their circular differences are zero.

When $c-k$ exceeds $U$, their circular differences are confined to values within $U$: $(c-k)\pm U$.

Numerically, circular differences are depicted in Table 5.9 with the circumference $U = 10$:

| **Circular values:** | 0 | 1 | 2 | 3 | 4 | 5 | 6 | 7 | 8 | 9 |
|---|---|---|---|---|---|---|---|---|---|---|
| 0 | **0** | 0.10 | 0.35 | 0.65 | 0.90 | 1 | 0.90 | 0.65 | 0.35 | 0.10 |
| 1 | 0.10 | **0** | 0.10 | 0.35 | 0.65 | 0.90 | 1 | 0.90 | 0.65 | 0.35 |
| 2 | 0.35 | 0.10 | **0** | 0.10 | 0.35 | 0.65 | 0.90 | 1 | 0.90 | 0.65 |
| 3 | 0.65 | 0.35 | 0.01 | **0** | 0.10 | 0.35 | 0.65 | 0.90 | 1 | 0.90 |
| 4 | 0.90 | 0.65 | 0.35 | 0.10 | **0** | 0.10 | 0.35 | 0.65 | 0.90 | 1 |
| 5 | 1 | 0.90 | 0.65 | 0.35 | 0.10 | **0** | 0.10 | 0.35 | 0.65 | 0.90 |
| 6 | 0.90 | 1 | 0.90 | 0.65 | 0.35 | 0.10 | **0** | 0.10 | 0.35 | 0.65 |
| 7 | 0.65 | 0.90 | 1 | 0.90 | 0.65 | 0.35 | 0.10 | **0** | 0.10 | 0.35 |
| 8 | 0.35 | 0.65 | 0.90 | 1 | 0.90 | 0.65 | 0.35 | 0.10 | **0** | 0.10 |
| 9 | 0.10 | 0.35 | 0.65 | 0.90 | 1 | 0.90 | 0.65 | 0.35 | 0.10 | **0** |

**TABLE 5.9**
A Matrix Exemplifying Circular Differences With Circumference $U = 10$

### 5.3.7 Other Difference Functions

The nominal, ordinal, interval, ratio, and polar metric differences are by far the most common ones. However, ${}_c\alpha_{metric}$ is not limited to the six functions defined above. It accepts any difference function that conforms to the rather minimal conditions of (5.3) and (5.4), (5.5) being optional. Defining difference functions for special purposes opens ${}_c\alpha_{metric}$ to very many applications, for example to the graphs of networks, pattern recognition, and facial identification.

Chapter 7 will introduce a difference function for multi-valued coding of units of analysis, applying it as in (5.6) but to sets of values, capable of embracing any of the single-valued difference functions defined above.

## 5.4 Comparative Properties of Metric Difference Functions

Stevens (1946) proposed and investigated the typology of the first four metrics defined above. Other typologies notwithstanding, his basic insights have endured. Nominal data are the weakest data, followed by ordinal data, interval data, and ratio data which are considered the strongest kind. His rank ordering of these four scale types is based on the fact that each stronger scale responds to a quality that a weaker one ignores. For example, the nominal difference function recognizes only whether two categories are same or different. The ordinal metric difference function assigns differences to neighborhood relations. The interval metric difference function quantifies differences between values, not just between neighboring ones. It has no reference point and is of infinite extension. The ratio metric difference function recognizes a zero point to which all differences refer. The polar metric difference recognizes two end points. Near them it behaves more like the ratio metric. In its center it behaves as the interval metric does. This allows one to rank the metrics according to their strength, how many differences they include. According to Stevens, e ratio metric is the strongest and the nominal metric is the weakest. One can conclude that:

*Each weaker metric omits qualities that a stronger one includes.*

Any difference functions that satisfy condition (5.3) and (5.4) can enter the calculation of ${}_c\alpha_{metric}$ provided the data justify their use and the analytical techniques to be applied can handle them as well.

Computing α-agreements with difference functions that are too weak for the nature of the data and the analytical methods to be employed omits qualities that might well contribute to the reliability of data but are not taken advantage of. For example, it is always possible to compute the ${}_c\alpha_{nominal}$ coefficient on ordered data but its result would not omit the reliabilities inherent in the rank ordering of the units or the numerical differences between scale points.

Omissions are undesirable. However the complement of applying an α-coefficient whose difference function exceeds the strength of the data and/or of the analytical method of choice leads to outright distortions. The classical example is to use a ${}_c\alpha_{interval}$-coefficient on the numbers on the jerseys of football players which defy addition and subtraction – just as when applying an analysis of variance to telephone numbers. The result would be uninterpretable, gibberish, and certainly misleading by assuming a quantitative relationship between them that does not exist.

When the off-diagonal observed coincidences exhibit no metric-related pattern, are unordered, randomly distributed, or totally absent, the applied metric cannot discern their differences and ${}_c\alpha$-reliabilities for different metrics are indistinguishable:

$${}_c\alpha_{\text{nominal}} = {}_c\alpha_{\text{ordinal}} = {}_c\alpha_{\text{interval}} = {}_c\alpha_{\text{ratio}} = {}_c\alpha_{\text{polar}} = {}_c\alpha_{\text{circular}}$$

This equality holds for any $\alpha$, including when $\alpha = 1$.

Provided the reliability data contain observed disagreements, the nominal ${}_c\alpha_{\text{nominal}}$, which makes no assumptions about any ordering of values can serve as a reference for presence or absence of a metric in the coded data. Reliability data that conform to a particular metric yield agreements ${}_c\alpha_{\text{metric}} > {}_c\alpha_{\text{nominal}}$. Conversely, reliability data that violate a particular metric yield metric-specific $\alpha$s smaller than for the nominal $\alpha$. This property is depicted in Figure 5.2, which compares the results of nominal and interval coefficients on three kinds of data. Unfortunately, applying $\alpha$-agreements with different metrics to the same data is the more informative about the prevalence of a metric the more disagreements are observed or the smaller the $\alpha$-agreements are.

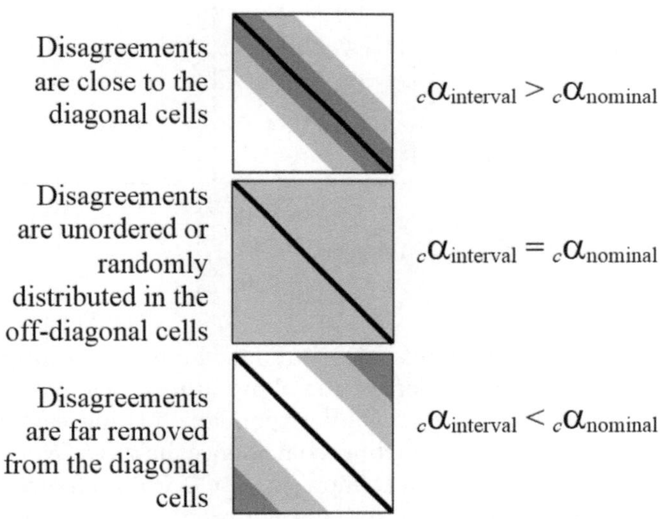


**FIGURE 5.2**
${}_c\alpha$-Agreements, Characteristics and Uncharacteristic of Interval Data

## 5.5 A Numerical Example for Two Metric Differences

The numerical example used to illustrate (2.10) and (2.3) is adapted here by replacing the icons 🕮, 🖂, ☎, 🖳, and 🗁 with numerical values 1, 2, 3, 4, and 5 respectively:

| | Units $u$: | 1 | 2 | 3 | 4 | 5 | 6 | 7 | 8 | 9 | 10 | 11 | 12 | |
|---|---|---|---|---|---|---|---|---|---|---|---|---|---|---|
| **Observers**: | Peter | 1 | 2 | 3 | 3 | 2 | 1 | 4 | 1 | 2 | . | . | . | |
| | Paul | 1 | 2 | 3 | 3 | 2 | 2 | 4 | 1 | 2 | 5 | . | . | |
| | Mary | . | 3 | 3 | 3 | 2 | 3 | 4 | 2 | 2 | 5 | 1 | 3 | |
| | Bill | 1 | 2 | 3 | 3 | 2 | 4 | 4 | 1 | 2 | 5 | 1 | . | |
| **Sums of pairable values** $m_u > 1$: | | 3 | 4 | 4 | 4 | 4 | 4 | 4 | 4 | 4 | 3 | 2 | 0 | 40 |

As noted in Chapter 2, of the 4×12 = 48 cells in this canonical form of reliability data 41 are occupied, 7 are missing. Moreover, the lone value 3 assigned by Mary to unit 12 is not pairable with other values in that unit, cannot contribute to assessments of the reliability of these data, and its $m_{12} = 0$ by definition. Therefore, these reliability data consist of a total of 40 pairable values.

Applying the definitions (5.1) to obtain the observed coincidence and (5.2) the expected ones yields matrices, already presented in Chapter 2, to which the above Table 5.2 added the matrix of perfect agreement for comparison:

| **Values:** | 1 | 2 | 3 | 4 | 5 | | 1 | 2 | 3 | 4 | 5 | | 1 | 2 | 3 | 4 | 5 | |
|---|---|---|---|---|---|---|---|---|---|---|---|---|---|---|---|---|---|---|
| 1 | 9 | . | . | . | . | | 7 | 1.33 | 0.33 | 0.33 | . | | 1.85 | 3 | 2.31 | 1.15 | 0.69 | 9 |
| 2 | . | 13 | . | . | . | | 1.33 | 10 | 1.33 | 0.33 | . | | 3 | 4 | 3.33 | 1.67 | 1 | 13 |
| 3 | . | . | 10 | . | . | | 0.33 | 1.33 | 8 | 0.33 | . | | 2.31 | 3.33 | 2.31 | 1.28 | 0.77 | 10 |
| 4 | . | . | . | 5 | . | | 0.33 | 0.33 | 0.33 | 4 | . | | 1.15 | 1.67 | 1.28 | 0.51 | 0.38 | 5 |
| 5 | . | . | . | . | 3 | | . | . | . | . | 3 | | 0.69 | 1 | 0.77 | 0.38 | 0.15 | 3 |
| | 9 | 13 | 10 | 5 | 3 | | 9 | 13 | 10 | 5 | 3 | | 9 | 13 | 10 | 5 | 3 | 40 |

$_c\alpha_{\text{metric}} = 1.000$ $\quad$ $_c\alpha_{\text{nominal}} = 0.743$, $_c\alpha_{\text{interval}} = 0.849$ $\quad$ $_c\alpha_{\text{metric}} = 0.000$

In ideally reliable data, all pairs of values are matching and occupy the diagonal of a coincidence matrix, shown in the left of the above three. Observed disagreements are absent. $D_o = 0$ which renders $_c\alpha_{\text{metric}} = 1$. The zero point of and α-agreement is demonstrated by the matrix to the right of the above three coincidence matrices. When the observed and expected coincidences are equal, $D_o = D_e$, and $_c\alpha_{\text{metric}} = 0$. Both reference points of α are so for any metric, which makes the data in the center matrix the focal point of all reliability considerations.

Attending to the matrix of actually observed coincidence, because the expected coincidences are already tabulated above, $_c\alpha_{\text{nominal}}$ is calculated with its computationally simplest expression in (2.8):

$$_c\alpha_{\text{nominal}} = 1 - \frac{n_{..} - \sum_{c=1}^{v} n_{cc}}{n_{..} - \sum_{c=1}^{v} e_{cc}} = 1 - \frac{40-(7+10+8+4+3)}{40-(1.85+4+2.31+0.51+0.15)} = 0.743$$

Now assuming the values 1 through 5 in this example constitute an interval scale, to calculate $_c\alpha_{\text{interval}}$ according to (5.5) requires reference to the interval metric difference function

$_{\text{interval}}\delta^2_{ck}$ in (5.13) whose differences are tabulated as in Table 5.6 but here only for the five present values:

| | **Values:** | 1 | 2 | 3 | 4 | 5 |
|---|---|---|---|---|---|---|
| | 1 | **0** | 0.06 | 0.25 | 0.56 | 1 |
| | 2 | 0.06 | **0** | 0.06 | 0.25 | 0.56 |
| $_{\text{interval}}\delta^2_{ck}$ | 3 | 0.25 | 0.06 | **0** | 0.06 | 0.25 |
| | 4 | 0.56 | 0.25 | 0.06 | **0** | 0.06 |
| | 5 | 1 | 0.56 | 0.25 | 0.06 | **0** |

With reference to the already available expected coincidences, simplifying the second form of (5.8) to minimize the number of additions by attending only to one of the two identical off-diagonal triangles of these matrices, and summing all coincidences with the same interval difference, yields:

$$_c\alpha_{\text{interval}} = 1 - \frac{_{\text{interval}}D_o}{_{\text{interval}}D_e} = 1 - \frac{\sum_{c=1}^{v}\sum_{k=1}^{v} n_{ck\ \text{interval}}\delta^2_{ck}}{\sum_{c=1}^{v}\sum_{k=1}^{v} e_{ck\ \text{interval}}\delta^2_{ck}} = 1 - \frac{\sum_{c-k=1}^{v-1} n_{ck\ \text{interval}}\delta^2_{ck}}{\sum_{c-k=1}^{v-1} e_{ck\ \text{interval}}\delta^2_{ck}}$$

$$= 1 - \frac{(1.33+1.33+0.33)0.0625+(0.33+0.33)0.25+0.33\cdot 0.5625}{(3+3.33+1.28+0.38)0.0625+(2.31+1.67+0.77)0.25+(1.15+1)0.5625+0.69\cdot 1}$$

$$= 0.849$$

Evidently,

$$_c\alpha_{\text{nominal}} = 0.743 < 0.849 = {}_c\alpha_{\text{interval}}$$

This difference can be explained by the distribution of mismatches in the matrix of observed coincidences. 6 mismatches concern neighboring values, 1.33 two values apart, 0.67 three values apart and none four values apart. It fits one upper-left pattern depicted in Figure 5.2. The largest possible interval difference of 1 for the 5–1 cell in this matrix is not observed, $n_{51} = 0.00$, yet, expected $e_{51} = 0.69$. Had this distribution been reversed, with the larger number of coincidences occurring where the interval differences are large, the inequality would have been reversed. As already mentioned, the nominal metric difference function does not respond to the location of the mismatching coincidences in the off-diagonal cells and therefore cannot distinguish degrees of differences due to any ordering of values.

# 6

# *The Quadrilogy for Single-Valued Predefined Units and Big Data*

The idea of a quadrilogy of reliability coefficients for big data emerged in response to an inquiry by Julien Cornebise, a researcher of machine learning from the UK. He faced the problem of measuring the reliability of 11,289,371 crowd-coded binary judgments – the presence or absence of a concern – made by a total of 3,712 coders of 2,653,322 units of analysis, between one an 56 per unit. Moreover, two experts coded 4,187 of these units to get a sense of the quality of the crowdsourced judgments.[1]

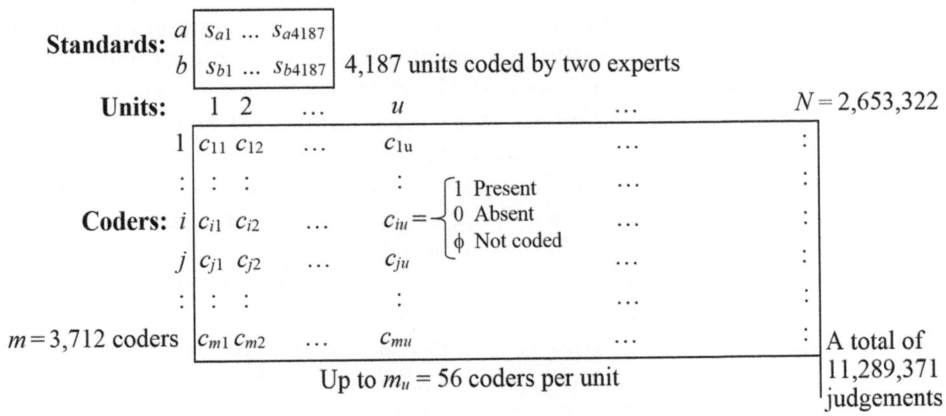


**TABLE 6.1**
Canonical form of Cornebise's reliability data: A matrix of 3.7 thousand unique coders-by-2.7 million units judged, occupied by 11.3 million judgements but 99.89% empty.

In social research, most evidence used to establish the reliability of data are generated by two coders. 56 is far beyond what commonly available reliability coefficients can handle. Most agreement coefficients, some of them reviewed in Chapter 4, are designed for only two. Fleiss' coefficient (Fleiss, 1971) can handle more than two coders but requires that all of them code all units of analysis. In Cornebise reliability data, the number of coders varied considerably. He recognized that Krippendorff's ${}_{c}\alpha$-coefficient (for coding predefined units of analysis) does not exhibit such limitations. It measures the agreement among variable numbers of interchangeable coders by pairing the values they assigned to units. This took care of one challenge that Cornebise's data posed, but there were more.

[1] Julien Cornebise, personal communications between 2017.2.2 to 2017.4.8. Data updated on 2020.9.11.

Another was the enormous size of his reliability data. Its canonical form of an $m$ coders-by-$N$ units matrix is sketched in Table 6.1.

To compute the α-agreement coefficient by its traditional definition, pairing all 11.3 million judgments with each other to compute the expected disagreement alone, already exceeds the practical limits of computation. No currently available software was able to handle such numbers. Fortunately, the binary coding of these data and the use of sums offered a simplification that paved the way to computationally feasible algorithms. Cornebise managed to write software in R that took advantage of these possibilities. However, obtaining meaningful reliabilities of these data posed more challenges.

The easiest one to tackle was extending α, originally conceived of as a measure of the replicability of data, to accept a trusted standard against which the accuracy of coding could be accessed, unit by unit. Whereas replicability takes all coders as interchangeable and pairs the values they assign to units of analysis with each other, regardless of how numerous they are, a standard, whether in the form of the consensus of a panel of experts' judgments or an objective measure, has to be regarded as a privileged "coder," and treated unlike the interchangeable coders. This calls for enumerating the contingencies between the standard and the values assigned by coders, not the coincidences needed in establishing replicability. Accuracy measures are important when the quality of data is the issues regardless of how they came about.

The final challenge was to present methodologically convincing alternatives in the face of the widely shared conviction among big data analysts, that the ease with which massive amounts of data support statistical hypotheses by traditional criteria renders reliability concerns moot as claimed by Mayer-Schönberger and Cukier's (2013) influential book. The assumption that big data provide more reliable insights into what is going on in the world has energized popular social science literature. However, in the social sciences, merely harvesting textual fragments and behavioral traces that people left behind unintentionally largely ignores their social origins and meanings. Especially short character strings – words, sentences, tweets – do not literally "contain" anything, least of all the roles they play in the social world. The practice of relying on data that are big and easily found has led McFarland and McFarland (2015) to conclude that big (and I would qualify uninterpreted) data can be and mostly are "precisely inaccurate." I agree with their largely sampling-based assessment but want to add two further concerns to theirs.

The belief that large sample sizes ensure reliability is not without a history of celebrating sociological abstractions. The latter have encouraged even content analysts to propose reliability measures that take majority judgments as the gold standard (Landis & Koch, 1977; Hubert, 1977; Craig, 1981; and more recently Fretwurst, 2015). This may well be justified in extreme cases, but it ignores an implied uncertainty. For example, when 99% of a large number of coders agree on what they have seen, read, and recorded, one may well come to trust the data they generate. However, as the proportion of majorities declines, conclusions based on them become increasingly doubtful. When they converge to, say, 51 percent, minor random fluctuations can easily flip one majority into its opposite, signaling the omnipresence of an unattended uncertainty in relying on majority judgments. Because majorities can be more or less decisive, majorities by themselves cannot provide sufficient assurances that data reliably represent the phenomena of interest.

Crowd coding of textual matter by large numbers of albeit unknown readers is a response to the admission that Artificial Intelligence (AI) fails to make sense of character strings and markers, easily harvested by algorithmic searches of social media, electronically accessible data bases, and economic transactions, whether data are big or small. What such usually short texts or recorded human actions mean to those who live with them escapes these data, and claims to be able to identify these meanings computationally are largely unsubstantiated. Crowd coding adds the human intelligence that AI is lacking. This is not to deny that big data analysts find lots of interesting pattern in big data, but the relationships of their findings to the social world are questionable without literary and socially competent interpreters on the ground. Cornebise did not face crowd-coded texts but satellite images of a political concern that humans were able to reliably identify but optical devices could not.

The reliance on overwhelming majorities by big data analysts is often justified by reference to the *Wisdom of Crowds*. Surowieki (2005) convincingly answers his question of "*Why the Many Are Smarter than the Few and How Collective Wisdom Shapes Business, Economies, Societies, and Nations*" by pointing to the underlying interactions among the members of crowds. Of course, such interactions are not free of shared biases, but they give members of crowds the ability to weed out individual idiosyncrasies and devious judgments, resulting in collectively more reliable data. The processes that constitute crowds are ignored when data are obtained by interviewing individuals as in public opinion research, or collecting tweets, emails, internet searches, or economic transactions by big data analysts. Such data may well contain collective flavors but are not vetted by the crowds of analytical concerns. Amazon's Mechanical Turk, a cost-effective way to get quick responses from large numbers of individual internet users to specific questions regarding the readings or interpretations of provided texts, is called "crowdsourcing." However, in Surowieki's conception this would be a misnomer that has misled researchers to equate bigness with wisdom. Majorities or averages are calculated measures that have little to do with wisdom.

There is much methodological work to be done before raw big data can be trusted. The following is concerned with assessing the reliability of their readings and recording as analyzable data by qualified coders, judges, or interpreters.

This chapter formalizes the distinctions among four out of five kinds of reliability outlined conceptually in Chapter 1. Of these, stability is not only irrelevant when replications are generated by multiple observers, but it is also computationally indistinguishable from the agreement of replicability. Thus, it focuses on the basic quadrilogy of reliability coefficients: replicability, accuracy, surrogacy, and decisiveness. As it was initially developed for measuring the reliability of big data, this chapter takes advantage of the conceptual simplifications afforded by binary data and generalizes the four measures to larger variables after the basic ideas are settled.

## 6.1 Big Binary Reliability Data

We start with the original problem of calculating reliabilities for binary judgments made by a variable number of observers. Table 6.2 deviates from the canonical form of reliability data in Tables 6.1 and 2.1 by representing the observers who judge units 0 or 1 not individually but by their frequencies. Table 6.2 provides the possibility of multiple standards against which multiple binary judgments on units can be checked for their accuracy.

| | | | | | | | |
|---|---|---|---|---|---|---|---|
| **Standard** or proposed **Surrogate's values:** | 0 | $s_{01}$ | $s_{02}$ | ... | $s_{0u}$ | ... | $s_{0N}$ |
| | 1 | $s_{11}$ | $s_{12}$ | ... | $s_{1u}$ | ... | $s_{1N}$ |
| **Units:** | | 1 | 2 | ... | $u$ | ... | $N$ = a large number |
| **Coders' values:** | 0 | $r_{01}$ | $r_{02}$ | ... | $r_{0u}$ | ... | $r_{0N}$ |
| | 1 | $r_{11}$ | $r_{12}$ | ... | $r_{1u}$ | ... | $r_{1N}$ |
| $\boldsymbol{r_{0u} + r_{1u}} =$ | | $m_1$ | $m_2$ | ... | $m_u$ | ... | $m_N$ |

**TABLE 6.2**
Binary Reliability Data and a Trusted Standard or Proposed Surrogate

## 6.2 Basic Binary Conceptions of the Quadrilogy

The graphical depictions of the pair comparisons that go into each of the four reliabilities were already introduced in Chapter 1, Figure 1.3, but deserve more detailed attention in this chapter.

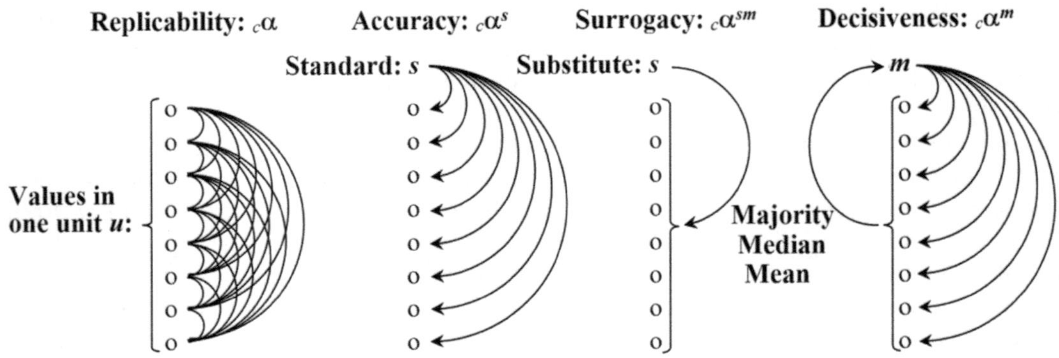


FIGURE 6.1
The Quadrilogy of Reliability Coefficients and the Values They Respectively Pair Within Any One Unit

### 6.2.1 The Replicability of the Process of Generating Data

Replicability is the ability of independently working observers to apply given coding instructions and only these to the same set of diverse phenomena and yield ideally identical data. The agreement measured between these replications is intended to give researchers the needed assurance that:

- The distinctions among the generated data represent the differences among the phenomena that observers faced while coding
- The coding instructions are reproducible elsewhere, and
- The extent of the observed agreement among replications justifies inferences that the generated data are sufficiently reliable to be analyzed.

The pair comparisons that the replicability measure requires within any one unit $u$ are depicted in the left image of Figure 6.1. Each unit $u$ in Table 6.2 is coded $m_u$ times which involves $m_u(m_u-1)$ possible pairs of values, which either match or they do not. This number of possible $c$-$k$ pair comparisons in $u$ will have to be weighted by $1/(m_u-1)$ to ensure that each value counts as 1 in the observed coincidences:

$$n_{ck} = \sum_{u=1}^{N} \frac{1}{m_u - 1} \sum_{c_u=0}^{1} r_{cu} \sum_{k_u=0}^{1} \begin{Bmatrix} r_{ku} - 1 & \text{iff } c_u = k_u \\ r_{ku} & \text{iff } c_u \neq k_u \end{Bmatrix} \text{ iff } m_u > 1 \tag{6.1}$$

The numbers of $c$–$k$ coincidences $n_{ck}$ are tabulated in the symmetrical 2-by-2 matrix of coincidences in Table 6.3. Its total $n_{..}$ equals the total number of pairable values within all units of the reliability data and its marginal sums distinguish their kinds, here 0s and 1s:

| **Coders' values *c* or *k*:** | 0 | 1 | |
|---|---|---|---|
| 0 | $n_{00}$ | $n_{01}$ | $n_{0\cdot} = n_{00} + n_{01} = n_{\cdot 0}$ |
| 1 | $n_{10}$ | $n_{11}$ | $n_{1\cdot} = n_{10} + n_{11} = n_{\cdot 1}$ |
| | $n_{\cdot 0}$ | $n_{\cdot 1}$ | $n_{\cdot\cdot} = n_{00} + n_{01} + n_{10} + n_{11}$ |

**TABLE 6.3**
The 2-by-2 Matrix of Coincidences $n_{ck}$ for Assessing the Replicability of Data Making

As discussed in Chapter 3, coincidences are symmetrically distributed: $n_{01} = n_{10}$, $n_{0\cdot} = n_{\cdot 0}$, and $n_{1\cdot} = n_{\cdot 1}$.

The observed disagreement ${}_cD_o = n_{01} + n_{10}$ is the sum of all mismatching pairs of values, whereas the expected disagreement ${}_cD_e = (n_0.n_{\cdot 1} + n_1.n_{\cdot 0})/(n_{\cdot\cdot} - 1)$ is the number of mismatches that could be observed if the assignment of values to units were unrelated to the phenomena that the data are expected to represent. Accordingly, the ${}_c\alpha$-agreement coefficient for binary data (2.9), reduces to:

$$ {}_c\alpha_{\text{binary}} = 1 - \frac{{}_cD_o}{{}_cD_e} = 1 - \frac{n_{01} + n_{10}}{\frac{1}{n_{\cdot\cdot} - 1}(n_0.n_{\cdot 1} + n_1.n_{\cdot 0})} = 1 - (n_{\cdot\cdot} - 1)\frac{n_{01}}{n_0.n_{\cdot 1}} \qquad (6.2) $$

Its range is: $-1 \cdot \le {}_c\alpha_{\text{binary}} \le +1$

If reliability data satisfy the preparatory conditions discussed in Section 1.7, according to Table 1.1, the two sources of unreliability that reduce ${}_c\alpha$ are intra-coder inconsistencies or instabilities of individual human coders and the inter-coder disagreements on interpreting the coding instructions in the face of the phenomena that observers are to turn into analyzable data.

The replicability coefficient ${}_c\alpha$, designed to indicate whether data can be trusted to represent the phenomena of analytical interest is well established, has been discussed in the preceding chapters, and deserves no further comment except to note that its binary version can indeed assume its negative extreme of ${}_c\alpha = -1$. Its simplification in the third expression of (6.2), already discussed in (2.9), is due to the symmetry of coincidences. As visualized in Figure 6.1, it pairs all values within units and sums them into the coincidences in terms of which ${}_c\alpha_{\text{binary}}$ is defined.

### 6.2.2 The Accuracy of Generated Data

The second image of Figure 6.1 depicts the pair comparisons that constitute the accuracy of generated data. It pairs all replications of a unit with a trusted standard. Assessing the accuracy of data is the most obvious application of this measure. However, accuracy runs into the paradox of requiring a trusted and hence known standard for the very same units of analysis of uncertain reliability which would make the coding superfluous. This paradox is resolved when testing the competencies of new coders before hiring them to generate the data for a larger project, to monitor the quality of coding of a larger project by measuring it at different intervals, or to test the seriousness and qualifications of potential crowd coders by including questions for which the analyst knows the answer. In practice, as exemplified in Cornebise reliability data, accuracies tend to be established on subsets of the eventually analyzed data, which pose the problem of representative sampling.

The motivation of publishing the accuracies of data is to ensure that readers or stakeholders of a research project understand the extent to which data mean what the chosen standard implies. By contrast, publishing the replicability of data gives its readers a sense how well the written coding instructions are communicable and would yield identical data when applied elsewhere. It follows that accuracy is a more stringent form of reliability.

When an applied standard is factually true, accuracy becomes a measure of the validity of generated data.

The sum of all *c*-*k*-contingencies $x_{ck}$ between the values *c* of replications and the values *k* of the standard is:

$$x_{ck} = \sum_{u=1}^{N} \sum_{c_u=0}^{1} r_{cu} \sum_{k_u=0}^{1} \frac{s_{ku}}{s_{0u}+s_{1u}} \text{ iff } m_u(s_{0u}+s_{1u}) > 0 \quad (6.3)$$

We use the notation $x_{ck}$ to distinguish contingencies from the coincidences called $n_{ck}$.

The ideal standard is unambiguous and single-valued, $s_{0u} + s_{1u} = 1$. When it is composed from multiple sources, the standard may include uncertainties. When it is unanimous, the proportion in (6.3) identifies that one value $k_u$, which is paired with all $r_{cu}$ in $m_u$. Under these ideal conditions, $r_{0u}$ would be added to $x_{0k}$ and $r_{1u}$ would be added to $x_{1k}$. However, when more than one standard is given, for example the two experts that Cornebise used to check on the quality of the crowdsourced data, there is the possibility that these standards disagree, leaving uncertain whether coders were correct or not. The possibility of an ambiguous standard is accounted for by the fraction in (6.3). It pairs the frequency of $k_u$ in the proportion of their occurrence with each $r_{cu}$.

The quality of a standard deserves separate attention. If it contains too many ambiguities, it could not be trusted would not deserve being called a standard worthy of evaluating the accuracy of coder-generated data. Before applying a standard derived from several sources it is advisable to measure the replicability of that standard by means of $\alpha_{binary}$ defined in (6.2) using the frequencies $s_{ku}$ in accounting for equation (6.1) in place of $r_{cu}$.

It should be noted that the accuracy of data can rarely exceed the replicability of the applied standard.

The frequencies defined by (6.3) give rise to the 2-by-2 matrix in Table 6.4, containing these *c*-*k* contingencies:

| | | | |
|---|---|---|---|
| **Standard's values *k*:** | 0 | 1 | |
| **Coders' values *c*:** 0 | $x_{00}$ | $x_{01}$ | $x_{0\cdot} = x_{00} + x_{01} \neq x_{\cdot 0}$ |
| 1 | $x_{10}$ | $x_{11}$ | $x_{1\cdot} = x_{10} + x_{11} \neq x_{\cdot 1}$ |
| | $x_{\cdot 0}$ | $x_{\cdot 1}$ | $x_{\cdot\cdot} = x_{00} + x_{01} + x_{10} + x_{11}$ |

**TABLE 6.4**
The 2-by-2 Matrix of Contingencies $x_{ck}$ for Assessing the Accuracy of Data

Unlike coincidence matrices, containing $n_{ck}$, Table 6.4 is a contingency matrix. Contingencies are not symmetrically distributed: $x_{01} \neq x_{10}$, $x_{\cdot 0} \neq x_{0\cdot}$ and $x_{\cdot 1} \neq x_{1\cdot}$. This asymmetry is easily recognized by the row sums enumerating the coders' use of values $c = 0$ and 1, and the column sums by enumerating how often the values *k* of the trusted standard are paired with values *c* assigned by coders.

Notationally, contingencies are referred to by $x$, The reference to a standard is indicated by using the superscript $s$ in the definitions of the observed disagreement ${}_cD_o^s = x_{01} + x_{10}$, of the expected disagreement ${}_cD_e^s = \frac{x_0.x._1 + x_1.x._0}{x..}$, and of the measure of the accuracy of data:

$$ {}_c\alpha_{\text{binary}}^s = 1 - \frac{{}_cD_o^s}{{}_cD_e^s} = 1 - \frac{x_{01} + x_{10}}{\frac{1}{x..}(x_0.x._1 + x_1.x._0)} \tag{6.4}$$

Its range is as for replicability: $-1 \leq {}_c\alpha_{\text{binary}}^s \leq +1$

According to Table 1.1, ${}_c\alpha_{\text{binary}}^s$ deviates from measuring 1.000 not only due to intra-observer inconsistencies and inter-observer disagreements, but most importantly, in response to how observers' values deviate from the trusted standard(s). For example, when all observers share the same prejudices regarding the phenomena they code, or have the same cognitive limitations, ${}_c\alpha_{\text{binary}}$-replicability cannot respond to these coder biases, but accuracy does. Thus as measures of reliability, accuracy is stronger than replicability.

Although the total sums $n..$ in Table 6.3 and $x..$ in Table 6.4 are numerically equal whenever each unit has at least two values $c_u$ and one standard $k_u$, Figure 6.1 reveals that accuracy ${}_c\alpha_{\text{binary}}^s$ involves far fewer pair comparisons than go into the measure ${}_c\alpha_{\text{binary}}$ for replicability.

Also, the definitions of their expected disagreements ${}_cD_e$ in (6.2) and ${}_cD_e^s$ in (6.4) are identical in form except for the denominators. Whereas in the denominator of ${}_cD_e$, $(n..-1)$ corrects for not pairing values with themselves, accuracy does not encounter this issue, and hence, the denominator of ${}_cD_e^s$ is $x..$ .

Majorities of binary judgments are inevitable whenever the number $n_{0u}+n_{1u}$ of coders judging unit $u$ is uneven and >2. When that number is even, ties are possible which neither confirm nor disconfirm the values of a proposed surrogate.

### 6.2.3 The Surrogacy of an Alternative for Coding of Data

When measuring accuracy, the standard is trusted; while the quality of observers' coding is in question and the target of evaluations. Surrogacy inverts the trust regarding these two variables by asking whether a device, preferred coding method, or proposed algorithm, (computationally in the position of the standard) is sufficiently close to the trusted observers' collective judgments to serve as a substitute for them. The measure of accuracy would not be able to determine the substitutability of a proposed alternative for generating data because it includes the diversity of individual judgments. Proposed single-valued substitutes need to be evaluated in terms of the trusted coding by human observers they are hoped to replace, as depicted in the third image of Figure 6.1.

A perfect single-valued surrogate for the coding of phenomena by multiple observers cannot be better than predicting a single value of what characterizes observers' collective judgments. In binary and nominal data, it needs to match or predict the values of the majorities of coders. This is formalized in the parentheses of the definition of its contingencies for surrogacy in (6.5):

$$x_{\bar{c}k} = \sum_{u=1}^{N} m_u \sum_{\bar{c}_u=0}^{1} \left\{ \begin{array}{l} 1 \;\text{ iff } r_{\bar{c}u} > r_{cu} \\ \left.\begin{array}{l} 0.5 \text{ and } \bar{c}_u = 0 \\ 0.5 \text{ and } \bar{c}_u = 1 \end{array}\right\} \text{iff } r_{0u} = r_{1u} \end{array} \right\} \sum_{k_u=0}^{1} \frac{s_{ku}}{s_{0u} + s_{1u}} \text{ iff } m_u > 1 \text{ and } s_{0u} + s_{1u} > 0 \tag{6.5}$$

Within these parentheses, the inequality $r_{\bar{c}u} > r_{cu}$ in unit $u$ establishes the value $\bar{c}_u$ as of the majority, paired with the value $k_u$ in $s_{ku}$, omitting the minority valued $c_u$ = not $(\bar{c}_u)$. Ties, $r_{0u} = r_{1u}$, add ½ of $m_u$ to $x_{\bar{c}k} = x_{0k}$ and ½ of $m_u$ to $x_{\bar{c}k} = x_{1k}$. As the data become large, ties are less likely. The proportion in (6.5) takes care of any disagreements among possible substitutes, just as in (6.3).

The frequencies $x_{\bar{c}k}$ of $\bar{c}$ - $k$ contingencies, defined by (6.5), constitute the 2-by-2 contingency matrix in Table 6.5. Just as for accuracy, the values $k$ of the candidate for substitution label its columns, the values $\bar{c}$ of the majorities among coders label its rows:

| **A potential surrogate's values $k$:** | | 0 | 1 | |
|---|---|---|---|---|
| **Majorities $\bar{c}$ of coders' values:** | 0 | $x_{00}$ | $x_{01}$ | $x_{0\cdot} = x_{00} + x_{01} \neq x_{\cdot0}$ |
| | 1 | $x_{10}$ | $x_{11}$ | $x_{1\cdot} = x_{10} + x_{11} \neq x_{\cdot1}$ |
| | | $x_{\cdot0}$ | $x_{\cdot1}$ | $x_{\cdot\cdot} = x_{00} + x_{01} + x_{10} + x_{11}$ |

**TABLE 6.5**
The 2-by-2 Matrix of Contingencies $x_{\bar{c}k}$ for Assessing the Surrogacy of Alternatives

Notationally, the additional reference to majorities is indicated by adding an $m$ to the superscript $s$ which stands now for the potential substitute for coded data. With the observed disagreement ${}_cD_o^{sm} = x_{01} + x_{10}$, the expected disagreement ${}_cD_e^{sm} = \dfrac{x_{0\cdot}x_{\cdot1} + x_{1\cdot}x_{\cdot0}}{x_{\cdot\cdot}}$ ′ the measure of surrogacy of a proposed substitute for the replications of human coders is:

$$ {}_c\alpha^{sm}_{\text{binary}} = 1 - \frac{{}_cD_o^{sm}}{{}_cD_e^{sm}} = 1 - \frac{x_{01} + x_{10}}{\frac{1}{x_{\cdot\cdot}}(x_{0\cdot}x_{\cdot1} + x_{1\cdot}x_{\cdot0})} \quad \text{where all are } x_{\bar{c}k} \tag{6.6} $$

Its range is as above: $-1 \cdot \le {}_c\alpha^{sm}_{\text{binary}} \le +1$

In sum, the measure of surrogacy ${}_c\alpha^{sm}_{\text{binary}}$ indicates the extent to which majorities of observers agree with the values of the candidate for their replacement above what can be expected when the two are unrelated. ${}_c\alpha^{sm}_{\text{binary}} = 1$ indicates its best possible fit, even when coders are not perfectly replicable. Whenever alternatives to the human coding of data are proposed – measuring devices, algorithms, and particular coders – ${}_c\alpha^{sm}_{\text{binary}}$ is indispensable indeed. It is sensitive to an important source of unreliability, the failure of a surrogate to match or predict the collective judgments of coders.

Earlier in this chapter, I articulated my misgivings of relying on conceptions of majorities among observers for indicating the replicability of the process of generating data. This does not apply to surrogacy. It is important to note that ${}_c\alpha^{sm}_{\text{binary}}$ does not respond to whether majorities are overwhelming or slim or how big the minorities actually are. Therefore, unlike measures of replicability and accuracy, surrogacy is not a measure of the reliability of data. Rather, it assesses the reliability of possible substitutes for human coding. This purposeful limitation motivates the fourth measure of this quadrilogy, the decisiveness of majorities. Decisiveness adds another dimension to decisions on whether a proposed surrogate could represent the individually coded data.

### 6.2.4 The Decisiveness of Majorities

A majority is decisive when all coders are in unanimous agreement on the valuation of a unit. It loses its decisiveness as the proportion of agreeing coders converges to the point of being equally distributed or tied. The pairing of values in the fourth image of Figure 6.1 suggests that decisiveness computationally resembles the measures of accuracy ${}_c\alpha^s_{\text{binary}}$ and surrogacy ${}_c\alpha^{sm}_{\text{binary}}$ except that it does not rely on an external standard. Instead, it compares the data generated by human coders with their own majorities. Majorities provide a standard that is internal to the data. Its disagreement within units resembles variance measures but for nominal data to start.

As the existence of majorities requires at least two coders, just as in (6.5), the contingencies $x_{c\bar{c}}$ for calculating decisiveness are defined by:

| | | | | |
|---|---|---|---|---|
| **Coders' majorities** $\bar{c}$ **:** | | 0 | 1 | |
| **Coders' values** $c$ **:** | 0 | $x_{00}$ | $x_{01}$ | $x_{0\cdot} = x_{00} + x_{01} \neq x_{\cdot 0}$ |
| | 1 | $x_{10}$ | $x_{11}$ | $x_{1\cdot} = x_{10} + x_{11} \neq x_{\cdot 1}$ |
| | | $x_{\cdot 0}$ | $x_{\cdot 1}$ | $x_{\cdot\cdot} = x_{00} + x_{01} + x_{10} + x_{11}$ |

**TABLE 6.6**
The 2-by-2 Matrix of Contingencies $x_{c\bar{c}}$ for Assessing the Decisiveness of Majorities

$$x_{c\bar{c}} = \sum_{u=1}^{N} r_{cu} \sum_{\bar{c}_u=0}^{1} \left\{ \begin{matrix} 1 \ \text{ iff } r_{\bar{c}u} > r_{cu} \\ \left. \begin{matrix} 0.5 \ \text{and} \ \bar{c}_u = 0 \\ 0.5 \ \text{and} \ \bar{c}_u = 1 \end{matrix} \right\} \text{iff } r_{0u} = r_{1u} \end{matrix} \right\} \text{iff } m_u > 1 \qquad (6.7)$$

The parentheses in (6.7) are identical to (6.5). The contingencies $x_{c\bar{c}}$ yield the 2-by-2 matrices of contingencies in Table 6.6, only formally resembling the contingencies $x_{ck}$ and $x_{\bar{c}k}$.

Notationally, we replace the reference to the standard in the superscript $s$ of ${}_c\alpha^s_{\text{binary}}$ by the superscript $m$ that refers to majorities in the observed disagreements ${}_cD^m_o = x_{01} + x_{10}$, the expected disagreements ${}_cD^m_e = \dfrac{x_0.x_{\cdot 1} + x_1.x_{\cdot 0}}{x_{\cdot\cdot}}$, and the agreement measure of the decisiveness of majorities:

$${}_c\alpha^m_{\text{binary}} = 1 - \frac{{}_cD^m_o}{{}_cD^m_e} = 1 - \frac{x_{01} + x_{10}}{\frac{1}{x_{\cdot\cdot}}(x_0.x_{\cdot 1} + x_{\cdot 0}x_1.)} \quad \text{where all are } x_{c\bar{c}} \qquad (6.8)$$

Its range $0 \leq {}_c\alpha^m_{\text{binary}} \leq +1$ is unlike that of the other binary agreement coefficients.

In Figure 6.1 one may recognize that the pair comparisons of accuracy between a standard $s$ and the individually coded replications are computationally split into two parts, those of surrogacy, relating a surrogate $s$ to the majorities $m$ of coders' judgments, and those of decisiveness, relating these majority judgments $m$ to the individually coded phenomena $c$. It follows that surrogacy and decisiveness are independent of each other, responding to disagreements that are indistinguishable in the measure of accuracy.

Measures of decisiveness serve at least four methodologically different indications. The most obvious one is as a measure of the extent to which the majorities of coders represents the phenomena of interest, evident in individual coders' judgments.

Decisiveness sets limits for replicability and accuracy. When coders' decisiveness is zero, replicability and accuracy measure are near zero as well.

Decisiveness is also a measure of the extent to which a substitute for individual coders' judgments is feasible. This interpretation becomes important, for example, in evaluating processes of machine learning, where it could measure the extent to which an algorithm failed to predict the coded data. Whether one is evaluating a substitute for individual coding, a model of data, or a theory of certain phenomena, decisiveness indicates how close this representation can possibly become given the diversity it cannot account for.

Finally, decisiveness, being negatively affected by everything that could go wrong in the process of generating data – lack of coder qualification, ambiguous coding instructions or misreading them, encountering unexpected complexities, breakdown of recording devices – the lack of decisiveness outlines the work that the designers of the process have do to improve the reliability of data. Conversely, decisiveness can also assess whether changes introduced in the coding instructions, in the choice of coders, or recording devices opened up the possibility of hope for substitutes for human coding to cover more ground.

Surrogacy and decisiveness work in tandem. Where one stops, the other starts. A substitute of perfect surrogacy combined with a dangerously low decisiveness measure is as worthless as an imperfect one facing perfect decisiveness. Both surrogacy and decisiveness should be considered in decisions on whether a proposed substitute for the human coding of data should be accepted or rejected. Surrogacy indicates how close a potential substitute is to what the data allow it to represent, decisiveness how reliable the best surrogate would be in the face of coded data. Cornebise could have used these two measures: surrogacy to see how well his two experts approximated the majorities among coders, and decisiveness to see what is outside of his control.

### 6.2.5 A Minimal Numerical Example for Comparing the Quadrilogy

The main benefit of the quadrilogy presented so far is its ability to handle huge data. However, the conceptual value of demonstrating the computations of these coefficients with a numerically realistic example could be easily drowned by large numbers of units and very many replications. Therefore, I offer a ridiculously simple example of binary coding just five units with the most typical disagreements. According to Table 6.2:

| | | | | | | | |
|---|---|---|---|---|---|---|---|
| **Standard** or proposed **Surrogate's values:** | 0 | | 1 | 1 | | | |
| | 1 | 2 | | 1 | 2 | | |
| **Units:** | | 1 | 2 | 3 | 4 | 5 = $N$ | |
| **Coders' values:** | 0 | 1 | 9 | | 5 | 8 | |
| | 1 | 8 | | 12 | 6 | 1 | |
| $r_{0u} + r_{1u} =$ | | 9 | 9 | 12 | 11 | 9 | 50 |

The following shows the coincidences $n_{ck}$ and contingencies $x_{ck}$, $x_{\bar{c}k}$, and $x_{c\bar{c}}$ separated for each unit:

| **Contributions from:** | Coincidences $n_{ck}$ | Contingencies $x_{ck}$ | Contingencies $x_{\bar{c}k}$ | Contingencies $x_{c\bar{c}}$ |
|---|---|---|---|---|
| **To:** | **Replicability** | **Accuracy** | **Surrogacy** | **Decisiveness** |
| **By:** Unit 1: | $\begin{matrix} & 0 & 1 \\ 0 & & 1 \\ 1 & 1 & 7 \end{matrix}$ | $\begin{matrix} & 1 \\ & 8 \end{matrix}$ | $\begin{matrix} & \\ & 9 \end{matrix}$ | $\begin{matrix} & 1 \\ & 8 \end{matrix}$ |
| Unit 2: | $\begin{matrix} 9 & \\ & \end{matrix}$ | $\begin{matrix} 9 & \\ & \end{matrix}$ | $\begin{matrix} 9 & \\ & \end{matrix}$ | $\begin{matrix} 9 & \\ & \end{matrix}$ |
| Unit 3: | $\begin{matrix} & \\ & 12 \end{matrix}$ | $\begin{matrix} & \\ 6 & 6 \end{matrix}$ | $\begin{matrix} & \\ 6 & 6 \end{matrix}$ | $\begin{matrix} & \\ & 12 \end{matrix}$ |
| Unit 4: | $\begin{matrix} 2 & 3 \\ 3 & 3 \end{matrix}$ | $\begin{matrix} & 5 \\ & 6 \end{matrix}$ | $\begin{matrix} & \\ & 11 \end{matrix}$ | $\begin{matrix} & 5 \\ & 6 \end{matrix}$ |
| Unit 5: | $\begin{matrix} 7 & 1 \\ 1 & \end{matrix}$ | N/A | N/A | $\begin{matrix} 8 & \\ 1 & \end{matrix}$ |
| **All units:** | $\begin{matrix} 18 & 5 \\ 5 & 22 \end{matrix}$ 50 | $\begin{matrix} 9 & 6 \\ 6 & 20 \end{matrix}$ 41 | $\begin{matrix} 9 & \\ 6 & 26 \end{matrix}$ 41 | $\begin{matrix} 17 & 6 \\ 1 & 26 \end{matrix}$ 50 |
| **Yields:** | $\alpha_{\text{binary}}=0.605$ | $\alpha^{s}_{\text{binary}}=0.369$ | $\alpha^{sm}_{\text{binary}}=0.656$ | $\alpha^{m}_{\text{binary}}=0.714$ |

Observe that for replicability, units 2 and 3 are in perfect agreement in the context of the values of all other units. Adding unit 5 to it reduces ${}_c\alpha_{\text{binary}}$ to 0.869. Adding unit 1 to it as well reduces ${}_c\alpha_{\text{binary}}$ to 0.799. Evidently, the main source of lacking replicability is unit 4 which brings the measure of replicability down to ${}_c\alpha_{\text{binary}} = 0.605$.

For decisiveness the overall assessment is similar. However, decisiveness responds to agreements of all values and their majorities, not with each other, which brings all contingencies in one column. Without unit 4, ${}_c\alpha^{m}_{\text{binary}}$ would be 0.900. All included yield ${}_c\alpha^{m}_{\text{binary}} = 0.714$.

Note that a standard or surrogate for unit 5 is missing. It contributes nothing to the accuracy and surrogacy measures. Therefore, the sum of the contingencies for accuracy and surrogacy is 41, not 50 as for replicability and decisiveness. Also, the measure of replicability applied to the standard is 0.400. The main culprit of this low value is due to unit 3, which will be seen in the contingencies for that unit below.

Although the replications in units 2 and 3 agree perfectly, the ambiguity of the standard or proposed alternative in unit 3 adds disagreements that negatively affect the accuracy and surrogacy alike. But because the values they exhibit in units 1 and 4 agree with the majorities in replications, for surrogacy, unit 3 is the only source of disagreement which measures ${}_c\alpha^{sm}_{\text{binary}} = 0.656$ . Accuracy on the other hand is reduced further by the disagreements in unit 1 and 4, yielding the lowest measure of the quadrilogy: ${}_c\alpha^{s}_{\text{binary}} = 0.369$.

Unit 4 deserves comparative attention. For replicability this unit is a perfect chance event. For surrogacy it amounts to a perfect match between the value of the substitute and the majority of replications. This is a microscopic demonstration that accuracy and decisiveness are equal if the standard's values correspond to the majorities of the replications.

Units 3 and 4 prove the claim articulated earlier that surrogacy and decisiveness are independent measures. Unit 3 exemplifies the case of decisiveness being perfect while surrogacy is low. Unit 1 and 4 are two cases of surrogacy being perfect while decisiveness is low. Unit 2 exemplifies the case of surrogacy and decisiveness being perfect. These miniature reliability data do not include the case of both being low.

Although these numerical data are extremely small, the suggestion that surrogacy and decisiveness partition the disagreements to which accuracy responds, made in discussing Figure 6.1, may be observed in the rows of the above example. In all four units the inaccuracies are always equal or larger than the disagreements within surrogacies and decisivenesses.

## 6.3 Extension to Other Than Binary Data of Medium Size

The preceding sections operationalized the quadrilogy for big data reliabilities. The advantage of Cornebise's data was that they involved only binary distinctions; the presence or absence of a concern. Binary reliability data rendered the computation of large volumes of reliability data feasible where all other coefficients failed. They also enabled defining the quadrilogy in its simplest form.

There was another justification for starting this chapter with binary data. The metric of binary data, concerning one category and its opposite, yes or no, 0 or 1 with no values in between, is that special case to which all metric differences – rankings, linear orderings and proportions, introduced in Chapter 5 – reduce to whenever only two values are observed. Thus, the explanations developed so far are fundamental to their generalization to what the remainder of this chapter adds: reliability data with larger numbers of ordered values.

We proceed much as in the above but rely on the canonical form of reliability data of Tables 2.1, 5.1 and 6.1 to which we add the reference to a single-valued standard for measuring the accuracy of data, or a possible alternative for human coding measured by its surrogacy. Its columns contain not only individual values, but they imply what they have

| | | | | | |
|---|---|---|---|---|---|
| **Standard** or **Surrogate**: | $k_1$ . . . . | $k_u$ . . . . . . . . | $k_{N_u}$ | if available |
| **Units**: | 1 2 3 . . | $u$ . . . . . . . . | $N_u$ | |
| **Replications**: 1 | $c_{11}$ . . . . | $c_{1u}$ . . . . . . . . | $c_{1N_u}$ | |
| : | : | : | : | |
| $i$ | $c_{i1}$ . . . . | $c_{iu}$ . . . . . . . . | $c_{iN_u}$ | |
| $j$ | $c_{j1}$ . . . . | $c_{ju}$ . . . . . . . . | $c_{jN_u}$ | |
| : | : | : | : | |
| $m$ | $c_{m1}$ . . . . | $c_{mu}$ . . . . . . . . | $c_{mN_u}$ | |
| **Sums** $m_u$ $\begin{cases} >0 \text{ for } {}_c\alpha^s \text{ and } k_u \\ >1 \text{ for } {}_c\alpha, {}_c\alpha^{sm}, {}_c\alpha^m \end{cases}$ | $m_1$ . . . . | $m_u$ . . . . . . . . | $m_{N_u}$ | $n_{..} = \sum_{u=1}^{N_u} m_u$ |

**TABLE 6.7**

Canonical Form of Reliability Data for Replicability, Including for Accuracy, Surrogacy, and Decisiveness

in common, their majorities or modes, medians, or means, needed for answering questions about the surrogacy of substitutes and the decisiveness of coding. Settling on a single-valued standard and surrogate, unlike in the binary reliability data in Table 6.2. This is not a principal limitation but a convenient one that can be overcome if necessary, but not in this section.

The reliability data in Table 6.7 differ from Tables 2.1 and 5.1 for replicability and Table 6.2 for the quadrilogy of binary data in in the following ways:

- Standards or surrogates for human coding are represented by single values $k_u$, assigned to units $1, 2, \ldots, u, \ldots, N_u$
- For accuracy, the number of values $c_u$ in $u$ that are pairable with values $k_u$ is $m_u > 0$
- For surrogacy, the number of values $c_u$ in $u$ that are pairable with values $k_u$ is $m_u > 1$
- For accuracy and surrogacy, the total number of pairable values is $x.. = \sum_{u=1}^{N_u} m_u \text{ iff } k_u$
- For replicability and decisiveness the number of values $c_u$ in $u$ have to be $m_u > 1$
- For replicability and decisiveness, the total number of pairable values is $n.. \text{ or } x.. = \sum_{u=1}^{N_u} m_u$

The possibility that $m$ observers do not code all $N_u$ units means that the number of values in units $u$ is $m_u \leq m$. Moreover, the condition of units containing pairable values may not apply to all $N_u$ units, wherefore the total number of pairable values is $\leq mN_u$.

### 6.3.1 The Replicability of the Process of Generating Data: *c*-Alpha Metric

Chapters 2 to 5 focused on measuring replicability, the classical concerns of most reliability analyses. The αs they developed were limited to measuring the extent to which the data-making process is replicable elsewhere. This chapter started by expanding the notion of reliability to four kinds, of which replicability is just one. Although the extension of the definition of replicability from binary to variables with $v \geq 2$ number of values has been treated previously, we repeat its definitions to set the stage of the definitions of accuracy, surrogacy and decisiveness. We define the observed coincidences just like (5.1) in terms of Table 5.1 by (6.9) in terms of Table 6.7, continuing naming them $n_{ck}$ in conformity to the binary definitions in (6.2):

$$n_{ck} = \sum_{u=1}^{N} \frac{1}{m_u - 1} \sum_{i=1}^{m} \sum_{j \neq i}^{m} 1 \text{ iff } c_{iu} \text{ and } k_{ju} \text{ and } m_u > 1 \tag{6.9}$$

These observed coincidences $n_{ck}$ are tabulated as in $v$-by-$v$ matrices, comparatively presented in Table 5.2, here reproduced as Table 6.8.

The expected coincidences are obtained from the marginal sums of the observed coincidences of Table 6.8 according to (5.2):

$$e_{ck} = \frac{n_c}{n..-1} \begin{cases} n_k & \text{iff } c \neq k \\ n_k - 1 & \text{iff } c = k \end{cases} \tag{6.10}$$

| Values of replications by coders: | 1 | . | . | $k$ | . | $v$ | |
|---|---|---|---|---|---|---|---|
| 1 | $n_{11}$ | . | . | $n_{1k}$ | . | $n_{1v}$ | $n_1$ |
| . | . | . | . | . | . | . | . |
| $c$ | $n_{c1}$ | . | . | $n_{ck}$ | . | $n_{cv}$ | $n_c = \sum_{k=1}^{v} n_{ck}$ |
| . | . | . | . | . | . | . | . |
| . | . | . | . | . | . | . | . |
| $v$ | $n_{v1}$ | . | . | $n_{vk}$ | . | $n_{vv}$ | $n_v$ |
| | $n_1$ | . | . | $n_k$ | . | $n_v$ | $n_{..} = \sum_{c=1}^{v} \sum_{k=1}^{v} n_{ck}$ |

**TABLE 6.8**
The $v$-by-$v$ Matrix of Observed Coincidences for Assessing Replicability

With generic references to the difference functions introduced in Chapter 5, and in terms of the observed and expected coincidences, the agreement coefficient for replicability becomes as in (5.5):

$$ {}_{c}\alpha_{\text{metric}} = 1 - \frac{D_o}{D_e} = 1 - \frac{\sum_{c=1}^{v} \sum_{k \neq c}^{v} n_{ck} \ {}_{\text{metric}}\delta_{ck}^2}{\sum_{c=1}^{v} \sum_{k \neq c}^{v} e_{ck} \ {}_{\text{metric}}\delta_{ck}^2} \quad (6.11) $$

From here on, we simplify the notations by omitting the subscripts for the metric of observed and expected disagreements leaving both decided by their context of use.

As the ${}_{c}\alpha_{\text{metric}}$ for replicability has been discussed previously and at lengths, we move on.

### 6.3.2 The Accuracy of Generated Data: *c*-Alpha-*s*

According to Figure 6.1, the observed contingencies for accuracy involve pairing all values $c_{iu}$ of coders' replications with available values $k_u$ of a trusted standard:

$$ x_{ck} = \sum_{u=1}^{N_u} \sum_{i=1}^{m} 1 \ \text{ iff } c_{iu} \text{ and } k_u \quad (6.12) $$

Its contingencies $x_{ck}$ are presented in the $v$-by-$v$ matrix of Table 6.9:

| | **Values of a standard:** | 1 | . | . | $k$ | . | $v$ | |
|---|---|---|---|---|---|---|---|---|
| | 1 | $x_{11}$ | . | . | $x_{1k}$ | . | $x_{1v}$ | $x_{1.}$ |
| | . | . | . | . | . | . | . | . |
| **Values of replications:** | $c$ | $x_{c1}$ | . | . | $x_{ck}$ | . | $x_{cv}$ | $x_{c.} = \sum_{k=1}^{v} x_{ck}$ |
| | . | . | . | . | . | . | . | . |
| | . | . | . | . | . | . | . | . |
| | $v$ | $x_{v1}$ | . | . | $x_{vk}$ | . | $x_{vv}$ | $x_{v.}$ |
| | | $x_{.1}$ | . | . | $x_{.k}$ | . | $x_{.v}$ | $x_{..} = \sum_{c=1}^{v} \sum_{k=1}^{v} x_{ck} = \sum_{u=1}^{N_u} m_u \text{ iff } k_u$ |

**TABLE 6.9**
Matrix of Observed Contingencies Between Assigned Values and a Standard

Unlike coincidence matrices in Tables 5.2 and 6.8, contingencies $x_{ck}$:

- Are asymmetrically distributed: $x_{ck} \neq x_{kc}$, $x_{c\cdot} \neq x_{\cdot c}$, and $x_{\cdot k\cdot} \neq x_{k\cdot}$
- Their totals $x_{\cdot\cdot}$ equal the number of values pairable with values of the standard.
- The column sums $x_{\cdot k}$ represent how often the standard's value $k$ is paired with observers' values $c$ whereas the row sums $x_{c\cdot}$ represent how often observers used pairable values $c$.

As in coincidence matrices for replicability, in contingency matrices for assessing accuracy, all perfect agreements occupy its diagonal, $x_{cc}$ and $x_{kk}$. However, as already stated, this similarity does no longer apply to the off-diagonal cells.

Whereas the expected coincidences represent the hypothetical condition of exhibiting no statistical relationship between values used by observers and the predefined units of phenomena of analytical interest, the expected contingencies for accuracy represent the condition under which the values used by observers are statistically unrelated to the values of the standard. The fact that the values paired are from two non-overlapping sets of values, renders the expected contingencies for accuracy simpler than (6.10) and implied in (6.2):

$$\varepsilon_{ck} = \frac{x_{c\cdot} x_{\cdot k}}{x_{\cdot\cdot}} \tag{6.13}$$

Just as we use the letter $x$ for contingencies and $n$ for coincidences, (6.13) uses the epsilon '$\varepsilon$' the expected contingencies in place of the letter '$e$' for the expected coincidences.

The form of the accuracy coefficient ${}_c\alpha^s$ for metric data is essentially like the one for replicability (6.11) but expressed in terms of the contingencies $x_{ck}$ and $\varepsilon_{ck}$:

$${}_c\alpha^s_{\text{metric}} = 1 - \frac{D^s_o}{D^s_e} = 1 - \frac{\sum_{c=1}^{v}\sum_{k\neq c}^{v} x_{ck\ \text{metric}}\delta^2_{ck}}{\sum_{c=1}^{v}\sum_{k\neq c}^{v} \varepsilon_{ck\ \text{metric}}\delta^2_{ck}} \tag{6.14}$$

------------------------ x ------------------------

Incidentally, when data for accuracy are nominal and all units $u$ are assigned $m_u = 1$ value $c_u$ and have one standard $k_u$ each, there is an interesting equivalence between Cohen's (1960) infamous κ (kappa) and the accuracy $\alpha^s_{\text{nominal}}$ applied to that simple case. Consider:

For nominal data, agreements are the complements of disagreements: $A^s = x_{\cdot\cdot} - D^s$

κ's proportion of observed agreement is: $P_a = \dfrac{\sum_{c=1}^{v} x_{cc}}{x_{\cdot\cdot}} = \dfrac{A^s_o}{x_{\cdot\cdot}} = \dfrac{x_{\cdot\cdot} - D^s_o}{x_{\cdot\cdot}}$

κ's proportion of expected agreement is: $P_e = \dfrac{\sum_{c=1}^{v} \frac{x_{c\cdot} x_{\cdot c}}{x_{\cdot\cdot}}}{x_{\cdot\cdot}} = \dfrac{A^s_e}{x_{\cdot\cdot}} = \dfrac{x_{\cdot\cdot} - D^s_e}{x_{\cdot\cdot}}$

In Cohen's conception of κ, limited to nominal coding by two coders:

$$\kappa = \frac{P_a - P_e}{1 - P_e} = \frac{A^s_o - A^s_e}{x_{\cdot\cdot} - A^s_e} = \frac{(x_{\cdot\cdot} - D^s_o) - (x_{\cdot\cdot} - D^s_e)}{x_{\cdot\cdot} - (x_{\cdot\cdot} - D^s_e)} = \frac{D^s_e - D^s_o}{D^s_e} = 1 - \frac{D^s_o}{D^s_e} = \alpha^s_{\text{nominal}}$$

Evidently, Cohen's so-called chance-corrected reliability coefficient κ does not measure the replicability of data. Its mathematical equivalence to a very limited ${}_c\alpha^S_{\text{nominal}}$ reveals κ to be a measure of the accuracy of one coder's categorizations compared with another coder as the trusted standard – and vice versa. The biases when trying to interpret κ as a measure of replicability have been discussed in Chapter 4 and Section 4.2.2 in particular although in other terms. This equivalence merely adds another interpretation to κ's oddities: Cohen's probably unrecognized conceptual confusion of replicability with accuracy, albeit on a small scale of two coders.

------------------------ x ------------------------

### 6.3.3 Majorities, Modes, Medians, and Means

The simplicity of binary data made the development of the quadrilogy easy to define and explain. For binary judgements among multiple observers majorities are obvious, including their absence in case of ties. This section generalizes the concept of majority as a characterization of a set of binary judgements obtained by multiple observers to data that exhibit various metrics: nominal, ordinal, interval, and ratio scales of measurements. These generalizations are needed in definitions of surrogacy and decisiveness, compatible with measures in replicability and accuracy of the quadrilogy of reliability coefficients.

Notationally, majorities or modes, medians, or means within a unit $u$ will be designated by $\overline{c}_u$ of sets $C_u$ of values $c_u$.

#### 6.3.3.1 *For Nominal Data*

For nominal data with more than four categories, single majorities or modes are not the only representations of a diversity of values in need of attention. A *mode* is the most frequently occurring category among a set of nominal categories.

To identify a mode, it is helpful to represent the $m_u$ values in any one unit $u$

$$c_{1u}, c_{2u}, \ldots, c_{iu}, \ldots, c_{mu}$$

in terms of their frequencies $r_{cu}$ of values $c_u$ in $u$ – just as in Table 6.2 but now for any number of values.

In the absence of ties, a single majority $\overline{c}_u$ in nominal data is the value $c_u$ with the largest frequency:

$$\overline{c}_u \mid r_{\overline{c}_u} = max_u(r_{c_u}) \tag{6.15a}$$

However, it may happen that the highest frequency in $u$ is shared among several values. Ties are observed when $h_u$ different values share the highest frequency $r_{c_{1u}} = r_{c_{2u}} = \ldots = r_{c_{h_u}}$. For example, among $m_u = 4$ categories there could be:

| | | |
|---|---|---|
| $h_u = 1$: @@@@; | @@@□; @@□⌘ | One mode of decreasing decisiveness |
| $h_u = 2 < m_u$: | @@□□ | Two tied modes |
| $h_u = 4 = m_u$: | @□⌘◆ | $h_u = m_u =$ four tied "modes" of distinct values. |

To embrace the possibility of a set of $h_u$ modes, i.e., distinct categories occurring with the same and largest frequency in unit $u$, (6.15a) is generalized to:

$$\left\{ \overline{c}_{u_1}, \cdots \overline{c}_{u_g}, \cdots \vec{c}_{u_{h_u}} \right\} \mid (r_{\overline{c}_{u1}} = \cdots = r_{\overline{c}_{ug}} = \cdots = r_{\overline{c}_{uh_u}}) = max_u(r_{c_u}) \tag{6.15b}$$

While the number of tied modes can range between $1 \le h_u \le m_u$, ties become less likely as the number $m_u$ of categories becomes large.

#### 6.3.3.2 For Ordinal or Ranked Data

For ordinal data, the *median* rank $\bar{c}_u$ of $m_u$ ranks in unit $u$ occupies the center of the frequency distribution of ordered ranks listed between the $c_{\text{lowest}}$ and $c_{\text{highest}}$ rank:

$$\bar{c}_u = \textit{The} \left(\frac{m_u + 1}{2}\right)^{\text{th}} \text{ rank between the } c_{\text{lowest}} \text{ and } c_{\text{highest}} \text{ rank} \tag{6.16}$$

When $m_u$ is uneven, $\bar{c}_u$ is the rank in the center of the distribution of all ranks. When $m_u$ is even, $\bar{c}_u$ may not be an observed rank. If not, a value between the two neighboring ranks serves as the median.

#### 6.3.3.3 For Interval, Ratio, and Polar Data

Sets of interval, ratio, and polar scale values have averages. The average of a set of values is the value from which the members of the set collectively minimally deviate. That value, the mean $\bar{c}_u$ within units $u$, is defined by:

$$\bar{c}_u = \frac{1}{m_u} \sum_{i=1}^{m_u} c_{iu} \tag{6.17}$$

### 6.3.4 The Surrogacy of an Alternative to Coding of Data: *c*-Alpha-*sm*

Surrogacy is the degree to which a proposed alternative agrees with the majorities, modes, medians, or means of replications.

The contingencies between $\bar{c}_u$ and a surrogate's values $k_u$ are:

$$x_{\bar{c}k} = \sum_{u=1}^{N_u} m_u \;\text{ for }\; \bar{c}_u \;\text{ and iff }\; k_u \tag{6.18a}$$

And for $h_u$-modal values consisting of a set $\left\{ \bar{c}_{u_1}, \ldots \bar{c}_{u_g}, \ldots, \bar{c}_{u_{h_u}} \right\}$ of tied modes:

$$x_{\bar{c}k} = \sum_{u=1}^{N} \frac{m_u}{h_u} \sum_{g=1}^{h_u} 1 \;\text{ for }\; \bar{c}_{u_g} \;\text{ and iff }\; k_u \tag{6.18b}$$

Contingencies $x_{\bar{c}k}$ are tabulated in a matrix as in Table 6.10 which may no longer be square as its row names consist of computed values $\bar{c}_u$ which may not occur among the recorded values $c_u$. See the following Table 6.10:

Its expected contingencies are defined just as for accuracy (6.13):

$$\varepsilon_{\bar{c}k} = \frac{x_{\bar{c}} \cdot x_{\cdot k}}{x_{\cdot\cdot}} \tag{6.19}$$

When tabulated, (6.18) and (6.19) form matrices with asymmetrically distributed contingencies $x_{\bar{c}k}$ and $\varepsilon_{\bar{c}k}$ just as for accuracy. However, unlike the contingency matrices for

| | | | | | | | | |
|---|---|---|---|---|---|---|---|---|
| **Values of a possible surrogate:** | | 1 | . | . | $k$ | . | $v$ | |
| | 1 | $x_{11}$ | . | . | $x_{1k}$ | . | $x_{1v}$ | $x_{1\cdot}$ |
| | . | . | . | . | . | . | . | . |
| **Coders' majorities,** | . | . | . | . | . | . | . | . |
| **medians,** or **means:** | $\bar{c}$ | $x_{\bar{c}1}$ | . | . | $x_{\bar{c}k}$ | . | $x_{\bar{c}v}$ | $x_{\bar{c}\cdot} = \sum_{k=1}^{v} x_{\bar{c}}$ |
| | . | . | . | . | . | . | . | . |
| | . | . | . | . | . | . | . | . |
| | . | . | . | . | . | . | . | . |
| | . | . | . | . | . | . | . | . |
| | | $x_{\cdot 1}$ | . | . | $x_{\cdot k}$ | . | $x_{\cdot v}$ | $x_{\cdot\cdot} = \sum_{\bar{c}} \sum_{k=1}^{v} x_{\bar{c}k} = \sum_{u}^{N_u} m_u > 1$ and $k_u$ |

**TABLE 6.10**
Matrix of Observed Contingencies Between Modes, Medians, or Means $\bar{c}$ and an Alternative

nominal data, those for ordinal, interval, ratio and polar data are unlikely to be square matrices as rows are computed medians and means which may not be part of the values of the proposed alternative. It follows that the matrix does not have a diagonal of perfectly matching contingencies $x_{\bar{c}=k}$.

The agreement coefficient for surrogacy, defined in terms of the observed and expected disagreements (6.18) and (6.19) resembles (6.3) and (6.14) except for the reference to $\bar{c}$ rather than $c$:

$$ {}_{c}\alpha_{\text{metric}}^{sm} = 1 - \frac{D_o^{sm}}{D_e^{sm}} = 1 - \frac{\sum_{\bar{c}} \sum_{k=1}^{v} x_{\bar{c}k} \; {}_{\text{metric}}\delta_{\bar{c}k}^2}{\sum_{\bar{c}} \sum_{k=1}^{v} \varepsilon_{\bar{c}k} \; {}_{\text{metric}}\delta_{\bar{c}k}^2} \tag{6.20} $$

It is important to emphasize that surrogacy is a measure of the reliability of a possible alternative for usually expensive human coders. It is not a measure of the reliability of data. As previously discussed, ${}_{c}\alpha_{\text{metric}}^{sm} = 1$ when the proposed alternative is the best possible fit of its values with what the coded data have in common, which is their majorities, medians, or means.

Surrogacy does not say anything about the variance among coded data, which is the topic of the following section.

### 6.3.5 The Decisiveness of Majorities, Medians, and Means: *c*-Alpha-*m*

By dictionary definitions, decisiveness is the ability to conclude with confidence or to make firm decisions. Its antonyms include indecisiveness, inconclusiveness, vagueness, and uncertainty. As such, decisiveness correlates with reliability. When data report a concern decisively, conclusions drawn from their analysis are less likely questionable. A proposal supported by 99% of all stakeholders is more decisive than one passed by a 51% majority. This is so in politics, at board meetings, with witnesses' accounts in a court of law, and recognitions of a phenomenon to be of a certain kind. Decisiveness is a measure independent from surrogacy.

Figure 6.1 should make the relationship between accuracy, surrogacy, and decisiveness clear. Accuracy evaluates replications with an external standard. Surrogacy evaluates a possible substitute with what observers have in common. Decisiveness evaluates what replications have in common with how they vary. It evaluates them with an internal standard.

According to Figure 6.1, the coincidences for decisiveness are obtained by pairing the $m_u$ replications $c_{iu}$ of $u$ with their own majorities or modes, medians, or means $\bar{c}_u$.

$$\text{For single } \bar{c}_u \text{ s} : x_{c\bar{c}} = \sum\nolimits_{u=1}^{N_u} \sum\nolimits_{i=1}^{m} 1 \text{ for } c_{iu},\ \bar{c}_u \text{ and iff } m_u > 1 \tag{6.21a}$$

$$\text{For } h_u \left\{ \bar{c}_{1u}, \ldots \bar{c}_{gu}, \ldots, \bar{c}_{h_u u} \right\} \text{ s} : x_{c\bar{c}} = \sum\nolimits_{u=1}^{N_u} \sum\nolimits_{i=1}^{m} \sum\nolimits_{g=1}^{h_u} \frac{1}{h_u} \text{ for } c_{iu},\ \bar{c}_{u_g} \text{ and iff } m_u > 1 \tag{6.21b}$$

| | | | | | | | | | | |
|---|---|---|---|---|---|---|---|---|---|---|
| **Coders' majorities, medians,** or **means:** | | 1 | . | . | $\bar{c}$ | . | . | . | . | |
| | 1 | $x_{11}$ | . | . | $x_{1\bar{c}}$ | . | . | . | . | $x_{1\cdot}$ |
| | . | . | . | . | . | . | . | . | . | . |
| **Values of replications:** | $c$ | $x_{c1}$ | . | . | $x_{c\bar{c}}$ | . | . | . | . | $x_{c\cdot} = \sum_{\bar{c}} x_{c\bar{c}}$ |
| | . | . | . | . | . | . | . | . | . | . |
| | $v$ | $x_{v1}$ | . | . | $x_{v\bar{c}}$ | . | . | . | . | $x_{v\cdot}$ |
| | | $x_{\cdot 1}$ | . | . | $x_{\cdot\bar{c}}$ | . | . | . | . | $x_{\cdot\cdot} = \sum_{c\,1}^{v} \sum_{\bar{c}} x_{cc}$ |

**TABLE 6.11**
Matrix of Observed Contingencies Between Coders' Values $c$ and Their Own $\bar{c}$

As Table 6.11 tabulates all contingencies of $m_u$ values $c_{iu}$ with their own modes, medians, or means $\bar{c}_u$, the matrix in Table 6.11 is not likely square.

The expected coincidences are defined just as for accuracy (6.15) and surrogacy (6.23), distinguished by their subscripts:

$$\varepsilon_{c\bar{c}} = \frac{x_{c\cdot} x_{\cdot\bar{c}}}{x_{\cdot\cdot}} \tag{6.22}$$

In which terms the ${}_c\alpha^m$ for decisiveness becomes:

$${}_c\alpha^m_{\text{metric}} = 1 - \frac{D^m_o}{D^m_e} = 1 - \frac{\sum_{\bar{c}} \sum_{k=1}^{v} x_{c\bar{c}\ \text{metric}} \delta^2_{c\bar{c}}}{\sum_{\bar{c}} \sum_{k=1}^{v} \varepsilon_{c\bar{c}\ \text{metric}} \delta^2_{c\bar{c}}} \tag{6.23}$$

The above-discussed three purposes of the decisiveness measure are: Evaluating the extent to which individual observers agree with what they have in common, measuring the extent to which a perfectly fitting surrogate could possibly substitute for individual human coding, and signaling how much free the coding process is of deviations of what coders share. There is virtue in considering measuring the reliability of data by ${}_c\alpha^m_{\text{binary}}$ in preference to ${}_c\alpha_{\text{binary}}$. This suggestion is attractive not only because decisiveness measures zero when data are totally indecisive about what they represent and one when they are unanimous, but also because the interval disagreements in decisiveness basically are variance measures. This resonates with many subsequently applied variance-related analytical methods. It also generalizes the notion of variance to metrics other than interval.

## 6.4 A Numerical Example Beyond the Binary

The above can be demonstrated by means of the example from Section 2.3 – its icons are represented here numerically to take advantage of the difference functions defined in Section 5.5, to which a standard or potential surrogate is added as in Table 6.7, and the members of the group of observers are given names to explore their accuracy.

| | | | | | | | | | | | | |
|---|---|---|---|---|---|---|---|---|---|---|---|---|
| **Standard** or **Surrogate**: | . | 3 | 3 | 3 | 2 | 3 | 4 | 2 | 2 | 5 | 1 | 3 |
| **Units** $u$: | 1 | 2 | 3 | 4 | 5 | 6 | 7 | 8 | 9 | 10 | 11 | 12 |
| **Observers**: Peter | 1 | 2 | 3 | 3 | 2 | 1 | 4 | 1 | 2 | . | . | . |
| Paul | 1 | 2 | 3 | 3 | 2 | 2 | 4 | 1 | 2 | 5 | . | . |
| Bill | . | 3 | 3 | 3 | 2 | 3 | 4 | 2 | 2 | 5 | 1 | 3 |
| Mary | 1 | 2 | 3 | 3 | 2 | 4 | 4 | 1 | 2 | 5 | 1 | . |
| **Sums of values** $m_u$: | 3 | 4 | 4 | 4 | 4 | 4 | 4 | 4 | 4 | 3 | 2 | 1 |

These data are too small to have practical meanings, but they may reveal some of the differences, starting with the number of pairable values $m_u$ in the replications. Replicability requires $m_u > 1$ values to be pairable among themselves, $n.. = 40$. For majorities and modes to be possible, decisiveness requires the same number, $x.. = 40$. Accuracy and surrogacy require only $m_u > 0$ plus a standard $k_u$. The latter is not the case for unit 1, and hence both contingencies are $x.. = 38$. Units that do not contain pairable values cannot contribute to evaluations of reliability and are ignored. Therefore, the 41 values in this table are not utilizable by all four coefficients alike.

The calculations of their replicability were already demonstrated in Section 2.3 for nominal data ${}_c\alpha_{\text{nominal}} = 0.738$ and in Section 5.5 for interval data, ${}_c\alpha_{\text{interval}} = 0.849$ in which their differences were adequately explained.

For accuracy, the values $k_u$ of the standard are paired with the observers' values $c_{iu}$. Its matrices of observed and expected contingencies sum to 38, as a standard for unit 1 is missing:

According to (6.14):

$$
{}_c\alpha^S_{\text{nominal}} = 1 - \frac{D^S_o}{D^S_e} = 1 - \frac{\sum_{c=1}^{v}\sum_{k=1}^{v} x_{ck\ \text{nominal}}\delta^2_{ck}}{\sum_{c=1}^{v}\sum_{k=1}^{v} \varepsilon_{ck\ \text{nominal}}\delta^2_{ck}} = 1 - \frac{x.. - \sum_{c=1}^{v} x_{cc}}{x.. - \sum_{c=1}^{v} \varepsilon_{cc}}
$$

$$
= 1 - \frac{38 - (2+9+11+4+3)}{38 - (0.32+4.11+4.92+0.53+0.24)} = 1 - \frac{9}{27.85} = 0.677
$$

Because observed disagreements are close to the diagonal, when regarding these values as interval data, accuracy can again be expected to be larger than when values are considered unordered $\left(\text{using the non-standardized }{}_{\text{interval}}\delta^2_{ck} = (c-k)^2\right)$.

$$
{}_{c}\alpha^{s}_{\text{interval}} = 1-\frac{D^{s}_{o}}{D^{s}_{e}} = 1-\frac{\sum_{c=1}^{v}\sum_{k=1}^{v} x_{ck\ \text{interval}}\delta^{2}_{ck}}{\sum_{c=1}^{v}\sum_{k=1}^{v} \varepsilon_{ck\ \text{interval}}\delta^{2}_{ck}}
$$
$$
= 1-\frac{(3+4+1)\cdot 1^2+1\cdot 2^2}{(0.68+\cdots+0.32+1.89+\cdots+0.39)1^2+(0.58+\cdots+0.87)2^2+\cdots+(0.16+0.47)4^2}
$$
$$
= 1-\frac{12}{85.58} = 0.860
$$

To evaluate whether the variable used above as the standard for evaluating the accuracy of the four observers could serve as a surrogate of the majorities of the four observers' categories, we pair each $k_u$ with the observers' majority. These comparisons yield the following contingency matrices and subsequent computations:

Contingencies for nominal data: **Observed** / **Expected**

| **Surrogate's Values** $k$: **Observers' majorities** $\bar{c}$: | Obs. 1 | Obs. 2 | Obs. 3 | Obs. 4 | Obs. 5 | Exp. 1 | Exp. 2 | Exp. 3 | Exp. 4 | Exp. 5 | |
|---|---|---|---|---|---|---|---|---|---|---|---|
| 1 | 2 | 4 | 1 | . | . | 0.37 | 2.21 | 3.13 | 0.74 | 0.55 | 7 |
| 2 | . | 8 | 5 | . | . | 0.68 | 4.11 | 5.81 | 1.37 | 1.03 | 13 |
| 3 | . | . | 10 | . | . | 0.53 | 3.16 | 4.47 | 1.05 | 0.79 | 10 |
| 4 | . | . | 1 | 4 | . | 0.26 | 1.58 | 2.24 | 0.53 | 0.39 | 5 |
| 5 | . | . | . | . | 3 | 0.16 | 0.95 | 1.34 | 0.32 | 0.24 | 3 |
| | 2 | 12 | 17 | 4 | 3 | 2 | 12 | 17 | 4 | 3 | 38 |

According to (6.20), surrogacy turns out to be:

$$
{}_{c}\alpha^{sm}_{\text{nominal}} = 1-\frac{D^{sm}_{o}}{D^{sm}_{e}} = 1-\frac{\sum_{\bar{c}=1}^{v}\sum_{k=1}^{v} x_{\bar{c}k\ \text{nominal}}\delta^{2}_{\bar{c}k}}{\sum_{\bar{c}=1}^{v}\sum_{k=1}^{v} \varepsilon_{\bar{c}k\ \text{nominal}}\delta^{2}_{\bar{c}k}} = 1-\frac{x_{..}-\sum_{c=1}^{v} x_{\bar{c}c}}{x_{..}-\sum_{c=1}^{v} \varepsilon_{\bar{c}c}}
$$
$$
= 1-\frac{38-(2+8+10+4+3)}{38-(0.37+4.11+4.47+0.53+0.24)} = 1-\frac{11}{28.28} = 0.611
$$

Note that the modes of nominal data stay within the set of categories that give rise to them. This accounts for the above and the following contingency matrices to be square. When values are ordered, this may not be so.

Contingencies for nominal data: **Observed** / **Expected**

| **Observers' majorities** $\bar{c}$: **Observers' values** $c$: | Obs. 1 | Obs. 2 | Obs. 3 | Obs. 4 | Obs. 5 | Exp. 1 | Exp. 2 | Exp. 3 | Exp. 4 | Exp. 5 | |
|---|---|---|---|---|---|---|---|---|---|---|---|
| 1 | 8.25 | 0.25 | 0.25 | 0.25 | . | 2.25 | 2.93 | 2.03 | 1.12 | 0.67 | 9 |
| 2 | 1.25 | 11.25 | 0.25 | 0.25 | . | 3.25 | 4.22 | 2.92 | 1.63 | 0.98 | 13 |
| 3 | 0.25 | 1.25 | 8.25 | 0.25 | . | 2.50 | 3.25 | 2.25 | 1.25 | 0.75 | 10 |
| 4 | 0.25 | 0.25 | 0.25 | 4.25 | . | 1.25 | 1.63 | 1.13 | 0.62 | 0.37 | 5 |
| 5 | . | . | . | . | 3 | 0.75 | 0.97 | 0.67 | 0.38 | 0.23 | 3 |
| | 10 | 13 | 9 | 5 | 3 | 10 | 13 | 9 | 5 | 3 | 40 |

In the above observed contingencies, note the contributions of pairing the $h_6 = 4$ tied majorities in unit 6 with each of its four values associated with that unit. It yields 4-by-4 evenly distributed frequencies, here 0.25, excluding the value 5. The observed disagreements, contributed by the other units, are added to the 1-2 and 2-3 cells of this matrix. With the expected contingencies, calculated from the marginal frequencies as usual, the decisiveness of these data is far larger than their surrogacy:

$$_{c}\alpha^{m}_{\text{nominal}} = 1 - \frac{D^m_o}{D^m_e} = 1 - \frac{\sum_{\ddot{c}=1}^{v}\sum_{k=1}^{v} x_{c\bar{c}}\ {}_{\text{nominal}}\delta^2_{c\bar{c}}}{\sum_{\bar{c}=1}^{v}\sum_{k=1}^{v} \varepsilon_{c\bar{c}}\ {}_{\text{nominal}}\delta^2_{\bar{c}k}} = 1 - \frac{x_{..} - \sum_{c=1}^{v} x_{c\bar{c}}}{x_{..} - \sum_{c=1}^{v} \varepsilon_{c\bar{c}}}$$

$$= 1 - \frac{40 - (8.25 + 11.25 + 8.25 + 4.25 + 3)}{40 - (2.25 + 4.22 + 2.25 + 0.62 + 0.23)} = 1 - \frac{5}{30.43} = 0.836$$

However, when data are of the interval kind, means may not equal the values they average causing their contingency matrices not to be square. This can be seen in the contingency matrices of our example:

| Contingencies for interval data: | **Observed** | | | | | **Expected** | | | | | |
|---|---|---|---|---|---|---|---|---|---|---|---|
| **Surrogate's Values** $k$: | 1 | 2 | 3 | 4 | 5 | 1 | 2 | 3 | 4 | 5 | |
| **Observers' means** $\bar{c}$: 1 | 2 | . | . | . | . | 0.11 | 0.63 | 0.89 | 0.21 | 0.16 | 2 |
| 1.25 | . | 4 | . | . | . | 0.21 | 1.26 | 1.79 | 0.42 | 0.32 | 4 |
| 2 | . | 8 | . | . | . | 0.42 | 2.53 | 3.58 | 0.84 | 0.63 | 8 |
| 2.25 | . | . | 4 | . | . | 0.21 | 1.26 | 1.79 | 0.42 | 0.32 | 4 |
| 2.5 | . | . | 4 | . | . | 0.21 | 1.26 | 1.79 | 0.42 | 0.32 | 4 |
| 3 | . | . | 9 | . | . | 0.47 | 2.85 | 4.02 | 0.95 | 0.71 | 9 |
| 4 | . | . | . | 4 | . | 0.21 | 1.26 | 1.79 | 0.42 | 0.32 | 4 |
| 5 | . | . | . | . | 3 | 0.16 | 0.95 | 1.35 | 0.32 | 0.22 | 3 |
| | 2 | 12 | 17 | 4 | 3 | 2 | 12 | 17 | 4 | 3 | 38 |

$$_{c}\alpha^{sm}_{\text{interval}} = 1 - \frac{D^{sm}_o}{D^{sm}_e} = 1 - \frac{\sum_{\bar{c}}\sum_{k=1}^{v} x_{\bar{c}k}\ {}_{\text{interval}}\delta^2_{\bar{c}k}}{\sum_{\bar{c}}\sum_{k=1}^{v} \varepsilon_{\bar{c}k}\ {}_{\text{interval}}\delta^2_{\bar{c}k}} \left(\text{using the non-standardized } {}_{\text{interval}}\delta^2_{\bar{c}k} = (\bar{c} - k)^2\right)$$

$$= 1 - \frac{(4+4)0.75^2 + 4 \cdot 0.5^2}{(0.21+1.26)0.25^2 + (1.26+1.79)0.75^2 + (0.63+\cdots+0.32)1^2 + \cdots + (0.16+0.16)4^2} =$$

$$= 1 - \frac{5.5}{78.419} = 0.930$$

Because the surrogacy of the interval data is larger than that of their accuracy, it stands to reason that the decisiveness would be lower than that of the accuracy:

Contingencies for interval data:

**Observed**

| **Observers' means** $\bar{c}$ : | 1 | 1.25 | 2 | 2.25 | 2.5 | 3 | 4 | 5 | |
|---|---|---|---|---|---|---|---|---|---|
| **Observers' values** $c$ : 1 | 5 | 3 | . | . | 1 | . | . | . | |
| 2 | . | 1 | 8 | 3 | 1 | . | . | . | |
| 3 | . | . | . | 1 | 1 | 8 | . | . | |
| 4 | . | . | . | . | 1 | . | 4 | . | |
| 5 | . | . | . | . | . | . | . | 3 | |
| | 5 | 4 | 8 | 4 | 4 | 8 | 4 | 3 | |

**Expected**

| | 1 | 1.25 | 2 | 2.25 | 2.5 | 3 | 4 | 5 | |
|---|---|---|---|---|---|---|---|---|---|
| 1 | 1.13 | 0.90 | 1.80 | 0.90 | 0.90 | 1.80 | 0.90 | 0.68 | 9 |
| 2 | 1.63 | 1.30 | 2.60 | 1.30 | 1.30 | 2.60 | 1.30 | 0.98 | 13 |
| 3 | 1.25 | 1.00 | 2.00 | 1.00 | 1.00 | 2.00 | 1.00 | 0.75 | 10 |
| 4 | 0.63 | 0.50 | 1.00 | 0.50 | 0.50 | 1.00 | 0.50 | 0.37 | 5 |
| 5 | 0.38 | 0.30 | 0.60 | 0.30 | 0.30 | 0.60 | 0.30 | 0.23 | 3 |
| | 5 | 4 | 8 | 4 | 4 | 8 | 4 | 3 | 40 |

$$
{}_c\alpha^m_{\text{interval}} = 1 - \frac{D^m_o}{D^m_e} = 1 - \frac{\sum_{c=1}^{v}\sum_{\bar{c}} x_{c\bar{c}}\ {}_{\text{interval}}\delta^2_{c\bar{c}}}{\sum_{c=1}^{v}\sum_{\bar{c}} \varepsilon_{c\bar{c}}\ {}_{\text{interval}}\delta^2_{c\bar{c}}} \left(\text{using the non-standardized } {}_{\text{interval}}\delta^2_{c\bar{c}} = (c-\bar{c})^2\right)
$$

$$
= 1 - \frac{6\cdot 0.06 + 2\cdot 0.25 + 2\cdot 0.56 + 2(1.5)^2}{(.9+1.3)(.25)^2 + (1.3+1)(.5)^2 + (1.3+1)(.75)^2 + \cdots + (.68+.38)4^2}
$$

$$
= 1 - \frac{6.5}{104.6252} = 0.938
$$

In sum:

| | | |
|---|---|---|
| Replicability | ${}_c\alpha_{\text{nominal}} = 0.738 < 0.849 = {}_c\alpha_{\text{interval}}$; | $n.. = 40$ |
| Accuracy | ${}_c\alpha^s_{\text{nominal}} = 0.677 < 0.860 = {}_c\alpha^s_{\text{interval}}$; | $x.. = 38$ |
| Surrogacy | ${}_c\alpha^{sm}_{\text{nominal}} = 0.611 < 0.930 = {}_c\alpha^{sm}_{\text{interval}}$; | $x.. = 38$ |
| Decisiveness | ${}_c\alpha^m_{\text{nominal}} = 0.836 < 0.938 = {}_c\alpha^m_{\text{interval}}$; | $x.. = 40$ |

Not explicated are: Accuracy re standard surrogacy for the remaining three.

| | | |
|---|---|---|
| Peter's | ${}_c\alpha^s_{\text{nominal}} = 0.510;\ x.. = 8$ | ${}_c\alpha^{sm}_{\text{nominal}} = 0.848;\ x.. = 9$ |
| Paul's | ${}_c\alpha^s_{\text{nominal}} = 0.542;\ x.. = 9$ | ${}_c\alpha^{sm}_{\text{nominal}} = 0.867;\ x.. = 10$ |
| Bill's | ${}_c\alpha^s_{\text{nominal}} = 1.000;\ x.. = 11$ | ${}_c\alpha^{sm}_{\text{nominal}} = 0.612;\ x.. = 10$ |
| Mary's | ${}_c\alpha^s_{\text{nominal}} = 0.615;\ x.. = 10$ | ${}_c\alpha^{sm}_{\text{nominal}} = 0.882;\ x.. = 11$ |

When comparing each individual observers' categorizations with the given standard, Peter's accuracy is the worst, not much better than Paul's. Mary's accuracy is nearly average, while Bill's accuracy is perfect. This is easily verifiable by comparing each observer's rows with the standard. Note also that comparisons require pairs of categories, the total number of which varies for each observer. The accuracy of all four observers, 0.677 is about average for each.

When comparing each observer's categories with the majorities among the remaining members of the group, the ability of Peter, Paul or Mary to serve as the surrogate of the other three is virtually indistinguishable. In fact the only problem for all three of them is the disagreement in unit 6. Here, Bill turns out to be the outlier, the worst possible surrogate of the other three. This should not be surprising as Bill's accuracy already deviates quite drastically from Peter, Paul, and Mary's. I should point out what is quantitatively apparent; that the above measure of the surrogacy of an external device, algorithm, or method, here equal to the standard, could serve as a substitute for the four observers has little to do with whether any one member of the group could serve as a replacement of the other three.

# 7

# *Multi-Valued Coding of Predefined Units*

This chapter builds on the preceding Chapters 2, 5, and 6 by generalizing the $_c\alpha$ for single-valued coding of predefined units of analysis to multi-valued coding of such units. By multi-valuedness I mean units that are valued by sets $C$ of values $c \in C$. Single-valuedness, discussed so far, is that special case of multi-valuedness where the number of values assigned to units is at most one. To indicate the difference between these two approaches to valuing predefined units of analysis, the capital "C" is used in the subscript of $\alpha$: $_C\alpha$.

Especially in the social sciences, there are many situations in which units are naturally described in terms of more than one category or value. Literary theory has recognized for eons that all texts afford multiple interpretations, not one. Dictionaries list alternative meanings among which readers may choose depending on a word's syntactical function, context of use, or the discourse being pursued in a text. Computational linguistics encounters multiple interpretations and struggles to disambiguate sentence to the extent possible. Conversation analysts have recognized that any utterance may serve several functions, can direct conversations along one of several paths, and encourage analysts to consider more than one speech act type to each utterance. People play different roles in their lives. Actors can assume multiple dramaturgical personalities. Published work may have several co-authors and draw on multiple sources.

Teaching social science students to define their research questions so that the multi-valuedness of phenomena does not pose problems for data processing and theory development is a common recommendation. It leaves much of the social world unattended and worse, encourages dismissing multi-valued phenomena as unworthy of attention, ambiguous, or left to philosophers. Yet, multi-valuedness is real.

Avoiding multi-valued phenomena is deeply ingrained in methodological conceptions of data languages in terms of a set of variables, each defined in terms of mutually exclusive categories or values, although there is much to be said in favor of avoiding uncertainties within the design of data languages and related coding instructions. However, their logic should accommodate more complex situations. For reasons of logic, a person has only one age at a time, cannot be simultaneously present at physically distinct locations, or believe something and its logical opposite. Yet, such contradictions happen and can be resolved by adopting different logical frames. Although we may believe a physical object to have but one temperature, it is conceivable that objects have several temperatures at different locations or according to different ways of measuring or at different times. Not all phenomena are either + or −, true or false. Truly multi-valued phenomena may make their recording difficult when a data language does not accommodate that possibility. In English, personal pronouns are gendered: 'he' or 'she,' not both. The neutral 'it' is reserved for impersonal objects. However, it often is important to recognize that individuals may be one or the other, both or neither. If this is the case, English gendered singular pronouns make gender-neutral references cumbersome to express, for example, by reporting that "he or she hurt herself and himself." To get out of this binary, I once proposed a non-gendered pronoun 'e':

| | | | |
|---|---|---|---|
| 'e' | for | 'he' and 'she' | – subjective |
| 'em' | for | 'him' and 'her' | – objective |
| 'es' | for | 'his' and 'hers' | – possessive |
| 'emself' | for | 'himself' and 'herself' | – reflexive |

It worked well in some of my own writings, even in everyday talk. But a gender-neutral pronoun creates a new single category, that bypasses multi-valuedness – like creating the title 'Ms' which refuses to indicate whether a woman is married or not. Traditional interviews have dealt with such uncertainties by including such default categories as 'other,' 'both,' or 'none of the above,' including 'NA' or 'no answer.' In interviews as well as textual analyses, such nondescript categories often attract cases where coders or interviewees were uncertain, are generally unreliable, and end up being uninformative. However, gender- and marital status-neutrality should not be relegated to a 'catch all else' category which ignores socially important identifications. The issue is whether that category can be treated as a multi-valued set of values or a category that is part of a variable of increasingly common categories of identifications.

The tendency in everyday language to add new categories where the distinctions in a variable are no longer applicable is more radically embraced by the analytical techniques available, the preponderance of which treat meaningful multi-valued phenomena as multi-dimensional. They represent units of analysis as a point in a multi-dimensional vector space, one value for each dimension. Accounting for the multi-valuedness of many social phenomena in terms of vector spaces is driven by the logic of conveniently available analytical techniques, not by the nature of the phenomena of interest. Multi-valued phenomena need to be represented by multi-valued data and analytical techniques that treat them accordingly.

A largely failed strategy to create data whenever given phenomena have multiple meanings, play several roles, or can be seen from multiple perspectives, is to force coders to describe them from a preferred perspective, or to select the primary, dominant, or most relevant attribute for a current research project. We always play numerous roles, being a parent, an author, a driver, a voter, a taxpayer, a bank client, etc. Forcing single-valued descriptions on such multi-valued phenomena not only omits what may well be analytically relevant, the agreements obtained from such coding instruction tend to suffer greatly and the resulting data are obviously unreliable.

To increase the reliability of such choices among alternatives, coders have been given unambiguous criteria, like the one of several adjectives mentioned first, written with the shortest number of characters, or most commonly used in a language. The improved measure of agreement is deceptively purchased by shifting the meanings of data to conceptually meaningless criteria.

The dominant motivation of forcing multi-valued phenomena into single-valued characterizations is due largely to easily available statistical methods applicable to single-valued data. Figure 1.1 and Section 1.1.4 already discussed the importance of matching the complexities of phenomena of analytical interest with the nature of the data language used to record them.

The lack of adequate statistical methods for multi-valued data starts with suitable reliability measures. In 1992, Krippendorff proposed several reliability measures to embrace multi-valuedness. My proposals have evolved since. This chapter develops a generalization of the preceding chapters' focus on agreement measures for the single-valued coding of predefined units of phenomena to two kinds of multi-valued agreements.

- Multi-valued agreement measures that are applicable to situations in which sets of values are attributed to unitized phenomena. Because multi-valued data are not too likely to be involved in issues of surrogacy and decisiveness, they are not addressed. Sections 7.1 concerns the replicability of sets of values, Section 7.2 the replicability of individual values in sets, and Section 7.3 with their accuracy.
- Multi-valued agreement measures that focus on how multi-valued sets distinguish among the units of the phenomena described in such terms. These measures are not affected by the number of values in sets, measuring only the values they agree on or omit. They can be conceived of as standardizing the above which brings them closer to the ${}_c\alpha$-agreements of previous chapters. Section 7.4 will address this difference function and conclude this chapter.

We shall retain the basic structure of alpha and focus on the appropriate difference functions for multi-valued characterizations of unitized phenomena. Moreover, inasmuch as these difference functions include the single-valued differences from Section 5.3, the alpha for multi-valued data amounts to another generalization of ${}_c\alpha$.

## 7.1 Multi-Valued Reliability Data for Replicability

Whereas single-valued reliability data consist of replications by observers who assign at most one category or value to units of analysis, multi-valued reliability data allow units to be described by sets of any number of values:

$$C = \{c_1, c_2, \ldots, c_{|C|}\} \text{ or } K = \{k_1, k_2, \ldots, k_{|K|}\};$$

Where: $|C|$ is the number of values $c \in C$, $C$'s cardinality
$C$ could be single-valued, $C = \{c\}$, or be empty, $C = \{\}$ i.e., missing

Multi-valued data may have various origins:

- Most obvious is the need to characterize one unit of analysis by several simultaneously applicable categories. For example, lists of

  Someone's friends

  An individual's personality traits

  Responses to the image of a Rorschach test

  Associations a product elicited from customers

  Co-authors of a publication
- Sets of legitimate interpretations of a text obtained from

  Different readers

  Crowdsourcing

  Word dictionaries

Search engines

Alternative translations

- Other examples concern legitimate variances of accounts within the same semantic domain, especially when averages or majorities would not adequately represent what matters. For example,

  The prices of gasoline charged by stations in a locale

  The distribution of grades given by teachers to their students in a class

  Divergent estimates of costs, sizes, speeds, volumes, or consumers of products

  Clusters resulting from a cluster analysis

  Equally valid alternative measures of the same phenomena

In their canonical form the reliability data for multi-valued data conform to Table 6.7, except for replacing the references to single values $c_{iu}$ by sets $C_{iu}$ of values $c$ that observer $i$ assigned to unit $u$:

| | | | | | | |
|---|---|---|---|---|---|---|
| **Standard** or **Surrogate**: | $K_1$ | . . . . | $K_u$ | . . . . . . . | $K_{N_u}$ | if applicable |
| **Units**: | 1 2 | . . . | $u$ | . . . . . . . | $N_u$ | |
| **Replications**: 1 | $C_{11}$ | . . . . | $C_{1u}$ | . . . . . . . | $C_{1N_u}$ | |
| : | : | | : | | : | |
| $i$ | $C_{i1}$ | . . . . | $C_{iu}$ | . . . . . . . | $C_{iN_u}$ | |
| $j$ | $C_{j1}$ | . . . . | $C_{ju}$ | . . . . . . . | $C_{jN_u}$ | |
| : | : | | : | | : | |
| $m$ | $C_{m1}$ | . . . . | $C_{mu}$ | . . . . . . . | $C_{mN_u}$ | |
| **Sums** $m_u$ $\begin{cases} >0 \text{ for } {}_C\alpha^s \text{ and } K_u \\ >1 \text{ for } {}_C\alpha, {}_C\alpha^{sm}, {}_C\alpha^m \end{cases}$ | $m_1$ | . . . . | $m_u$ | . . . . . . . | $m_{N_u}$ | $n_{..} = \sum_{u=1}^{N_u} m_u$ |

**TABLE 7.1**

The Canonical Form for Multi-Valued Reliability Data

Much as in Tables 2.4 and 5.1, the canonical form of multi-valued data in Table 7.1 consists of:

- A predefined set of phenomena, labeled 1, 2, ..., $u$, ..., $N_u$, the units of analysis
- Up to $m$ replications, labeled 1, 2, ..., $i$, $j$, ..., $m$, observers, coders, tests, or devices
- Sets $C_{iu}$ of values $\{c_1, c_2, c_3, \ldots, c_{|C|}\}_{iu}$ assigned by observer $i$ to unit $u$.
- $|C|$, the number of values $c \in C$, the set $C$'s cardinality, including $|C| = 1$ of single values {c} and $|C| = 0$ for empty sets {} or missing values, also denoted $\phi$
- $m_u$ is the number of pairable sets of values with $|C_{\cdot u}| > 0$ assigned to unit $u$.

  For replicability, $m_u > 1$ and for accuracy, $m_u > 0$ plus a standard $K_u$
- For replicability and accuracy, $n_{..}$ is total number of pairable sets of values. As accuracy requires the availability of a standard it may have different totals $n_{..}$
- $|C||K|$ the number of pairs of values in two coinciding sets $C$ and $K$

### 7.1.1 Its Observed Coincidences

In terms of Table 7.1, the observed coincidences of sets *C* and *K* are obtained analogue to (6.1):

$$n_{CK} = \sum_{u=1}^{N_u} \frac{1}{m_u - 1} \sum_{i=1}^{m} \sum_{j \neq i}^{m} 1 \text{ for } C_{iu} \text{ and } K_{ju} \text{ iff } m_u > 1 \tag{7.1}$$

and tabulated in Table 7.2 together with the number of values in sets C and their sum:

| Sets of values: | 1 | . | . | $K$ | . | $v$ | Number of sets | Number of values |
|---|---|---|---|---|---|---|---|---|
| 1 | $n_{11}$ | . | . | $n_{1K}$ | . | $n_{1v}$ | $n_{1\cdot}$ | . |
| . | . | . | . | . | . | . | . | . |
| $C$ | $n_{C1}$ | . | . | $n_{CK}$ | . | $n_{Cv}$ | $n_{C\cdot} = \sum_{K=1}^{v} n_{CK}$ | $n_{C\cdot}\lvert C\rvert$ |
| . | . | . | . | . | . | . | . | . |
| . | . | . | . | . | . | . | . | . |
| $v$ | $n_{v1}$ | . | . | $n_{vK}$ | . | $n_{vv}$ | $n_{v\cdot}$ | . |
| | $n_{\cdot 1}$ | . | . | $n_{\cdot K}$ | . | $n_{\cdot v}$ | $n_{\cdot\cdot} = \sum_{C=1}^{v}\sum_{K=1}^{v} n_{CK}$ ; | $\sum_{C=1}^{v} n_{C\cdot}\lvert C \rvert$ |

**TABLE 7.2**
Matrix of Observed Multi-Valued Coincidences

Analogue to Tables 5.1 and 6.9, coincidence matrices:

- Are symmetrical: $n_{CK} = n_{KC}$ and their marginal sums are $n_{\cdot C} = n_{C\cdot}$ and $n_{\cdot K} = n_{K\cdot}$
- Tabulate the coincidences of a total of $v$ pairable sets C and K.
- Contain the matching coincidences of sets $C = K$ in their diagonal cells $n_{CC}$ and $n_{KK}$.

Mismatching coincidences of sets $C \neq K$ are in their off-diagonal cells

- Sum to the total of $n_{\cdot\cdot}$ pairable sets of values in the reliability data (not to be confused with the values they contain)

However, sets must not be confused with the values they contain:

- Each set C is composed of $\lvert C \rvert$ values $c \in C$ and of $\lvert K \rvert$ values $k \in K$
- There are $v$ different sets C of values c
- Each *C*–*K* cell in Table 7.2 contains $\lvert C \rvert \lvert K \rvert$ *c*–*k* pairs of values
- Rows and columns sum to $n_{C\cdot}$ and $n_{\cdot K}$ sets containing $n_{C\cdot}\lvert C \rvert$ and $n_{\cdot K}\lvert K \rvert$ values
- The total number of pairable values is: $\sum_{C=1}^{v} n_{C\cdot}\lvert C \rvert = \sum_{K=1}^{v} n_{\cdot K}\lvert K \rvert$

Note that, except for limiting the trusted standard to single sets $K_u$ in Table 7.1, the reliability data defined so far correspond to Table 6.7 and an earlier proposal for measuring multi-valued data (Krippendorff & Craggs, 2016). This proposal has been abandoned, however, because its software approached computational limits as soon as the number of values from which multi-valued sets were composed exceeded eleven. This limitation was due to computing the expected disagreements by randomizing not only the pairs of values that observers used when forming sets but also the sizes of the sets they used. The following acknowledges the sizes of sets but expresses expectations by pairing of values across them.

## 7.2 Multi-Valued Difference Functions

Multi-valued difference functions measure the difference between any two sets $C$ and $K$ of values:

$$C = \{c_1, c_2, c_3, \ldots, c_{|C|}\}$$

$$K = \{k_1, k_2, k_3, \ldots, k_{|K|}\}$$

When values in sets $C$ and $K$ are unordered categories, the multi-valued difference function for nominal data is defined by:

$$_{\text{nominal}}\Delta_{CK} = |C||K| - |C \cap K|^2 \quad (7.2)$$

Where: $|C|$ = $C$'s cardinality, the number of values $c \in C$ as defined above
$|C||K|$ = The total number of $c$–$k$ pairs between $C$ and $K$
= The number of single-valued differences to consider variously
$|C \cap K|^2$ = The number of $c$-$k$ pairs of values that $C$ and $K$ share

(7.2) can be visualized with Figure 7.1:

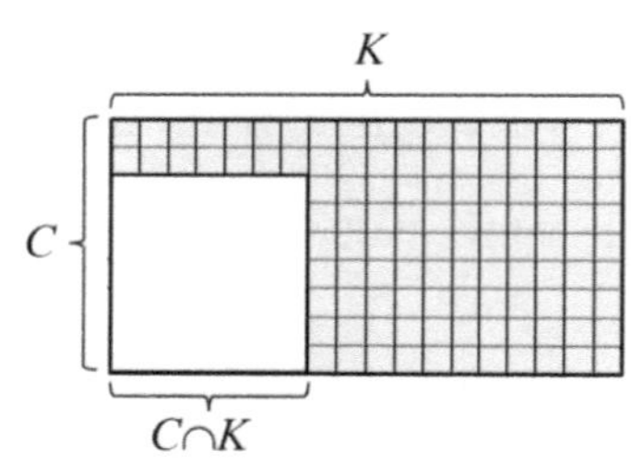


**FIGURE 7.1**
A Visualization of the Nominal Difference Between Two Multi-Valued Sets

According to Figure 7.1, (7.2), pairs all individual values in $C$ with all individual values in $K$, $|C||K|$ pairs in total. Of this total there would be exactly $|C \cap K|$ pairs of shared values. However, considering that $|C \cap K|$ are the values shared by two observers' judgments, any mismatching pairs within the $|C \cap K|^2$ pairs must be considered of zero difference by virtue of belonging to the paired sets. This operationalizes the motivation for multi-valued coding, which is grounded in the judgments by observers that the nature of the phenomena of interest may not be describable in terms of mutually exclusive values but afford several simultaneously appropriate and mutually supportive qualities, attributes, or descriptions. As Figure 7.1 illustrates, (7.2) accounts for all $|C||K|$ pairs of any two coinciding sets $C$ and $K$ of values and considers those in $|C \cap K|^2$ as matching.

When $C$ and $K$ are single-valued: $\quad {}_{\text{nominal}}\Delta_{\{c\}\{k\}} = {}_{\text{nominal}}\delta^2_{ck} = \begin{cases} 0 \text{ iff } c = k \\ 1 \text{ iff } c \neq k \end{cases}$

Rewriting (7.2) in terms of the sums of the nominal metric differences ${}_{\text{nominal}}\delta^2_{ck}$ of the $|C||K|$ individual cells yields:

$${}_{\text{nominal}}\Delta_{CK} = |C||K - |C \cap K|^2 = \sum_{c \in C} \sum_{k \in K} {}_{\text{nominal}}\delta^2_{ck} - \sum_{c \in C \cap K} \sum_{k \in K \cap C} {}_{\text{nominal}}\delta^2_{ck}$$

This offers the path to generalize (7.2) to all metric differences ${}_{\text{metric}}\delta^2_{ck}$ introduced in Section 5.3.

$${}_{\text{metric}}\Delta_{CK} = \sum_{c \in C} \sum_{k \in K} {}_{\text{metric}}\delta^2_{ck} - \sum_{c \in C \cap K} \sum_{k \in K \cap C} {}_{\text{metric}}\delta^2_{ck} \tag{7.3}$$

It should be noted that while ${}_{\text{nominal}}\Delta_{CK}$ ignores any ordering of values, treating all of them as categories that merely match or differ, the general multi-valued difference function ${}_{\text{metric}}\Delta_{CK}$ of (7.3) requires that all values in sets have the same metric. This is a severe limitation by comparison to (7.2) for nominal data. When a person is describable in terms of their profession, age, income, hobbies, and telephone numbers, a nominal difference function can handle this diversity at the expense of metric difference, for example one person 10 years of age would be considered to be of a different age than an 11 year old. The same difference would be noted between a 10 and an 80 year old. When an ordered metric is chosen, all values must share the same metric properties. Hence, (7.3) limits the use of multi-valued data to metrically compatible values, such as lengths, temperature readings, incomes, and the like, the multiplicity of which is considered more important than their single-valued majorities, medians or means. Sets of metrically equal values may result from different occasions at which something was measured, competing estimates of the same phenomena, alternative treatment results of the same illness. Given these restrictions, the following will be stated with the general form of the multi-valued difference ${}_{\text{metric}}\Delta_{CK}$ although the nominal ${}_{\text{nominal}}\Delta_{CK}$ may well be the more useful one when the values in sets are unordered or their ordering can be ignored.

Because single-valued differences are standardized, $0 \leq {}_{\text{metric}}\delta^2_{ck} \leq 1$, required in (5.5), by making single-valued differences functions of various metrics part of the definition (7.3) of ${}_{\text{metric}}\Delta_{CK}$, it follows that the range of ${}_{\text{metric}}\Delta_{CK}$ equals that of ${}_{\text{nominal}}\Delta_{CK}$. Stated generally, the range of the multi-valued difference functions is:

$$0 \leq {}_{\text{metric}}\Delta_{CK} \leq |C||K| \tag{7.4}$$

To get a sense of how the nominal and interval differences behave relative to each other Table 7.3 lists several numerical examples. Note that the interval values in these examples range between $c_{\text{smallest}} = 1$ and $c_{\text{largest}} = 10$, and hence the single-valued interval metric differences in this table are standardized according to (5.13): ${}_{\text{interval}}\delta^2_{ck} = ((c-k)(10-1))^2$.

The first six rows in Table 7.3 show single-valued sets of values whose differences were already discussed in the numerical example of Section 5.5. Whereas the nominal differences are either 0 or 1, the interval differences are nuanced but within the same range between 0 and 1. The 7th and last rows demonstrate how duplications of values are quantified. The 8th and 9th rows make clear that differences are not affected by the ordering of values in sets. In the 10th row, the value 1 of the 9th row is replaced by the largest value, here 10.

| $C$ | $K$ | $\|C\|\|K\|$ | ${}_{\text{nominal}}\Delta_{CK}$ | ${}_{\text{interval}}\Delta_{CK}$ |
|---|---|---|---|---|
| {} | {1} | NA | NA | NA |
| {1} | {1} | 1 | 0 | 0 |
| {2} | {1} | 1 | 1 | 0.0123 |
| {4} | {1} | 1 | 1 | 0.1111 |
| {8} | {1} | 1 | 1 | 0.6049 |
| {10} | {1} | 1 | 1 | 1 |
| {1,1} | {1} | 2 | 0 | 0 |
| {1,2} | {2} | 2 | 1 | 0.0123 |
| {2,1} | {2} | 2 | 1 | 0.0123 |
| {2,10} | {2} | 2 | 1 | 0.7901 |
| {2,10} | {1,2} | 4 | 3 | 1.8025 |
| {2,6,10} | {1,2} | 6 | 5 | 2.3086 |
| {2,4,6,8,10} | {1,2} | 10 | 9 | 3.5185 |
| {2,3,4,6,8,9,10} | {1,2,3} | 21 | 17 | 5.2469 |
| {2,3,4,5,6,7,8,9,10} | {1,2,3,4,5} | 35 | 19 | 9.3209 |
| {10,10,10,10,10,10} | {1,1,1,1,1,1} | 36 | 36 | 36 |

**TABLE 7.3**
Numerical Examples of Nominal and Interval Multi-Valued Differences

It amounts to a considerable increase in their interval differences but does not affect their nominal differences. The remaining rows add mismatching values, which increases both kinds of differences; for nominal data the count of mismatching pairs of values, for interval data by how far the mismatching values are apart. The last example consists of two sets of 6 duplications, unlike in each set. Their difference involves $6 \times 6 = 36$ pairs equal to 36 times the individual difference in row 6.

## 7.3 Multi-Valued Disagreements for Replicability

Note that the metric difference function for nominal data (7.2) and its extension (7.3) to all other metrics, discussed here, apply to predefined units of analysis. However, they are applicable more generally whenever data are conceived of in terms of multi-valued descriptions.

### 7.3.1 Observed Multi-Valued Disagreements

This section concerns measuring the agreement related to multiple valuations, using the multi-valued difference function (7.2) and (7.3) – not with its standardized version (7.19) to be discussed in Section 7.6.

The observed disagreement follows in the footsteps of (5.3). It is defined in terms of the coincidence matrices of sets, not of single values, applies the multi-valued difference function (7.3) instead of any of the single-valued metric differences, and is expressed as a proportion of the total number of pairs of values involved:

$$_{C}D_{o} = \frac{\sum_{C=1}^{v}\sum_{K=1}^{v} n_{CK} \, {}_{\text{metric}}\Delta_{CK}}{\sum_{C=1}^{v}\sum_{K=1}^{v} n_{CK} \, |C||K|} \tag{7.5}$$

Notationally, we omit references to the metric of the disagreements and take them to be as indicated in the metric of the multi-valued difference function and of ${}_C\alpha_{\text{metric}}$.

### 7.3.2 Expected Multi-Valued Disagreement

The expected coincidences among sets follow (5.2):

$$e_{CK} = \frac{n_C}{n..-1}\begin{cases} n_K & \text{iff } C \neq K \\ n_K - 1 & \text{iff } C = K \end{cases} \tag{7.6}$$

And the expected disagreement conforms to the logic of (7.5):

$${}_C D_e = \frac{\sum_{C=1}^{v}\sum_{K=1}^{v} e_{CK}\ {}_{\text{metric}}\Delta_{CK}}{\sum_{C=1}^{v}\sum_{K=1}^{v} e_{CK}\,|C||K|} \tag{7.7}$$

### 7.3.3 *C*-Alpha for Multi-Valued Replicability Data

$${}_C\alpha_{\text{metric}} = 1 - \frac{{}_C D_o}{{}_C D_e} \tag{7.8}$$

If all sets contain at most one value, the observed disagreement (7.5) reduces to $1/n..$ of (5.6) and the expected disagreement (7.7) becomes $1/n..$ of (5.7). It follows that the multi-valued agreement coefficient of (7.8) is a generalization to sets of values of the single-valued agreement coefficient (5.8) and (2.5).

### 7.3.4 A Numerical Example of *C*-Alpha

The computation of ${}_C\alpha_{\text{metric}}$ may be demonstrated by means of a simple example of nine units coded by three observers:

| **Units** $u$: | 1 | 2 | 3 | 4 | 5 | 6 | 7 | 8 | 9 | |
|---|---|---|---|---|---|---|---|---|---|---|
| **Observers:** F | {b,d} | {a,b,d} | {a} | {a} | {b,d} | {a,b,d} | | {a,b,d} | {a,b} | |
| G | {d} | {d} | {a} | {a} | {b,d} | {a,b,d} | {b} | {a} | {a,b} | |
| H | | {a,b} | {a} | | {b,d} | {a,b,d} | | {a} | {a,b} | |
| **Sums of pairable sets** $m_u > 1$: | 2 | 3 | 3 | 2 | 3 | 3 | 0 | 2 | 3 | $22 = n..$ |

Recall from the discussion in Chapter 2, that values, here sets of values, must be pairable within units. The lone {b}, identified by observer G for unit 7 is not pairable with any other value in unit 7 and has to be ignored. It follows that these reliability data have eight units with pairable sets, not nine, and they consist of a total of 40 categories, of 3 kinds distributed over $n.. = 22$ sets of values of three sizes in $3 \times 9 = 27$ cells.

(7.1) defines the entries in the matrix of observed coincidence in Table 7.2. They conform to definitions (2.10) and (5.1) for single-valued coincidences which weigh all possible pairs of sets so that each set contributes exactly one to that coincidence matrix. Accordingly, in the canonical form of this example, units 3 through 6 and 9 contain perfectly matching sets which occupy the diagonals in the following coincidence matrix for $n_{CK}$. There are 7 {a}s,

4 {a,b}s and {b,d}s and 5 {a,b,d}s. Units 2 is coded by three different sets. Its six possible pairs contribute 0.5 to six cells in the following coincidence matrix. While mismatching sets are found in the off-diagonal cells, the multi-valued difference function gives credit for shared values. For example, all sets in unit 8 have {a} in common and the two sets in unit 1 share {d}:

$n_{CK}$

| Sets C: | {a} | {b} | {d} | {a,b} | {b,d} | {a,b,d} | $n_{C\cdot}$ |
|---|---|---|---|---|---|---|---|
| {a} | 6 | | | | | 1 | 7 |
| {b} | | 0 | | | | | 0 |
| {d} | | | 0 | 0.5 | 1 | 0.5 | 2 |
| {a,b} | | | 0.5 | 3 | | 0.5 | 4 |
| {b,d} | | | 1 | | 3 | | 4 |
| {a,b,d} | 1 | | 0.5 | 0.5 | | 3 | 5 |
| | 7 | 0 | 2 | 4 | 4 | 5 | 22 |

$_{\text{nominal}}\Delta_{CK} = |C||K| - |C \cap K|^2$

| Sets C: | {a} | {b} | {d} | {a,b} | {b,d} | {a,b,d} |
|---|---|---|---|---|---|---|
| {a} | **0** | 1 | 1 | 1 | 2 | 2 |
| {b} | 1 | **0** | 1 | 1 | 1 | 2 |
| {d} | 1 | 1 | **0** | 2 | 1 | 2 |
| {a,b} | 1 | 1 | 2 | **0** | 3 | 2 |
| {b,d} | 2 | 1 | 1 | 3 | **0** | 2 |
| {a,b,d} | 2 | 2 | 2 | 2 | 2 | **0** |

$|C||K|$

| Sets C: | {a} | {b} | {d} | {a,b} | {b,d} | {a,b,d} |
|---|---|---|---|---|---|---|
| {a} | 1 | 1 | 1 | 2 | 2 | 3 |
| {b} | 1 | 1 | 1 | 2 | 2 | 3 |
| {d} | 1 | 1 | 1 | 2 | 2 | 3 |
| {a,b} | 2 | 2 | 2 | 4 | 4 | 6 |
| {b,d} | 2 | 2 | 2 | 4 | 4 | 6 |
| {a,b,d} | 3 | 3 | 3 | 6 | 6 | 9 |

As the sets in this example contain unordered categories a, b, and d, only the nominal multi-valued difference (7.2) applies. They are tabulated in the matrix at the center of the above three. Multi-valued differences between matching sets in the diagonal are zero. Differences between mismatching the first three single values, e.g. {a}≠{b}, are $_{\text{nominal}}\Delta_{\{a\}\{b\}} = 1 = |C||K|$, which happens to be their maxima. {a,b,d} and {d} share {d} which is credited by $_{\text{nominal}}\Delta_{\{a,b,d\}\{d\}} = 2 < |C||K| = 3$. $_{\text{nominal}}\Delta_{\{a,b\}\{b,d\}} = 3 < 4$, etc. The matrix on the right of the above shows these maxima, relative to which the observed disagreement is defined.

In terms of these three tables, the observed disagreement (7.5) becomes:

$$_{C}D_{o} = \frac{\sum_{C=1}^{v}\sum_{K=1}^{v} n_{CK}\ {}_{\text{nominal}}\Delta_{CK}}{\sum_{C=1}^{v}\sum_{K=1}^{v} n_{CK}\ |C||K|}$$

$$= \frac{2(1\cdot2+0.5\cdot2+1\cdot1+0.5\cdot2+0.5\cdot2)}{6\cdot1+3\cdot4+3\cdot4+3\cdot9+2(1\cdot3+0.5\cdot2+1\cdot2+0.5\cdot3+0.5\cdot6)} = \frac{12}{78} = 0.154$$

The expected coincidences (7.6) are shown in the following table:

$e_{CK}$

| Sets C: | {a} | {b} | {d} | {a,b} | {b,d} | {a,b,d} | $n_{C\cdot}$ |
|---|---|---|---|---|---|---|---|
| {a} | 2.00 | | 0.67 | 1.33 | 1.33 | 1.67 | 7 |
| {b} | | 0 | | | | | 0 |
| {d} | 0.67 | | 0.10 | 0.38 | 0.38 | 0.48 | 2 |
| {a,b} | 1.33 | | 0.38 | 0.57 | 0.76 | 0.95 | 4 |
| {b,d} | 1.33 | | 0.38 | 0.76 | 0.57 | 0.95 | 4 |
| {a,b,d} | 1.67 | | 0.48 | 0.95 | 0.95 | 0.95 | 5 |
| | 7 | 0 | 2 | 4 | 4 | 5 | 22 |

And enumerated as by definition (7.7):

$$
\begin{aligned}
{}_{C}D_e &= \frac{\sum_{C=1}^{v}\sum_{K=1}^{v} e_{CK\ \text{metric}}\Delta_{CK}}{\sum_{C=1}^{v}\sum_{K=1}^{v} e_{CK}\,|C||K|} \\
&= \frac{2(0.67\cdot 1+1.33\cdot 1+1.33\cdot 2+1.67\cdot 2+0.38\cdot 2+0.38\cdot 1+0.48\cdot 2+0.76\cdot 3+0.95\cdot 2+0.95\cdot 2)}{2\cdot 1+0.1\cdot 1+0.57\cdot 4+0.57\cdot 4+0.95\cdot 9+2(0.67\cdot 1+1.33\cdot 2+1.33\cdot 2+1.67\cdot 3+\ldots+0.95\cdot 6)} \\
&= \frac{31.238}{72.095} = 0.433
\end{aligned}
$$

The multi-valued C-Alpha replicability (7.8) for our example becomes:

$$
{}_{C}\alpha_{\text{nominal}} = 1 - \frac{{}_{C}D_o}{{}_{C}D_e} = 1 - \frac{0.154}{0.433} = 0.644
$$

Visual examination of the reliability data in their canonical form renders ${}_{C}\alpha_{\text{nominal}}$ = 0.644 plausible. Mismatches are observed in three out of six units. However, not only do these three units contain small amounts of agreement in the form of shared categories; they also account for only 40% of all values. The three units with perfect agreement account for 60% of all values, obviously overwhelming the smaller number of observed mismatches.

## 7.4 $c\in C$-Alpha for Individual Values in Multi-Valued Sets of Values

The coincidence matrices for multi-valued data as defined in Table 7.1 aggregate the values that observers use in terms of the sets that unite them. Their row and column sums enumerate sets of values. To not confuse values with sets of values, Table 7.2 mentions, and the numerical example lists, not only the row and column sums of sets but also the number of values that rows implicitly represent. The reasons for defining ${}_{C}\alpha$ for replicability on a matrix of coincidences of sets of values rather than the customary matrices of individual values lie in the virtue of multi-valued difference functions discounting all differences among values that two paired sets share. Figure 7.1 visualizes the $|C||K|$ pairs of values, each occurring with a frequency of $n_{CK}$, of which the differences between pairs of values in $|C\cap K|^2$ are zero.

### 7.4.1 Its Observed Disagreements

Inasmuch as any one chosen value $c$ may or may not occur in various individual sets $C$ and $K$ and intersections $C \cap K$ and with frequencies $n_{CK}$ requires identifying that value's participation in any $C$–$K$-coincidences of sets. For one chosen value $c$, distinct from all values other than $c$, the observed disagreement can be stated as:

$$
{}_{c\in C}D_o = \frac{\sum_{c\in C}\sum_{K}\sum_{k\notin C\cap K} n_{CK}}{\sum_{c\in C}\sum_{K} n_{CK}\,|K|} \qquad (7.9)
$$

The first two sums in (7.9) focus on {c}–K pairs of values. Its numerator enumerates all mismatches among the set of pairs enumerated in the denominator whether matching or not. So, just as for the observed disagreement among all sets of values in (7.5), ${}_{c\in C}D_o$ is the proportion of mismatching pairs with one value $c$ of analytical interest.

To make this conception clear, I am continuing the use of the numerical example introduced in Section 7.3.4. Attending to individual values calls for expanding the 6-by-6 matrix of observed coincidences of sets to the following 10-by-10 matrix of values that form the observed sets of values. Its entries consist of the frequencies $n_{CK}$ of the $|C||K|$ observed pairs of values, boxed for the convenience of relating them to the matrix of coincidences of sets, shown above. Of the $|C||K|$ observed pairs in each box, the differences among the $|C\cap K|^2$ pairs of values are zero, and the remaining $|C||K|-|C\cap K|^2$ mismatches contribute to the observed disagreement. To visually differentiate them, the following values-by-values coincidences show pairs that contribute to the disagreements in bold; those that count as agreements are set in regular typeface and gray:

| **Multi-valued sets:** | | | | | {a,b} | | {b,d} | | {a,b,d} | | | ${}_{c\in C}D_o$'s numerator | |
|---|---|---|---|---|---|---|---|---|---|---|---|---|---|
| | **Individual values:** | a | b | d | a | b | b | d | a | b | d | ↓ | ${}_{c\in C}D_o$'s denominator |
| | a | 6 | | | | | | | 1 | **1** | **1** | **2** | 9 |
| | b | | | | | | | | | | | **0** | 0 |
| | d | | | | **½** | **½** | **1** | 1 | **½** | **½** | ½ | **3** | 4.5 |
| {a,b} | a | | | **½** | 3 | 3 | | | ½ | ½ | **½** | **1** | 8 |
| | b | | | **½** | 3 | 3 | | | ½ | ½ | **½** | **1** | 8 |
| {b.d} | b | | | **1** | | | 3 | 3 | | | | **1** | 7 |
| | d | | | 1 | | | 3 | 3 | | | | **0** | 7 |
| {a,b,d} | a | 1 | | **½** | ½ | ½ | | | 3 | 3 | 3 | **0.5** | 11.5 |
| | b | **1** | | **½** | ½ | ½ | | | 3 | 3 | 3 | **1.5** | 11.5 |
| | d | **1** | | ½ | **½** | **½** | | | 3 | 3 | 3 | **2** | 11.5 |
| | | | | | | | | | | | | **12** mismatches | 78 totals |

Whereas the observed disagreement for all sets of values in (7.5) can be recognized in the two sums: ${}_CD_o=\dfrac{12}{78}$, (7.8) separates these two sums into what individual values $c\in C$ contribute to the total. For example, the value "a" occurs in three rows and columns, mismatching in 3.5 out of 28.5 possible pairs. For all three values in the reliability data, the observed disagreements are:

$$ {}_{a\in C}D_o=\frac{2+1+0.5}{9+8+11.5}=0.123 $$

$$ {}_{b\in C}D_o=\frac{0+1+1+1.5}{0+8+7+11.5}=0.132 $$

$$ {}_{d\in C}D_o=\frac{3+0+2}{4.5+7+11.5}=0.217 $$

Obviously, the sum of the three numerators equals 12, which is the numerator of ${}_CD_o$ in ${}_C\alpha_{\text{nominal}}$ for sets of values, including all individual values, which ${}_{c\in C}D_o$ separates for each value. Also, the sum of the three denominators equals 78, which is the denominator of ${}_CD_o$ for the same reasons.

### 7.4.2 Its Expected Disagreement

Its expected disagreement covering all sets follows (7.8):

$$_{C}D_e = \frac{\sum_{C=1}^{v}\sum_{K=1}^{v} e_{CK\ \text{metric}}\Delta_{CK}}{\sum_{C=1}^{v}\sum_{K=1}^{v} e_{CK}\,|C||K|}$$

### 7.4.3 $c \in C$-Alpha for Individual Values in Sets of Values

With references to values $c \in C$ in the position of the standard, the expected disagreements $_{C}D_e$ from (7.8) the agreement $_{c\in C}\alpha$ for individual value $c$ becomes:

$$_{c\in C}\alpha_{\text{metric}} = 1 - \frac{_{c\in C}D_o}{_{C}D_e} \tag{7.10}$$

With the expected disagreement for all values $_{C}D_e = 0.433$, calculated above, the agreements of the three values in the example are:

$$_{a\in C}\alpha_{\text{metric}} = 1 - \frac{0.123}{0.433} = 0.716$$

$$_{b\in C}\alpha_{\text{metric}} = 1 - \frac{0.132}{0.433} = 0.695$$

$$_{d\in C}\alpha_{\text{metric}} = 1 - \frac{0.217}{0.433} = 0.499$$

Evidently, $_{a\in C}\alpha$ and is more reliable than $_{C}\alpha = 0.644$, $_{d\in C}\alpha$ is less so. The value *'d,'* implicated in all three units with mismatching sets, is what pulls $_{C}\alpha$ down. Due to the extremely small example, none of these differences can be considered statistically significant.

## 7.5 *C*-Alpha-s for Multi-Valued Accuracy

To obtain the multi-valued reliabilities for accuracy, we follow the script used to derive the accuracy measure from the definition of the replicability measure for single-valued coding of predefined units of analysis in Section 6.3.

The reliability data for the multi-valued C-Alpha coefficient for accuracy makes use of the standard in the reliability data of Table 7.1, replacing its individual values $k_u$ in Table 6.7 by sets of values $K_u$.

The observed contingencies between $C_{iu}$ assigned by observers $i$ to unit $u$ and the standard $K_u$ for the same unit $u$ amount to a mere replacement of the references to single values in (6.14) by references to multi-valued sets:

$$x_{CK} = \sum_{u=1}^{N_u}\sum_{i=1}^{m} 1 \ \text{ iff } C_{iu} \text{ and } K_u \tag{7.11}$$

These contingencies are entered into a matrix of observed contingencies as in Table 6.9, Its cells are asymmetrically distributed with the sets $K$ of the standard defining its columns and $C$ of the aggregated replications by observers defining its rows.

The asymmetry of contingency matrices does not prevent the use of the definition (7.5) for obtaining the observed disagreements for multi-valued accuracies:

$$_{C}D_{o}^{s} = \frac{\sum_{C=1}^{v}\sum_{K=1}^{v} x_{CK} \; {}_{\text{metric}}\Delta_{CK}}{\sum_{C=1}^{v}\sum_{K=1}^{v} x_{CK} \, |C||K|} \qquad (7.12)$$

The definition of the expected contingencies follows (6.15):

$$\varepsilon_{CK} = \frac{x_{C\cdot}x_{\cdot K}}{x_{\cdot\cdot}} \qquad (7.13)$$

The multi-valued accuracies are obtained just as in (7.7) but with the expected contingencies$\mu_{CK}$:

$$_{C}D_{e}^{s} = \frac{\sum_{C=1}^{v}\sum_{K=1}^{v} \varepsilon_{CK} \; {}_{\text{metric}}\Delta_{CK}}{\sum_{C=1}^{v}\sum_{K=1}^{v} \varepsilon_{CK} \, |C||K|} \qquad (7.14)$$

And the multi-valued accuracy for predefined units becomes:

$$_{C}\alpha_{\text{metric}}^{s} = 1 - \frac{_{C}D_{o}^{s}}{_{C}D_{e}^{s}} \qquad (7.15)$$

The multi-valued agreement for replicability $_{C}\alpha_{\text{metric}}$ and for accuracy $_{C}\alpha_{\text{metric}}^{s}$ differs only by being obtained from matrices of observed and expected coincidence matrices and of contingency matrices respectively.

## 7.6 $\bar{C}$-Alpha for the Replicability of Standardized Sets, Ignoring Their Sizes

It is not uncommon for an analysis to focus not on the values contained in sets but on the differences between the multi-valued phenomena, regardless of the sets' sizes. The agreement measure for such situations is obtained by standardizing the difference function applied to sets.

In a previous paper (Krippendorff & Craggs, 2016), we used the Dice coefficient of similarity $d = 2|C \cap K| / |C| + |K|$ (Dice, 1945) for this purpose. However, this effort ran into computational limits that had little to do with the coefficient but with the computational implications of getting to the expected disagreements. I therefore prefer to standardize the nominal version of (7.2) by:

$$\bar{\Delta}_{CK} = \frac{{}_{\text{nominal}}\Delta_{CK}}{|C||K|} = 1 - \frac{|C \cap K|^2}{|C||K|} \tag{7.16}$$

Its range corresponds to that of single-valued differences:

$$0 \le \bar{\Delta}_{CK} \le 1 \tag{7.17}$$

Because the standardized multi-valued difference function has the same properties as the single-valued metric differences (5.3), (5.4), and (5.5), the observed (5.6) and expected (5.7) disagreements are applicable to the coincidences among multi-valued sets (7.1) as in Table 7.2:

$$_{\bar{C}}D_o = \sum_{C=1}^{v} \sum_{K=1}^{v} n_{CK} \bar{\Delta}_{CK} \tag{7.18}$$

With the expected coincidences as defined by (7.6), the expected disagreement becomes:

$$_{\bar{C}}D_e = \sum_{C=1}^{v} \sum_{K=1}^{v} e_{CK} \bar{\Delta}_{CK} \tag{7.19}$$

And the $_{\bar{C}}\alpha$-coefficient that ignores the sizes of sets of values assigned to units becomes as usual:

$$_{\bar{C}}\alpha = 1 - \frac{_{\bar{C}}D_o}{_{\bar{C}}D_e} \tag{7.20}$$

### 7.6.1 The Numerical Example Continued

To illustrate how the foci on multi-valued sets and their standardized version differ computationally, we make use of the numerical example from Section 7.3.4. The following shows the matrix of the standardized difference function $\bar{\Delta}_{CK}$ of (7.16) for the multi-valued sets of values:

$\bar{\Delta}_{CK}$

| | {a} | {b} | {d} | {a,b} | {b,d} | {a,b,d} |
|---|---|---|---|---|---|---|
| {a} | 0 | 1 | 1 | 0.5 | 1 | 0.67 |
| {b} | 1 | 0 | 1 | 0.5 | 0.5 | 0.67 |
| {d} | 1 | 1 | 0 | 1 | 0.5 | 0.67 |
| {a,b} | 0.5 | 0.5 | 1 | 0 | 0.75 | 0.33 |
| {b,d} | 1 | 0.5 | 0.5 | 0.75 | 0 | 0.33 |
| {a,b,d} | 0.67 | 0.67 | 0.67 | 0.33 | 0.33 | 0 |

Note the range of $\bar{\Delta}_{CK}$ is between 0 and 1. The product $|C||K|$ in (7.16), which varies with the sizes of sets C and K, here merely affects the granularity of $\bar{\Delta}_{CK}$.

The two disagreements (7.18) and (7.19) are obtained by multiplying the observed and expected coincidences with $\bar{\Delta}_{CK}$ (for convenience the products for only one off-diagonal triangle is enumerated below):

$$_{\bar{C}}D_o = \sum_{C=1}^{v}\sum_{K=1}^{v} n_{CK}\bar{\Delta}_{CK} = 2(1\cdot 0.67+0.5\cdot 1+1\times 0.5+0.5\cdot 0.67+0.5\cdot 0.33) = 4.333$$

$$_{\bar{C}}D_e = \sum_{C=1}^{v}\sum_{K=1}^{v} e_{CK}\bar{\Delta}_{CK}$$
$$= 2\begin{pmatrix} 0.67\cdot 1+1.33\cdot 1+1.33\cdot 0.5+1.67\cdot 0.67+0.38\cdot 1+0.38\cdot 0.5 \\ +0.48\cdot 0.67+0.76\cdot 0.75+0.95\cdot 0.33+0.95\cdot 0.33 \end{pmatrix} = 11.460$$

and $_{\bar{C}}\alpha$ for sets of multi-valued units or (7.20) becomes:

$$_{\bar{C}}\alpha = 1-\frac{_{\bar{C}}D_o}{_{\bar{C}}D_e} = 1-\frac{4.333}{11.460} = 0.622$$

The computational difference between $_{C}\alpha_{\text{nominal}}$ of (7.8) and $_{\bar{C}}\alpha$ of (7.20) can be observed by comparing the two 6-by-6 matrices of their respective difference functions. $_{\text{nominal}}\Delta_{CK}$ responds to the sizes of the paired sets; $\bar{\Delta}_{CK}$ does not. For example, (7.8) considers $_{\text{nominal}}\Delta_{\{a\}\{b,d\}} = {}_{\text{nominal}}\Delta_{\{a\}\{a,b,d\}} = 2$ and of the same difference whereas (7.20) takes $\bar{\Delta}_{\{a\}\{b,d\}} = 1$ and its maximum difference, whereas $\bar{\Delta}_{\{a\}\{a,b,d\}} = 0.67$. The two difference function move further apart when the sizes of the sets become larger and of unequal sizes. Indeed, the two coefficients respond to different conceptions of differences and hence measure different agreements. The fact that the two coefficients, $_{C}\alpha_{\text{nominal}} = 0.644$ and $_{\bar{C}}\alpha = 0.622$, are virtually identical is due to the minimal variation in the sizes and numbers of sets in this example.

# 8

## *Partitioning Continua and Coding Relevant Segments*

Previous chapters addressed the reliability of coding predefined units of analysis – predefined because their autonomous and distinct nature is not in question. The independences of units are prerequisites of counting them and creating statistical accounts. However, social phenomena are rarely of this kind. Forcing researchers to impose units on social phenomena that occur in contexts or in continua is analytically convenient but often ignores socially meaningful phenomena. For example, although the very idea of public opinion assumes that opinions arise in conversations among individuals, yet, to obtain convenient statistical accounts, public opinion researchers interview individuals in the absence of acquaintances who tend to be part of the process of making up one's mind. Also, while articulated opinions do not come as countable entities, opinion researchers force them into enumerable answers to predefined questions.

The artificiality of imposing predefined units is especially problematic when it comes to continuous phenomena. Historical processes and moving images unfold in time. Texts are written and read sequentially. Everything said is said in response to something said previously and articulated in view of likely responses. In the 1960s, when we felt the need to measure the reliability of identifying violence on U.S. television, we faced the problem that coders disagreed on where violence began and where it ended. To cope with such disagreements, we explored a variety of filmic categories to narrow coders' choices and focused our attention on the categories of violent acts for our reliability assessments. This was inadequate, no doubt, but still better than imposing semantically arbitrary units on phenomena of analytical interest.

Unable to recognize simple units in continuous phenomena and obligated to provide evidence of the reliability of their data in compliance with the requirements of available analytical techniques, quantitatively oriented researchers tend to impose mechanically defined units on such phenomena. For example, Turow, Cappella, and Jamieson (1996) analyzed call-in radio talk shows in 30-second segments. A clock is reliable by definition and relying on its units obviates the reliability of unitizations. However, a 30-second segment may start in the middle of a question, include only part of an argument, or catch the end of a response without connection to what triggered it. Similarly, typographical units such as printed lines of text, column inches of newsprint, or pages of books are easily identifiable units, which may well be natural for typesetters but have nothing to do with how authors frame what they wish to convey, and readers interpret what they read. Even grammatical definitions of units – words, noun phrases, and sentences – or rhetorical ones – propositions, premises and conclusions, arguments, themes – can severely limit the analysis of videos, political speeches, unfolding social events, or military conflicts, whether they are tape-recorded, narrated, transcribed, or ordered.

Most phenomena of interest in the humanities and the social sciences have linear extensions in time and in space. For example, narratives, whether told, published, dramatized or performed have a natural beginning and an end. Listeners, readers, or viewers may well distinguish noticeable parts in such continua – actors, events, injustices, or victories – but they conceptualize and talk about them in their own narratives. In conversations, speakers take turns, partly for physiological reasons (it is difficult to listen to others while talking) and partly in compliance with politeness conventions. Participants know when it is their turn to speak, get upset when they are interrupted or not listened to, which attests to units that are natural to those particular occasions, not easily identifiable without their contexts. External observers and analysts of conversation have difficulties making the same distinctions that participants do. Working from transcripts of tape recordings allows them to distinguish patterns that the participants may not be aware of. However, they are the analysts' units.

Similarly, literary scholars commonly underline or annotate relevant sections of text irrespective of their lengths, quote them when needed, or assign them to categories of their interests. Qualitative content analysis software supports such processes by allowing users to highlight relevant segments of longer documents, apply codes to them, search for similar sequences in the remaining body of text, list them verbatim or enumerate them by the categories assigned to them. Television dramas, ongoing political negotiations, historical processes, and technological developments are all continuous processes in which researchers may distinguish parts that are relevant in searching for answers to their research questions from irrelevant matter between them. Their lengths depend on the meanings distributed in continua and the research questions pursued.

Identifying segments that are indigenous to continuous processes is a prerequisite for generating data that can serve as surrogates for phenomena that extend in time and in space. It deserves its own reliability measures. I developed three kinds of agreement coefficients applicable to textual continua. This chapter concerns the reliability of coding of segments of unequal lengths within partitioned continua. It assures researchers that the volume of coded matter is reliable. But by summing their lengths, it accepts a metric that is applicable to all of its segments, more important or less important. The latter is what Chapter 9 will address, leaving Chapter 10 to provide the reliability of distinctions within continua.

While the substance of this chapter was published in the third edition of my *Content Analysis* (Krippendorff, 2013), its fourth edition (Krippendorff, 2019) outlined the new approach presented here. Regrettably, I discovered one of its formulations was incomplete, evident in the computational example employed in this chapter, too late to be corrected in the first printing of the book but promised to be corrected in subsequent printings. Readers should follow this chapter and not be surprised that it deviates slightly from pages 345–346 of the fourth edition. I hope the present chapter withstands the test of time.

## 8.1 Replicability of Single-Valued Partitions of Continua

### 8.1.1 Its Reliability Data

The canonical form of continuous reliability data is illustrated in Figure 8.1.

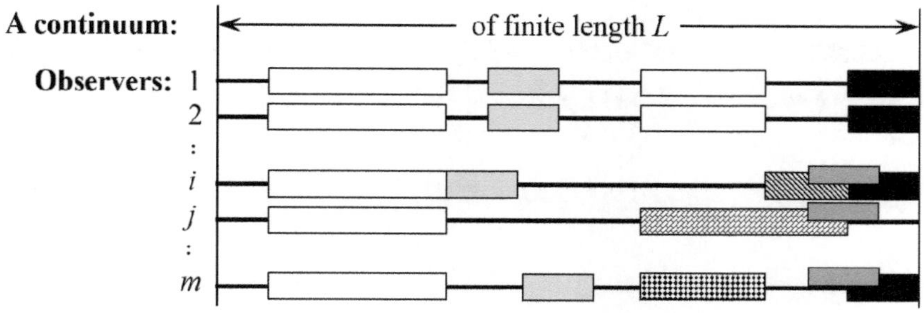


**FIGURE 8.1**
A Continuum Partitioned into Variously Valued Segments

Reliability data:

- Result from $m > 1$ *observers, coders, judges, or annotators*, labeled: 1, 2, …, $i$, $j$, …, $m$.
- Each independently *partitions* a continuum of length $L$ into any number of non-overlapping *segments* $S$ of variable lengths.
- Segments $S_{ig}$ and $S_{jh}$ are *consecutively numbered*:

  For observer $i$: 1, 2, …, $g$, …, and for observer $j$: 1, 2, …, $h$, … .
- Segments are *categorized or single-valued*. Segments of relevant matter, the identified units, are valued $c \neq \phi$, where $c$ could be a category, name, description, or numerical value possessing any common metric properties (nominal, ordinal, interval, ratio, etc.). Segments of unmarked or of irrelevant matter, the gaps between observer-identified and valued units, are identified $c = \phi$.
- *Lengths* are expressed in integers, representing the *smallest indivisible unit* in the continuum. Indivisibility depends on the research question and the nature of the continuum being unitized. When unitizing texts, the arguably smallest indivisible units are characters, including blanks, but they could be taken also to be words or even larger textual units. When unitizing movies, the smallest indivisible unit might be the individual frames as in 35mm tapes, or conventional units of time (seconds, minutes, hours, …) and lengths (millimeters, inches, miles, …). When unitizing digitized matter, the smallest indivisible units might be binary digits or bits.
- Within that continuum, segments $S_{ig}$ and $S_{jh}$ are *located* by *their endings*: $End_{ig}$ and $End_{jh}$. Their *beginnings* are the endings of the preceding segments: $End_{ig-1}$ for $S_{ig}$ and $End_{jh-1}$ for $S_{jh}$. The first segment begins at $End_{i1-1} = End_{j1-1} = 0$, the last ends at $End_{ilast} = End_{jlast} = L$.
- The *lengths of segments* are denoted by $L(S_{ig}) = End_{ig} - End_{ig-1}$ and $L(S_{jh}) = End_{jh} - End_{jh-1}$.
- The *length of the continuum*, shared by all observers, is the sum of the lengths of its segments: $L = \sum_g L(S_{ig}) = \sum_h L(S_{jh})$ . See Figure 8.2.

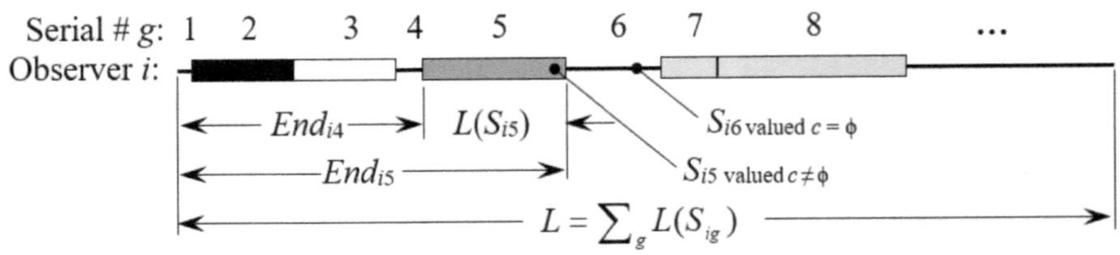


**FIGURE 8.2**
Lengths in One Observer's Partition of a Continuum

## 8.1.2 Its Observed Coincidences

Reliability is perfect when all segments pairwise agree among replications regarding their lengths, locations in the continuum, and values assigned to them. In Figure 8.1, this is exemplified by the seven segments $S_{i2}$. Disagreements arise when segments partly intersect, and/or are unequally valued.

To capture the coinciding intersections of segments, valued $c$ and $k$ respectively, $l_{ck}$ is defined as the sum of all pairs of intersections of segments, adjusted so that the sum of all coincidences equals the $mL$ of the $m$ independent replications:

$$\ell_{ck} = \sum_{i}^{m} \sum_{g} \frac{\sum_{j \neq i}^{m} \sum_{h} L\left(S_{ig \text{ valued } c} \cap S_{jh \text{ valued } k}\right)}{m-1} \tag{8.1}$$

The matrix of the observed coincidences tabulates them, much as for predefined units:

| **Values:** | $\phi$ | 1 | . | $k$ | . | . | $v$ | |
|---|---|---|---|---|---|---|---|---|
| $\phi$ | $\ell_{\phi\phi}$ | $\ell_{\phi 1}$ | . | $\ell_{\phi k}$ | . | . | $\ell_{\phi v}$ | $\ell_{\phi\cdot}$ |
| 1 | $\ell_{1\phi}$ | $\ell_{11}$ | . | $\ell_{1k}$ | . | . | $\ell_{1v}$ | $\ell_{1\cdot}$ |
| . | . | . | . | . | . | . | . | . |
| . | . | . | . | . | . | . | . | . |
| $c$ | $\ell_{c\phi}$ | $\ell_{c1}$ | . | $\ell_{ck}$ | . | . | $\ell_{cv}$ | $\ell_{c\cdot} = \sum_{k=\phi}^{v} \ell_{ck}$ |
| . | . | . | . | . | . | . | . | . |
| $v$ | $\ell_{v\phi}$ | $\ell_{v1}$ | . | $\ell_{vk}$ | . | . | $\ell_{vv}$ | $\ell_{v\cdot}$ |
| | $\ell_{\cdot\phi}$ | $\ell_{\cdot 1}$ | . | $\ell_{\cdot k}$ | . | . | $\ell_{\cdot v}$ | $mL = \sum_{c=\phi}^{v} \sum_{k=\phi}^{v} \ell_{ck}$ |

**TABLE 8.1**
Observed Coincidences Among All Intersecting Segments

The observed coincidences in Table 8.1 resemble of course the observed coincidences for predefined units in Table 2.2. They differ by representing the sum of the lengths of intersecting segments of the continuum, instead of the frequencies of predefined units. Note that the sums in (8.1) enumerates all $m(m-1)$ pairs of $m$ observers' partitions, the sum $j \neq i$ prevents pairing observed intersection with themselves. The fraction $1/(m-1)$ ensures that all segments are counted only once so that all cells in Table 8.1 sum to $mL$.

### 8.1.3 Its Observed Disagreements

The observed disagreement for segmenting the whole continuum follows the logic established by (2.2). However, the uncoded or irrelevant gaps between identified segments exhibit no structure and are not ordered, which prevents the use of any metric difference function other than nominal. Section 8.1.6 will overcome this limitation, but only for where a metric is applicable.

Notationally, the following uses the subscript $u$ for unitizing. Accordingly, the observed disagreement of partitioning continua into segments, including coded and uncoded segments $c$ and $k$, recognizes the lengths of all coinciding intersections $\ell_{ck}$ is:

$$ {}_uD_o = \sum_{c=\phi}^{v} \sum_{k=\phi}^{v} \ell_{ck\ \text{nominal}} \delta_{ck}^2 \tag{8.2}$$

### 8.1.4 Its Expected Coincidences

Just as in the αs for the replicability of coding predefined units, the expected coincidences among segments of various lengths needs to exclude the pairing of segments with themselves. Whereas predefined units bear the same weight and discounting their being paired with themselves can be expressed by their frequencies, identified segments of various lengths have their own contiguity, need to be taken whole where possible, and pairing them to obtain expected contingencies needs to take accounts of their unequal lengths. To do so, one has to distinguish two kinds of segments. The uncoded and irrelevant gaps between observer-identified and coded segments have no contiguity. Their sum can be treated just as are predefined and logically separate units of measurement. By contrast, the valued and hence relevant segments have meaningful lengths. It follows that the segments $L(S_{ig\ \text{valued}\ =\ c})$ summed to $\ell_{c\cdot}$ have to be taken as whole and contiguous. Thus, in expressing the expected coincidences, the space of all possible pairs of units, $(mL)^2$ contains all $\ell_{\varphi\cdot}^2$ pairs of uncoded matter of which $\ell_{\varphi\cdot}$ measuring units are paired with themselves and should not be counted. $(mL)^2$ also contains all intersections of coded units of which all pairs of segments paired with themselves, $(L(S_{ig\ \text{valued}\ \neq\varphi}))^2$ needs to be excluded from $(mL)^2$. This leads to the expression of the expected coincidences which treats these two kinds of segments separately:

| **Values:** | $\phi$ | 1 | . | $k$ | . | . | $v$ | |
|---|---|---|---|---|---|---|---|---|
| $\phi$ | $e_{\phi\phi}$ | $e_{\phi 1}$ | . | $e_{\phi k}$ | . | . | $e_{\phi v}$ | $e_{\phi\cdot}$ |
| 1 | $e_{1\phi}$ | $e_{11}$ | . | $e_{1k}$ | . | . | $e_{1v}$ | $e_{1\cdot}$ |
| . | . | . | . | . | . | . | . | . |
| . | . | . | . | . | . | . | . | . |
| $c$ | $e_{c\phi}$ | $e_{c1}$ | . | $e_{ck}$ | . | . | $e_{cv}$ | $e_{c\cdot} = \sum_{k=\phi}^{v} e_{ck}$ |
| . | . | . | . | . | . | . | . | . |
| $v$ | $e_{v\phi}$ | $e_{v1}$ | . | $e_{vk}$ | . | . | $e_{vv}$ | $e_{v\cdot}$ |
| | $e_{\cdot\phi}$ | $e_{\cdot 1}$ | . | $e_{\cdot k}$ | . | . | $e_{\cdot v}$ | $mL = \sum_{c=\phi}^{v} \sum_{k=\phi}^{v} e_{ck}$ |

**TABLE 8.2**
Expected Coincidences Among All Possible Intersecting Segments

$$e_{ck} = \frac{\ell_{c\cdot}\ell_{\cdot k} - \begin{cases} \ell_{\phi\cdot} & \text{iff } c = k = \phi \\ \sum_i^m \sum_g L(S_{ig \text{ valued} = c=k\neq\phi})^2 & \text{iff } c = k \neq \phi \end{cases}}{mL - \frac{\ell_{\phi\cdot}}{mL} - \frac{1}{mL}\sum_i^m \sum_g L(S_{ig \text{ valued} \neq\phi})^2} \tag{8.3}$$

They are tabulated in Table 8.2.

Whenever segments are of unequal lengths, not pairing them with themselves has the effect of the marginal sums of the matrix of expected coincidences deviating slightly from those in the matrix of observed coincidences in Table 8.1.

$$e_{c\cdot} = \sum_{k=\varphi}^{v} e_{ck} \approx \ell_{c\cdot}$$

$$e_{\cdot k} = \sum_{c=\varphi}^{v} e_{ck} \approx \ell_{\cdot k}$$

$$\text{but } e_{\cdot\cdot} = \sum_{c=\phi}^{v}\sum_{k=\phi}^{v} e_{ck} = \sum_{c=\phi}^{v}\sum_{k=\phi}^{v} \ell_{ck} = mL$$

However, with increasing lengths $L$ of the partitioned continuum, increasing number of identified segments, and decreasing diversity of their lengths, which can be anticipated in reliability data of practical research projects, this discrepancy disappears.

### 8.1.5 Its Expected Disagreements

Its expected disagreement, expressed with and without reference to separately established expected coincidences $\varepsilon_{ck}$ becomes:

$${}_uD_e = \sum_{c=\phi}^{v}\sum_{k=\phi}^{v} e_{ck\ \text{nominal}}\delta_{ck}^2 = \frac{\sum_{c=\phi}^{v}\ell_{c\cdot}\sum_{k=\phi}^{v}\ell_{\cdot k\ \text{nominal}}\delta_{ck}^2}{mL - \frac{\ell_{\phi\cdot}}{mL} - \frac{1}{mL}\sum_i^m \sum_g L(S_{ig \text{ valued} \neq\phi})^2} \tag{8.4}$$

### 8.1.6 *u*-Alpha for Partitioned Continua

The form of the agreement coefficient ${}_u\alpha$, stated in general terms, and in their simpler form for nominal data becomes:

$${}_u\alpha_{\text{nominal}} = 1 - \frac{{}_uD_o}{{}_uD_e} = 1 - \frac{\sum_{c=\phi}^{v}\sum_{k=\phi}^{v}\ell_{ck\ \text{nominal}}\delta_{ck}^2}{\sum_{c=\phi}^{v}\sum_{k=\phi}^{v} e_{ck\ \text{nominal}}\delta_{ck}^2} = 1 - \frac{mL - \sum_{c=\phi}^{v}\ell_{cc}}{mL - \sum_{c=\phi}^{v} e_{cc}} \tag{8.5}$$

Ideally, all metrics (0.4), (0.5), (5.10), and (5.12) through (5.16) are applicable to ${}_u\alpha$. However, as articulated in Section 5.4, the applicability of any metric is limited to the weakest metric of its values. It follows that whenever a partitioned continuum includes gaps between valued segments, only the nominal metric difference (0.4) and (5.10) is applicable to all segments, relevant and irrelevant, and hence (8.5) defines ${}_u\alpha_{\text{nominal}}$.

### 8.1.7 A Numerical Example

As a numerical example, consider these reliability data consisting of three partitions of a continuum:

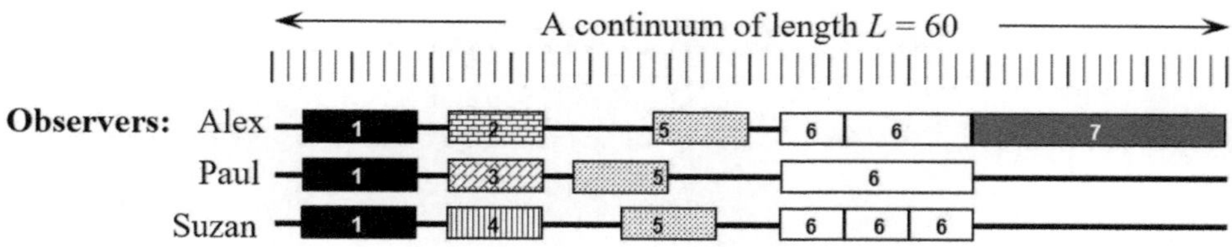


The two coincidence matrices obtained after applying (8.1) and (8.3) to these partitions are:

**Observed coincidences**

| | φ | 1 | 2 | 3 | 4 | 5 | 6 | 7 | |
|---|---|---|---|---|---|---|---|---|---|
| φ | 45 | . | . | . | . | 10 | . | 16 | 71 |
| 1 | . | 21 | . | . | . | . | . | . | 21 |
| 2 | . | . | 0 | 3 | 3 | . | . | . | 6 |
| 3 | . | . | 3 | 0 | 3 | . | . | . | 6 |
| 4 | . | . | 3 | 3 | 0 | . | . | . | 6 |
| 5 | 10 | . | . | . | . | 8 | . | . | 18 |
| 6 | . | . | . | . | . | . | 36 | . | 36 |
| 7 | 16 | . | . | . | . | . | . | . | 16 |
| | 71 | 21 | 6 | 6 | 6 | 18 | 36 | 16 | 180 |

**Expected coincidences**

| φ | 1 | 2 | 3 | 4 | 5 | 6 | 7 | |
|---|---|---|---|---|---|---|---|---|
| 28.20 | 8.42 | 2.41 | 2.41 | 2.41 | 7.22 | 14.43 | 6.41 | 71.91 |
| 8.42 | 1.66 | 0.71 | 0.71 | 0.71 | 2.13 | 4.27 | 1.90 | 20.51 |
| 2.41 | 0.71 | 0 | 0.20 | 0.20 | 0.61 | 1.22 | 0.54 | 5.89 |
| 2.41 | 0.71 | 0.20 | 0 | 0.20 | 0.61 | 1.22 | 0.54 | 5.89 |
| 2.41 | 0.71 | 0.20 | 0.20 | 0 | 0.61 | 1.22 | 0.54 | 5.89 |
| 7.22 | 2.13 | 0.61 | 0.61 | 0.61 | 1.68 | 3.66 | 1.63 | 18.15 |
| 14.43 | 4.27 | 1.22 | 1.22 | 1.22 | 3.66 | 6.32 | 3.25 | 35.59 |
| 6.41 | 1.90 | 0.54 | 0.54 | 0.54 | 1.63 | 3.25 | 1.35 | 16.16 |
| 71.91 | 20.51 | 5.89 | 5.89 | 5.89 | 18.15 | 35.59 | 16.16 | 180 |

This example is constructed to highlight a few properties easily identifiable in the graphical representation of the above reliability data. Note that the identified units in the first group match in location, length and kind and their sum ends up in the $\ell_{11}$-cell of the matrix of observed coincidences. The units in the second group match in location, length but not in kind. They occupy the mismatching cells of the three categories 2, 3, and 4. The units in the third group match in length, kind but not in their location. Their matching intersections turn up in the $\ell_{55}$-cell and their intersections with blanks in the $\ell_{5\varphi}$ and $\ell_{\varphi 5}$ -cells. The units in the fourth group occupy the same stretch on the continuum, are of the same kind but of different lengths. It is important to recognize that (8.1) accounts for all intersections, relevant and irrelevant, summing their lengths by kind but not their location on the continuum. Therefore, all units in the fourth group turn up in the $\ell_{66}$-cell of the matrix of observed coincidences. The lone unit valued 7 pairs only with blanks left by Paul and Susan which is represented in the two cells $\ell_{\varphi 7}$ and $\ell_{7\varphi}$. Finally, the total sum $mL = 3{\times}60 = 180$, is the sum of the lengths of all intersections, each counting only once.

Regarding the expected coincidences, we recognize that the somewhat complicated form of (8.3) is meant to prevent identified units from being paired with themselves. Doing so would underestimate the expected disagreements. For example, the sum of the three units valued 1 in the first group is $\ell_{1\cdot} = 3 \times 7 = 21$. Its square, $21^2$, is the sum of 9 squares, $7^2$, one for each possible pair of segments of length 7. It includes $3 \times 7^2$ due to units paired with themselves, which should not be counted as agreement. Therefore, the denominator of $\varepsilon_{11}$ subtracts $3 \times 7^2$ from $21^2$ in order to count only the results of pairing segments other than with themselves. Dividing the difference by the common denominator, which happens to be $180 - (962/180) = 174.66$, yields the expected coincidence $\varepsilon_{11} = 1.6833$. The off-diagonal cells of the expected coincidences do not need such corrections.

According to the third expression in the definition of ${}_{u}\alpha_{\text{nominal}}$ in (8.5), which expresses ${}_{u}\alpha$ as a function of the observed and expected coincidences already available above, the reliability of all segments in the continuum becomes:

$$ {}_{u}\alpha_{\text{nominal}} = 1 - \frac{mL - \sum_{c=\phi}^{v} \ell_{cc}}{mL - \sum_{c=\phi}^{v} e_{cc}} = 1 - \frac{180-(45+21+8+36)}{180-(28.46+1.68+1.24+5.80)} = 0.510 $$

## 8.2 The Family of *u*-Alpha Coefficients

The general limitation of ${}_{u}\alpha$ to treating the segments of partitions as categorical or nominal, even if identified segments are coded with values of higher order metrics, can be lifted by splitting its disagreements into two parts; computing a binary agreement measure to evaluate the distinction between identified segments and the gaps between them, or between relevant and irrelevant segments, and an agreement measure able to apply more powerful metrics to the valued intersections, found in a *v*-by-*v* matrix nominally part of Table 8.1. To start, we break the reliability data into two mutually exclusive kinds and define the following reliability measures:

${}_{u}\alpha_{\text{nominal}}$

↙ ↘

The reliability of relevance = ${}_{\phi|u}\alpha_{\text{binary}}$ taken up in the following section.

${}_{cu}\alpha_{\text{metric}}$ = The reliability of valued segments. taken up in Section 8.2.3.

### 8.2.1 $\phi \mid u$-Alpha for Distinguishing Relevant and Irrelevant Matter

This binary agreement measure is important for two reasons. First, it answers the question of whether coders can reliably distinguish between relevant and irrelevant matter, which is the reliability of the data that inform a research question. It responds to errors of omission and commission.

Second, it frees the analyst to apply more powerful metric differences on the valued intersections, as appropriate. The latter is the topic of Section 8.3.2.

The distinction between relevant and irrelevant matter is a binary one. Notationally, the reliability of such a distinction will be signified by "$\phi \mid u$" in place of the subscript *u* of ${}_{u}\alpha$.

The ${}_{\phi|u}\alpha$-coefficient is obtained by applying the definitions of ${}_{u}\alpha$ in (8.2), (8.4) and (8.5) to coincidence matrices that ignore the distinctions among all identified units by summing the four quadrants in Table 8.1 along the dotted lines. It effectively sums all relevant matter into one category $+ \neq \phi$ and reduces the matrix of observed coincidences in Table 8.1 to the following 2-by-2 matrix:

| **Values**: | $\phi$ | $+$ | = Identified segments |
|---|---|---|---|
| $\phi$ | $\ell_{\phi\phi}$ | $\ell_{\phi+}$ | $\ell_{\phi .}$ |
| $+$ | $\ell_{+\phi}$ | $\ell_{++}$ | $\ell_{+.}$ |
| | $\ell_{.\phi}$ | $\ell_{.+}$ | $\ell_{..} = mL$ |

**TABLE 8.3**
Observed Coincidences of the Distinction between Identified Segments and Blanks

$$\text{Where: } \ell_{\phi+} = \ell_{+\phi} = \sum_{c=1}^{v} \ell_{c\phi} = \sum_{k=1}^{v} \ell_{\phi k} \tag{8.6}$$
$$\ell_{++} = \sum_{c=1}^{v} \sum_{k=1}^{v} \ell_{ck}$$

Its expected coincidences can be obtained either by summing the entries in the four quadrants in Table 8.2 or by applying (8.3) to the marginal sums in Table 8.1, considering the lengths $L(S_{ig})$ of segments in the canonical form of the reliability data:

| **Values**: | $\phi$ | $+$ | = Identified segments |
|---|---|---|---|
| $\phi$ | $e_{\phi\phi}$ | $e_{\phi+}$ | $e_{\phi .}$ |
| $+$ | $e_{+\phi}$ | $e_{++}$ | $e_{+.}$ |
| | $e_{.\phi}$ | $e_{.+}$ | $\ell_{..} = mL$ |

**TABLE 8.4**
Expected Coincidences of the Distinction between Identified Segments and Blanks

Where:
$$e_{\phi+} = e_{+\phi} = \sum_{c=1}^{v} e_{c\phi} = \sum_{k=1}^{v} e_{\phi k} \tag{8.7}$$
$$e_{++} = \sum_{c=1}^{v} \sum_{k=1}^{v} e_{ck}$$

As (8.6) and (8.7) suggest, its observed and expected coincidences can be obtained by summing the entries in the two $v + 1$-by-$v + 1$ coincidence matrices in Tables 8.1 and 8.2 to the two 2-by-2 coincidence matrices in Tables 8.3 and 8.4. Taking again advantage of the algebraic properties of dichotomous data, discussed in Chapter 2, the simplest form of the binary ${}_{\phi|u}\alpha$-coefficient is:

$$_{\phi|u}\alpha_{\text{binary}} = 1 - \frac{_{\phi|u}D_o}{_{\phi|u}D_e} = 1 - \frac{\sum_{c\neq\phi}^{v} \ell_{\phi c}}{\sum_{c\neq\phi}^{v} \varepsilon_{\phi c}} = 1 - \frac{\ell_{+\phi} + \ell_{\phi+}}{e_{+\phi} + e_{\phi+}} = 1 - \frac{\ell_{+\phi}}{e_{+\phi}} \tag{8.8}$$

### 8.2.2 The Numerical Example Continued

For the numerical example, the 2-by-2 observed and expected coincidence matrices, obtained by applying (8.8) to the already available coincidence matrices, are:

Observed coincidences

| Values: | $\phi$ | + | |
|---|---|---|---|
| $\phi$ | 45 | 26 | 71 |
| + | 26 | 83 | 109 |
| | 71 | 109 | 180 |

Expected coincidences

| Values: | $\phi$ | + | |
|---|---|---|---|
| $\phi$ | 28.5 | 44.3 | 72.8 |
| + | 44.3 | 62.9 | 107.2 |
| | 72.8 | 107.2 | 180 |

Regarding this example, (8.8) yields:

$$_{\phi|u}\alpha_{\text{binary}} = 1 - \frac{_{\phi|u}D_o}{_{\phi|u}D_e} = 1 - \frac{\ell_{\phi+}}{e_{\phi+}} = 1 - \frac{26}{44.31} = 0.413$$

### 8.2.3 *cu*-Alpha for Intersections of Relevant Segments With Diverse Metrics

The advantage of separating the reliability of distinguishing between relevant and irrelevant matter from the reliability of coding relevant matter has already been discussed above. It calls for reformulating (8.2), (8.4), and (8.5) to be applicable to relevant matter which excludes all intersections with irrelevant matter for the benefit of applying all available metrics to these data.

Notationally, a "*c*" is added to $_u\alpha$'s subscript, indicating that $_{cu}\alpha$ addresses the coding of freely unitized segments, exclusive of all intersections with irrelevant matter, labeled $\phi$

The observed coincidences for applying $_{cu}\alpha$ can be obtained by enumerating them from the reliability data in Figure 8.1, excluding all intersections involving irrelevant matter labeled $\phi$, and considering the possibility that each valued segment $S_{ig}$ could be paired with fewer than $m$–1 other valued segments $S_{jh}$. In the case of the latter, reference to the $m$–1 pairs for each segment in the denominator of (8.1) has to be qualified and replaced by the actual number of valued segments with which $S_{ig}$ is being paired. This number varies between 0 and $m$–1:

$$0 \le \left(\sum_{j\neq i}^{m} 1 \text{ iff} \sum_h L\left(S_{ig \text{ valued } c\neq\phi} \cap S_{jh \text{ valued } \neq\phi}\right) > 0\right) \le m-1$$

which replaces the $m$–1 in (8.1) whenever only coded segments are considered. The lengths of *c*-*k*-coded coincidences is $\ell^*_{ck}$:

$$\ell^*_{ck} = \sum_{i=1}^{m}\sum_g \frac{\sum_{j\neq i}^{m}\sum_h L\left(S_{ig \text{ valued } c\neq\phi} \cap S_{jh \text{ valued } k\neq\phi}\right)}{\sum_{j\neq i}^{m} 1 \text{ iff } \left(\sum_h L(S_{ig \text{ valued } c\neq\phi} \cap S_{jh \text{ valued } \neq\phi})\right) > 0} \tag{8.9}$$

These coincidences form a *v*-by-*v* coincidence matrix, nominally similar to the submatrix in Table 8.1 but with its own marginal sums and coincidences:

$$\ell^*_{c\cdot} = \sum_{k=1}^{v} \ell^*_{ck} \neq \ell_{c\cdot}$$

$$\ell^*_{\cdot k} = \sum_{c=1}^{v} \ell^*_{ck} \neq \ell_{\cdot k}$$

$$\ell^*_{\cdot\cdot} = \sum_{c=1}^{v}\sum_{k=1}^{v} \ell^*_{ck}$$

Its observed disagreement omits uncoded matter but follows (8.2) otherwise:

$$_{cu}D_o = \sum_{c=1}^{v}\sum_{k=1}^{v} \ell^*_{ck}\ {}_{\text{metric}}\delta^2_{ck} \tag{8.10}$$

Its expected coincidences are obtained from the marginal sums of the $v$-by-$v$ matrix of observed coincidences $\ell^*_{ck}$. Its definition follows (8.3) except for omitting all references to irrelevant matter $\phi$ and acknowledging that coded segments $S_{ig}$ may not intersect with $m$–1 other segments:

$$e^*_{ck} = \frac{\ell^*_{c\cdot}\ell^*_{\cdot k} - \sum_{i=1}^{m}\sum_{g}\dfrac{\sum_{j\neq i}^{m}\left(\sum_{h} L(S_{ig\text{ valued } c=k\neq\phi} \cap S_{jh\text{ valued }\neq\phi})\right)^2}{\sum_{j\neq i}^{m} 1 \text{ iff } \left(\sum_{h} L(S_{ig\text{ valued }\neq\phi} \cap S_{jh\text{ valued }\neq\phi})\right) > 0} \quad \text{iff } c=k}{\ell^*_{\cdot\cdot} - \dfrac{1}{\ell^*_{\cdot\cdot}}\sum_{i=1}^{m}\sum_{g}\dfrac{\sum_{j\neq i}^{m}\left(\sum_{h} L(S_{ig\text{ valued }\neq\phi} \cap S_{jh\text{ valued }\neq\phi})\right)^2}{\sum_{j\neq i}^{m} 1 \text{ iff } \left(\sum_{h} L(S_{ig\text{ valued }\neq\phi} \cap S_{jh\text{ valued }\neq\phi})\right) > 0}} \tag{8.11}$$

Uneven marginal sums $\ell^*_{c\cdot}$ account for uneven corrections of the expected coincidences, whose marginal sums $e^*_{c\cdot}$ may deviate slightly from those of the observed, especially when data are small:

$$e^*_{c\cdot} = \sum_{k=\phi}^{v} e^*_{ck} \approx \ell^*_{c\cdot}$$

$$e^*_{\cdot k} = \sum_{c=\phi}^{v} e^*_{ck} \approx \ell^*_{\cdot k}$$

$$\text{but}\, e^*_{\cdot\cdot} = \sum_{c=\phi}^{v}\sum_{k=\phi}^{v} e^*_{ck} = \ell_{++}$$

In these terms, the expected disagreement becomes:

$$_{cu}D_e = \sum_{c=1}^{v}\sum_{k=1}^{v} e^*_{ck}\ {}_{\text{metric}}\delta^2_{ck} \tag{8.12}$$

The $_{cu}\alpha$-coefficient, obtained from the $v$-by-$v$ observed and expected coincidences of valued intersections becomes:

$$_{cu}\alpha_{\text{metric}} = 1 - \frac{_{cu}D_o}{_{cu}D_e} = 1 - \frac{\sum_{c=1}^{v}\sum_{k=1}^{v} \ell^*_{ck}\ {}_{\text{metric}}\delta^2_{ck}}{\sum_{c=1}^{v}\sum_{k=1}^{v} e^*_{ck}\ {}_{\text{metric}}\delta^2_{ck}} \tag{8.13}$$

As already mentioned, while $_u\alpha$ and $_{\phi|u}\alpha$ cover the total length $L$ of all $m$ continua, the reliability of the distinction between relevant and irrelevant matter is obtained by $_{\phi|u}\alpha$. Unreliabilities of this distinction are manifest in the intersections of relevant and irrelevant matter. Under these circumstances, $_{cu}\alpha$ covers a smaller part of valued continua, only pairable coded segments. To account for the omission of irrelevant segments and non-pairable relevant segments, a useful indicator of its coverage is:

$$100\frac{\ell^*_{..}}{\ell_{..}}\% \text{ of the continuum covered by } {}_{cu}\alpha$$

### 8.2.4 The Numerical Example Continued

Turning to the numerical example in Figure 8.1 and the $v$-by-$v$ matrices of the observed and expected coincidences obtained by (8.9) and (8.11), these two matrices are:

$v$-by-$v$ Matrix of observed coincidences

| | ϕ | 1 | 2 | 3 | 4 | 5 | 6 | 7 | |
|---|---|---|---|---|---|---|---|---|---|
| ϕ | . | . | . | . | . | . | . | . | . |
| 1 | . | 21 | . | . | . | . | . | . | 21 |
| 2 | . | . | 0 | 3 | 3 | . | . | . | 6 |
| 3 | . | . | 3 | 0 | 3 | . | . | . | 6 |
| 4 | . | . | 3 | 3 | 0 | . | . | . | 6 |
| 5 | . | . | . | . | . | 8 | . | . | 8 |
| 6 | . | . | . | . | . | . | 36 | . | 36 |
| 7 | . | . | . | . | . | . | . | . | 0 |
| | . | 21 | 6 | 6 | 6 | 8 | 36 | 0 | 83 |

Matrix of expected coincidences

| | ϕ | 1 | 2 | 3 | 4 | 5 | 6 | 7 | |
|---|---|---|---|---|---|---|---|---|---|
| ϕ | . | . | . | . | . | . | . | . | . |
| 1 | . | 3.86 | 1.66 | 1.66 | 1.66 | 2.20 | 9.93 | . | 20.97 |
| 2 | . | 1.66 | 0 | 0.47 | 0.47 | 0.63 | 2.84 | . | 6.07 |
| 3 | . | 1.66 | 0.47 | 0 | 0.47 | 0.63 | 2.83 | . | 6.07 |
| 4 | . | 1.66 | 0.47 | 0.47 | 0 | 0.63 | 2.83 | . | 6.07 |
| 5 | . | 2.20 | 0.63 | 0.63 | 0.63 | 0.25 | 3.78 | . | 8.13 |
| 6 | . | 9.93 | 2.84 | 2.83 | 2.83 | 3.78 | 13.45 | . | 35.69 |
| 7 | . | . | . | . | . | . | . | . | . |
| | . | 20.89 | 6.07 | 6.07 | 6.07 | 8.40 | 35.69 | . | 83 |

In this example, the observed coincidences $\ell^*_{ck}$ happen to equal the coincidences $\ell_{ck}$ in the $v$-by-$v$ submatrix of the matrix of all observed coincidences in Table 8.1. As already mentioned, this may not be so when valued segments are intersecting with fewer that ($m$–1) other valued segments. The extreme example of this condition is the lone unit valued 7, which is absent from the above.

Just as ${}_{c}\alpha_{\text{metric}}$ becomes ${}_{c}\alpha_{\text{nominal}}$ when the nominal difference function ${}_{\text{nominal}}\delta^2_{ck}$ is inserted to account for categorical data, so does (8.13) becomes simplified:

$${}_{cu}\alpha_{\text{nominal}} = 1 - \frac{{}_{cu}D_o}{{}_{cu}D_e} = 1 - \frac{\ell^*_{..} - \sum_{c=1}^{v} \ell^*_{cc}}{\ell^*_{..} - \sum_{c=1}^{v} e^*_{cc}} = 1 - \frac{83-(21+8+36)}{83-(3.86+0.25+13.45)} = 0.725$$

Here $100\frac{\ell^*_{..}}{\ell_{..}} = 100\frac{83}{180} = 46\%$ of the continuum is covered by ${}_{cu}\alpha$.

Numerically, ${}_{cu}\alpha_{\text{nominal}} = 0.725$ is significantly larger than both, ${}_{u}\alpha_{\text{nominal}} = 0.510$ and ${}_{\phi|u}\alpha_{\text{binary}} = 0.413$. One can conclude from these differences that the unreliabilities in this example are located prior to the valuing of identified segments, in the distinction between relevant and irrelevant matter ${}_{\phi|u}\alpha_{\text{binary}}$, part of ${}_{u}\alpha_{\text{nominal}}$.

With ${}_{cu}\alpha$, representing all intersections among valued segments, covering 46% of all intersection, the fact that 54% are due to disagreements with and among irrelevant segments is another indication of where unreliabilities originate. ${}_u\alpha$ averages all unreliabilities. ${}_{\phi|u}\alpha$ covers all unreliabilities of the distinction between relevant and irrelevant segments, which qualifies reports of the frequencies of relevant units of analysis. ${}_{cu}\alpha_{\text{metric}}$ focuses on coding relevant segments.

The most important advantage of ${}_{cu}\alpha$ over ${}_u\alpha$ is its ability to make use of metrics more powerful than the nominal metric. Assuming the metric of values assigned to units are of the interval kind, ranging between $c_{\text{smallest}} = 1$ and $c_{\text{largest}} = 6$, inserting the standardized interval metric differences ${}_{\text{interval}}\delta^2_{ck} = \left( {}^{(c-k)}/_{(6-1)} \right)^2$ of (5.13) into (8.13) yields the ${}_{cu}\alpha_{\text{interval}}$-coefficient for the above example. For simplicity and taking advantage of the symmetry of coincidences, the following enumerates $\ell^*_{ck}\ {}_{\text{interval}}\delta^2_{ck}$ and $e^*_{ck}\ {}_{\text{interval}}\delta^2_{ck}$ only for cells in one off-diagonal triangle of the above two coincidence matrices, yields the ${}_{cu}\alpha_{\text{interval}}$-coefficient for the above example:

$$
\begin{aligned}
{}_{cu}\alpha_{\text{interval}} &= 1 - \frac{\sum_{c=1}^{v}\sum_{k=1}^{v} \ell^*_{ck}\ {}_{\text{interval}}\delta^2_{ck}}{\sum_{c=1}^{v}\sum_{k=1}^{v} e^*_{ck}\ {}_{\text{interval}}\delta^2_{ck}} = 1 - \frac{\sum_{(c-k)=1}^{5} \ell^*_{ck}\ {}_{\text{interval}}\delta^2_{ck}}{\sum_{(c-k)=1}^{5} e^*_{ck}\ {}_{\text{interval}}\delta^2_{ck}} \\
&= 1 - \frac{(3+3)0.2^2 + 3\cdot 0.4^2}{(1.66+0.47+0.47+0.63+3.78)0.2^2 + (1.66+0.47+0.63+2.83)0.4^2 + \cdots + 9.93\cdot 1^2} \\
&= 1 - \frac{0.72}{16.17} = 0.956
\end{aligned}
$$

From the distribution of its observed coincidences it is quite obvious that disagreements stem exclusively from the confusion of three neighboring values, 2, 3, and 4. The extreme difference between values 1 and 6, ${}_{\text{interval}}\delta^2_{16} = 1$, is not observed, $\ell^*_{16} = 0$, but is expected to be $e^*_{16} = 9.93$. As discussed in Section 5.4, such a distribution of coincidences suggests values to be ordered. Although our example is too small to support such conclusions, this observation is confirmed by the fact that ${}_{cu}\alpha_{\text{interval}} = 0.958$ is significantly higher than when values are regarded as unordered categories, ${}_{cu}\alpha_{\text{nominal}} = 0.725$. This difference is due largely to the distribution of mismatching valuations. When treated as nominal data, their location in the off-diagonal cells does not matter. But when considered interval data, it turns out that they conform with how the interval metric differences evaluate them. Here, all deviations are overwhelmingly close to the diagonal of perfect matches and far removed from the extreme of the disagreement in the 1–6 cell.

## 8.3 Relations between the *u*-Alpha Family and *c*-Alpha

The ${}_u\alpha$-family of agreement coefficients accepts variations in the lengths of segments, identified units and gaps between them in a continuum. The ${}_c\alpha$-coefficient, by contrast, is insensitive to such variations among units, taking all units to be of uniform size, whether naturally so or by predefinition. Yet in the case of single-valued coding their logic is similar. This may be seen by contrasting their general forms, (2.13) above with (8.13) below:

$$_{c}\alpha_{\text{metric}} = 1 - (n_{..}-1)\frac{\sum_{c=1}^{v}\sum_{k=1}^{v} n_{ck\,\text{metric}}\,\delta_{ck}^{2}}{\sum_{c=1}^{v} n_{c.}\cdot\sum_{k=1}^{v} n_{\cdot k\,\text{metric}}\,\delta_{ck}^{2}}$$

$$_{cu}\alpha_{\text{metric}} = 1 - \left(\ell^{*}_{..} - \frac{1}{\ell^{*}_{..}}\sum_{i}^{m}\sum_{g}\frac{\sum_{j\neq i}^{m}\left(\sum_{h} L(S_{ig\ \text{valued}\neq\phi}\cap S_{jh\ \text{valued}\neq\phi})\right)^{2}}{\sum_{j\neq i}^{m} 1 \text{ iff } \left(\sum_{h} L(S_{ig\ \text{valued}\neq\phi}\cap S_{jh\ \text{valued}\neq\phi})\right) > 0}\right)\frac{\sum_{c=1}^{v}\sum_{k=1}^{v}\ell^{*}_{ck\ \text{metric}}\,\delta_{ck}^{2}}{\sum_{c=1}^{v}\ell^{*}_{c\cdot}\cdot\sum_{k=1}^{v}\ell^{*}_{\cdot k\ \text{metric}}\,\delta_{ck}^{2}}$$

Note that the expressions on the right sides of (2.13) and (8.13) are isomorphic. The former refers to the frequencies $n_{ck}$ of coincidences among predefined and valued units and the latter to the sum of the lengths of single-valued intersections of unequal lengths. They differ quite radically regarding the contents of their parentheses. When the lengths of all single-valued segments are equal to 1, the expected disagreement sums all $n_{..}(n_{..}-1)$ pairs of $n_{..}$ segments, and to grant each equal weight without pairing segments with themselves, (3.13) divides that sum of evpectations by $(n_{..}-1)$. The certainly more complex expression in the parentheses of (8.13) does the same except for ensuring that when pairing segments of unequal lengths, intersections of segments are not paired with themselves. The latter would be inconsistent with how the observed disagreements among segments of unequal lengths are enumerated and deflate the expected disagreement, which is all the more significant as (8.13 ) concerns segments of lengths usually much larger than predefined units.

Another relationship worthy of note is that while $_{u}\alpha$ applies to all segments of partitions of whole continua, whether segments are relevant and coded as such or the irrelevant and considered uncoded gaps between relevant segments, $_{cu}\alpha$ omits the latter. This omission is a direct analogue to $_{c}\alpha$ which ignores missing data, i.e., missing codes in units not coded by all observers. For example, the segment valued 7 in our numerical example is not pairable with any other identified segment. It is excluded from the $v$-by-$v$ coincidence matrix of observed and expected coincidences leading to $_{cu}\alpha$. Had the continuum been regarded as a set of 60 predefined units, the 16 units valued 7 would have been ignored as well. Thus, $_{cu}\alpha$ ignoring intersections with irrelevant segments $\phi$ equals $_{c}\alpha$'s applicability to so-called missing codes.

## 8.4 The Replicability of Multi-Valued Partitions of Continua

The virtues of multi-valued coding have been articulated in the preceding Chapter 7. Here, its conception is applied to segments of continua.

### 8.4.1 Its Reliability Data

Multi-valuedness is evident whenever segments of a continuum have multiple and equally valid interpretations, signaling that these qualities are not mutually exclusive. For example, a lengthy story may feature actors, motivations, actions, effects, interpersonal relations, benefits and disasters in combination and in different parts. Allowing some of them to be simultaneously developed and narrated implies multi-valued accounts distributed over stretches of texts, times and spaces. Multi-valued accounts of identified segments in a partitioned continuum do not significantly alter the form of reliability data in Figure 8.1. It merely accepts sets of values as descriptors of segments:

$$C = \{c_1, c_2, c_3, \ldots, c_{|C|}\}$$

$$K = \{k_1, k_2, k_3, \ldots, k_{|K|}\}$$

Multi-valuedness occurs whenever observers decide on a set of values appropriate to characterize relevant segments, as in accounts of complex narratives or the qualities that define an individual of interest. Each observer thereby partitions a continuum into mutually exclusive multi-valued segments. Multi-valuedness can also come about when observers identify overlapping and separately valued segments from which a partition of the continuum is constructed, one for each observer, as in Figure 8.3.

Except for the multi-valuedness of its segments, the partitions of $m > 1$ observers are treated just as depicted in Figure 8.1 and described in terms of Figure 8.2.

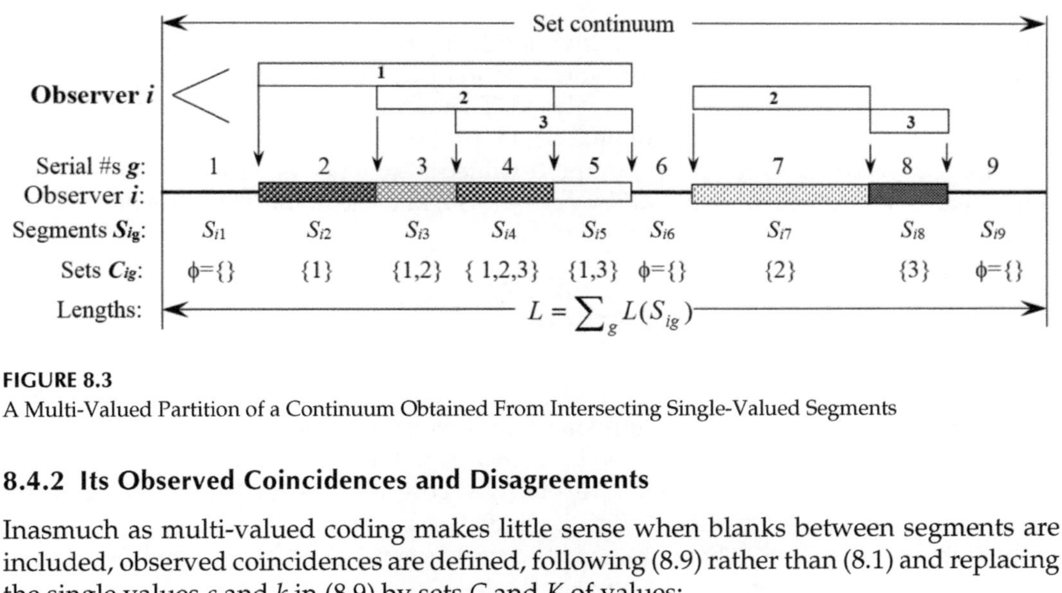


**FIGURE 8.3**
A Multi-Valued Partition of a Continuum Obtained From Intersecting Single-Valued Segments

### 8.4.2 Its Observed Coincidences and Disagreements

Inasmuch as multi-valued coding makes little sense when blanks between segments are included, observed coincidences are defined, following (8.9) rather than (8.1) and replacing the single values $c$ and $k$ in (8.9) by sets $C$ and $K$ of values:

$$\ell^*_{CK} = \sum_{i=1}^{m} \sum_{g} \frac{\sum_{j \neq i}^{m} \sum_{h} L\left(S_{ig \text{ valued } C \neq \phi} \cap S_{jh \text{ valued } K \neq \phi}\right)}{\sum_{j \neq i}^{m} 1 \text{ iff } \left(\sum_{h} L(S_{ig \text{ valued } C \neq \phi} \cap S_{jh \text{ valued } K \neq \phi})\right) > 0} \tag{8.14}$$

Table 8.5 tabulates the observed $C$–$K$-coincidences of multi-valued intersections of segments:

Note that coincidences $\ell^*_{CK}$ refer to sets of values and their marginal sums show their joint lengths. Each implicitly refers to the values they respectively contain, explicitly enumerated to the right of the matrix in Table 8.5.

| Sets of values: | 1 | 2 | . | . | $K$ | . | Lengths by sets | Lengths by values |
|---|---|---|---|---|---|---|---|---|
| 1 | $\ell^*_{11}$ | $\ell^*_{12}$ | . | . | $\ell^*_{1K}$ | . | $\ell^*_{1\cdot}$ | . |
| 2 | $\ell^*_{21}$ | $\ell^*_{22}$ | . | . | $\ell^*_{2K}$ | . | $\ell^*_{2\cdot}$ | . |
| . | . | . | . | . | . | . | . | . |
| $C$ | $\ell^*_{C1}$ | $\ell^*_{C2}$ | . | . | $\ell^*_{CK}$ | . | $\ell^*_{C\cdot} = \sum_K \ell^*_{CK}$ | $\ell^*_{C\cdot}\lvert C\rvert$ |
| . | . | . | . | . | . | . | . | . |
| . | . | . | . | . | . | . | . | . |
| | $\ell^*_{\cdot 1}$ | $\ell^*_{\cdot 2}$ | . | . | $\ell^*_{\cdot K}$ | . | $\ell^*_{\cdot\cdot} = \sum_C \sum_K \ell^*_{CK}$ | $\sum_C \ell^*_C \lvert C\rvert$ |

**TABLE 8.5**
Observed Coincidences Among Multi-Valued Segments

The multi-valued difference function (7.2):

$$_{\text{nominal}}\Delta_{CK} = \ |C||K| - |C \cap K|^2$$

does not enumerate the values shared by intersecting sets. As suggested in the examples given in Chapter 7, multi-valued sets rarely contain ordered values of the same metric, especially when textual continua are analyzed. Therefore, the most sensible multi-valued difference function is the one for nominal data reproduced in (7.2) and above.

In the observed disagreement for multi-valued segments of unequal lengths, the coincidences $n_{CK}$ in (7.5) are replaced by $\ell^*_{CK}$ of (8.14) and the nominal multi-valued difference function (7.2) is applied to them. It yields:

$$_{Cu}D_o = \frac{\sum_C \sum_K \ell^*_{CK} \ {}_{\text{nominal}}\Delta_{CK}}{\sum_C \sum_K \ell^*_{CK} \, |C||K|} \tag{8.15}$$

As discussed in Section 7.3.1, a multi-valued difference function evaluates $|C||K|$ pairs of values, giving credit for those shared by both sets. In the numerator of (8.15), the multi-valued difference function is applied to all observed coincidences. Its denominator consists of the number of pairs of values involved, which is the maximum possible disagreement in the numerator in (8.15). Therefore, the observed disagreement is a proportion between 0 and 1, and 0 when all agree and 1 when none does.

### 8.4.3 Its Expected Coincidences and Disagreements

The definition of the expected coincidences follows (8.11), omitting all references to irrelevant segments $\phi$ but replacing references to single values $c$ and $k$ by sets $C$ and $K$ of values, just as for (8.14):

$$e^*_{CK} = \frac{\ell^*_{C\cdot}\ell^*_{\cdot K} - \sum_{i=1}^{m}\sum_g \dfrac{\sum_{j\neq i}^{m}\left(\sum_h L(S_{ig\text{ valued } C=K\neq\phi} \cap S_{jh\text{ valued }\neq\phi})\right)^2}{\sum_{j\neq i}^{m} 1 \text{ iff } \left(\sum_h L(S_{ig\text{ valued }\neq\phi} \cap S_{jh\text{ valued }\neq\phi})\right) > 0} \text{ iff } C = K}{\ell^*_{\cdot\cdot} - \dfrac{1}{\ell^*_{\cdot\cdot}}\sum_{i=1}^{m}\sum_g \dfrac{\sum_{j\neq i}^{m}\left(\sum_h L(S_{ig\text{ valued }\neq\phi} \cap S_{jh\text{ valued }\neq\phi})\right)^2}{\sum_{j\neq i}^{m} 1 \text{ iff } \left(\sum_h L(S_{ig\text{ valued }\neq\phi} \cap S_{jh\text{ valued }\neq\phi})\right) > 0}} \tag{8.16}$$

Note that (8.16) conforms to (8.11) by replacing the references to individual values $c$ and $k$ by $C$ and $K$ of values jointly assigned to segments. This prevents the pairing of sets with themselves which would bias the expected disagreement. The use of the multi-valued difference function does much of the work:

$$_{Cu}D_e = \frac{\sum_C \sum_K e^*_{CK\ \text{nominal}} \Delta_{CK}}{\sum_C \sum_{K \neq C} e^*_{CK} |C||K|} \tag{8.17}$$

### 8.4.4 *Cu*-Alpha for the Replicability of Multi-Valued Coding of Segments of Unequal Lengths

$$_{Cu}\alpha_{\text{nominal}} = 1 - \frac{_{Cu}D_o}{_{Cu}D_e} \tag{8.18}$$

## 8.5 The Accuracy of Multi-Valued Partitions of Continua

### 8.5.1 Its Reliability Data, Numerically Exemplified

Reliability data for accuracy consist of $m$ partitions of a continuum, for example, created by $m$ independent observers, coders or judges, plus one partition of that continuum considered the trusted standard against which the $m$ partitions are evaluated for their accuracy.

We add to the terms defined in Section 8.1.1:

- A standard $s$ consisting of a trusted partition of the continuum against which the $m$ replications are evaluated for their accuracy.
- All contingencies that enumerate the intersections of multi-valued coding of segments of various lengths conforming to Section 8.4.1. Single-valued coding is that special case of multi-valued coding in which sets contain only one value.

Accuracy, defined in Section 6.3.2 for binary coefficients, in Section 6.6.2 for coefficients applicable to $v > 2$ values, plus different metrics, defined in Chapter 5, expands the range of disagreement measures to assess the accuracy of data. Observed disagreements among replications are numerically larger than their disagreements with a trusted standard but do not affect the total number of contingencies.

When allowing segments to be multi-valued, multiple sources contributing different partitions to the standard can be handled by a single partition constructed according to Figure 8.3.

Keeping the example simple, we add one partition as the standard to the numerical example, introduced in Section 8.1.7:

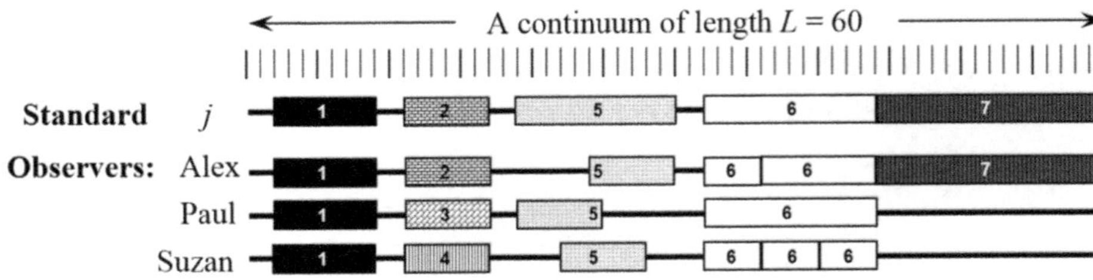


## 8.5.2 Its Observed Contingencies and Disagreements

Accuracy differs from the more common replicability in three respects:

- Segments $S_{ig}$ are not paired with each other but with the segments of the standard $S_{sh}$
- While the intersections of all segments in $m$ partitions with that standard are summed, the resulting contingencies form a matrix that preserves references to a standard's sets $K$ of values the columns of a contingency matrix and all observers' sets of $C$ values in the rows of this matrix. It follows that contingency matrices are asymmetrical and may not be square $v$-by-$v$ matrices.
- Because the segments of the trusted standard are not part of the segments identified by observers, there is no chance that they be paired with themselves, significantly simplifying the agreement coefficients for accuracy, starting with the contingencies for accuracy:

$$\ell^s_{CK} = \sum_i^m \sum_{g,h} L\left(S_{ig \text{ valued } C \neq \phi} \cap S_{sh \text{ valued } K \neq \phi}\right) \tag{8.19}$$

In which the standard $s$ assumes the place of coder $j$ in the definition (8.14) and its segments are parts of all intersections. These contingencies are entered in the asymmetrical contingency matrix in Table 8.6.

| **Standard's sets of values:** | | 1 | 2 | . | . | $K$ | . | **Lengths by sets** | **Lengths by values** |
|---|---|---|---|---|---|---|---|---|---|
| | 1 | $\ell^s_{11}$ | $\ell^s_{12}$ | . | . | $\ell^s_{1K}$ | . | $\ell^s_{1\cdot}$ | . |
| | 2 | $\ell^s_{21}$ | $\ell^s_{22}$ | . | . | $\ell^s_{2K}$ | . | $\ell^s_{2\cdot}$ | . |
| | . | . | . | . | . | . | . | . | . |
| **Coders' sets of values:** | $C$ | $\ell^s_{C1}$ | $\ell^s_{C2}$ | . | . | $\ell^s_{CK}$ | . | $\ell^s_{C\cdot} = \sum_K \ell^s_{CK} \neq \ell^s_{\cdot C}$ | $\ell^s_{C\cdot}\lvert C\rvert$ |
| | . | . | . | . | . | . | . | . | . |
| | . | . | . | . | . | . | . | . | . |
| | | $\ell^s_{\cdot 1}$ | $\ell^s_{\cdot 2}$ | . | . | $\ell^s_{\cdot K}$ | . | $\ell^s_{\cdot\cdot} = \sum_C \sum_K \ell^s_{CK}$ | $\sum_C \ell^s_{C\cdot}\lvert C\rvert$ |

**TABLE 8.6**
Observed Contingencies of Multi-Valued Segments Intersecting With a Standard

In these terms, the observed inaccuracies correspond to (8.15) in form:

$${}_{Cu}D^s_o = \frac{\sum_C \sum_K \ell^s_{CK} \; {}_{\text{nominal}}\Delta_{CK}}{\sum_C \sum_K \ell^s_{CK} \lvert C \rvert \lvert K \rvert} \tag{8.20}$$

By the convention introduced in Chapter 5, disagreements become inaccuracies and are indicated by the superscript *s*.

### 8.5.3 Its Expected Contingencies and Disagreements

Recall that all inaccuracies involve pairing segments provided by observers with a trusted standard. Inasmuch as they involve two distinct segments, the issue of self-pairing does not arise. This results in a very simple expression for the expected contingencies:

$$\varepsilon_{CK}^{s} = \frac{\ell_{C\cdot}^{s}\ell^{s}_{\cdot K}}{\ell^{s}_{\cdot\cdot}} \tag{8.21}$$

The expected disagreements or inaccuracies, compatible with (8.20), is also the proportion:

$$_{Cu}D_e^s = \frac{\sum_C \sum_K \varepsilon_{CK}^s \; {}_{\text{nominal}}\Delta_{CK}}{\sum_C \sum_K \varepsilon_{CK}^s \,|C||K|} = \frac{\sum_C \ell_{C\cdot}^s \sum_K \ell^s_{\cdot K} \; {}_{\text{nominal}}\Delta_{CK}}{\sum_C \ell_{C\cdot}^s \sum_K \ell^s_{\cdot K} \,|C||K|} \tag{8.22}$$

### 8.5.4 *Cu*-Alpha-s for the Accuracy of Multi-Valued Coding of Segments of Unequal Lengths

$$_{Cu}\alpha^s_{\text{nominal}} = 1 - \frac{_{Cu}D_o^s}{_{Cu}D_e^s} \tag{8.23}$$

### 8.5.5 The Numerical Example Concluded

According to the numerical example introduced in Section 8.5.1, which includes a standard, the two contingency matrices are:

| Standard's sets of values: | | Observed contingencies 1 | 2 | 3 | 4 | 5 | 6 | 7 | Expected contingencies 1 | 2 | 3 | 4 | 5 | 6 | 7 | |
|---|---|---|---|---|---|---|---|---|---|---|---|---|---|---|---|---|
| Coders' sets of values: | 1 | 21 | . | . | . | . | . | . | 4.05 | 3.47 | . | . | 3.47 | 6.94 | 3.08 | 21 |
| | 2 | . | 6 | . | . | . | . | . | 1.16 | 0.99 | . | . | 0.99 | 1.98 | 0.88 | 6 |
| | 3 | . | 6 | . | . | . | . | . | 1.16 | 0.99 | . | . | 0.99 | 1.98 | 0.88 | 6 |
| | 4 | . | 6 | . | . | . | . | . | 1.16 | 0.99 | . | . | 0.99 | 1.98 | 0.88 | 6 |
| | 5 | . | . | . | . | 18 | . | . | 3.47 | 2.97 | . | . | 2.97 | 5.95 | 2.64 | 18 |
| | 6 | . | . | . | . | . | 36 | . | 6.94 | 5.95 | . | . | 5.95 | 11.89 | 5.28 | 36 |
| | 7 | . | . | . | . | . | . | 16 | 3.08 | 2.64 | . | . | 2.64 | 5.29 | 2.35 | 16 |
| | | 21 | 18 | 0 | 0 | 18 | 36 | 16 | 21 | 18 | 0 | 0 | 18 | 36 | 16 | 109 |

Because the above example consists of single-valued segments, and only for these, the following simplifications hold:

$$\sum_C \sum_K \ell^s_{CK} |C||K| = \sum_c \sum_k \ell^s_{ck} = \ell^s.. = \sum_c \sum_c \varepsilon^s_{ck} = \sum_C \sum_K \varepsilon^s_{CK} |C||K|$$

$$\sum_C \sum_K \ell^s_{CK\ \text{nominal}} \Delta_{CK} = \ell^s.. - \sum_c \ell^s_{cc}$$

$$\sum_C \sum_K \varepsilon^s_{CK\ \text{nominal}} \Delta_{CK} = \ell^s.. - \sum_c \varepsilon^s_{cc}$$

$$_{\text{nominal}}\Delta_{\{c\}\{k\}} = {}_{\text{nominal}}\delta^2_{ck}$$

$$_{Cu}\alpha^s_{\text{nominal}} = 1 - \frac{_{Cu}D^s_o}{_{Cu}D^s_e} = 1 - \frac{\ell^s.. - \sum_c \ell^s_{CC}}{\ell^s.. - \sum_c \varepsilon^s_{CC}} \qquad (8.24)$$

Regarding this example, one can easily see that there are only two observed inaccuracies of segments valued 3 and valued 4 and the standard valued 2. Applying the simplified form (8.24) to these data yields the following:

$$_{Cu}\alpha^s_{\text{nominal}} = 1 - \frac{\ell^s.. - \sum_c \ell^s_{cc}}{\ell^s.. - \sum_c \varepsilon^s_{cc}}$$

$$= 1 - \frac{109 - (21 + 6 + 18 + 36 + 16)}{109 - (4.05 + 0.99 + 2.97 + 11.88 + 2.35)} = 1 - \frac{12}{86.76} = 0.862$$

Inasmuch as the standard embraces all segments valued 5 and counts segment valued 7 as a match, it can hardly be surprising that the accuracy $_{Cu}\alpha^s_{\text{nominal}} = 0.862$ of the coded data is larger than the replicability $_{Cu}\alpha_{\text{nominal}} = 0.725$ of the coding process.

# 9

# Preserving the Coherency of Identified Segments in Continua

Whereas the family of ${}_u\alpha$-coefficients, developed in Chapter 8, accounts for the lengths of intersecting segments of partitioned continua, there is a need for an agreement measure that preserves the contiguity of identified segments, not summing their intersections. After all, textual and visual segments tend to be meaningful wholes. Quotations that a scholar wants to use in support of an argument are coherent despite being composed of parts. Propositions have a natural beginning and an end.

The problem this chapter seeks to overcome can be recognized in the numerical example introduced in Section 8.1.7 of the previous chapter. It featured six segments, all valued 6. They collectively occupy the same part of the continuum but are of three different lengths. That there are no gaps between them is secondary to the fact that they end up summed in the 6–6 cell of the matrix of coincidences, irrespective of their varying lengths. Sums always hide the varying length of the summed segments. This is perfectly justifiable whenever researchers seek assurances that the volumes and kinds of relevant matter they are analyzing are reliable. With such research objectives in mind, it makes sense to focus on the sums of coincidences regardless of the distinctions among segments and consider the following pairs of segments indistinguishable:

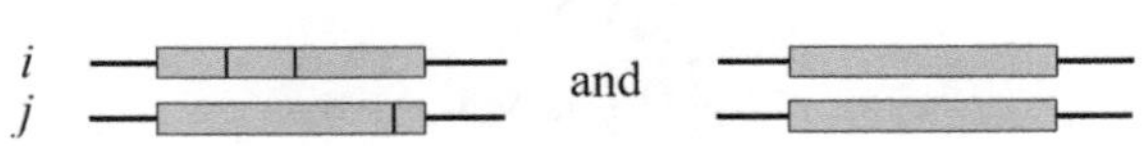


As well as:

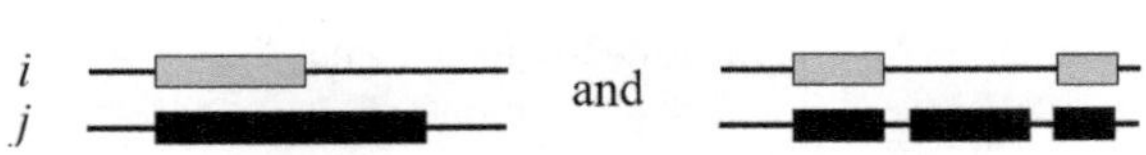


The ${}_u\alpha$-coefficient does not recognize their differences. Moreover, when the reliability of coding identified segments is the issue, ${}_{cu}\alpha$ cannot respond to intersections that are not valued and has to exclude intersections of identified and valued segments with irrelevant matter from a measure designed to apply various metric differences, for example:

These observations are not meant to discourage uses of agreement measures concerned with partitions of continua and summing their disagreements in terms of their lengths. They are to point out the virtues of recognizing the coherences of segments as significant data, and the need for an alternative agreement measure that is sensitive to these coherences. This chapter overcomes the limitation of ${}_u\alpha$ and ${}_{cu}\alpha$ by defining an agreement measure that respects, to the extent possible, the coherence of identified segments.

Notationally, to distinguish the difference function, disagreements and the α-coefficient from the family of ${}_u\alpha$-agreements for unitizing presented in Chapter 8 the capital "U" will be used in in their subscripts.

This chapter starts with developing an appropriate difference function. For the convenience of its presentation, it does not make use of the multi-valued coding of segments and is not concerned with reliabilities other than replicability.

## 9.1 Its Difference Function

The definition of its difference function relies on the terminology introduced in Section 8.1.1 but applies it differently. Whereas Section 8.1.2 defined differences among segments in terms of the lengths of their intersections (8.1), and summed and tabulated their intersections in the form of coincidence matrices as in Table 8.1, to overcome the above-mentioned omissions, its differences need to respond to any pair of intersecting segments, omitting only intersections among irrelevant segments valued ϕ. The above also suggested that its ${}_U\alpha$-agreement has to involve a difference function for all pairs of segments, single-valued, relevant but not coded, and intersecting with irrelevant matter in continua.

The three cases of differences that are depicted in Figure 9.1.

According to Figure 9.1, the difference ${}_U\delta_{igjh}$ between any one or two single-valued segments

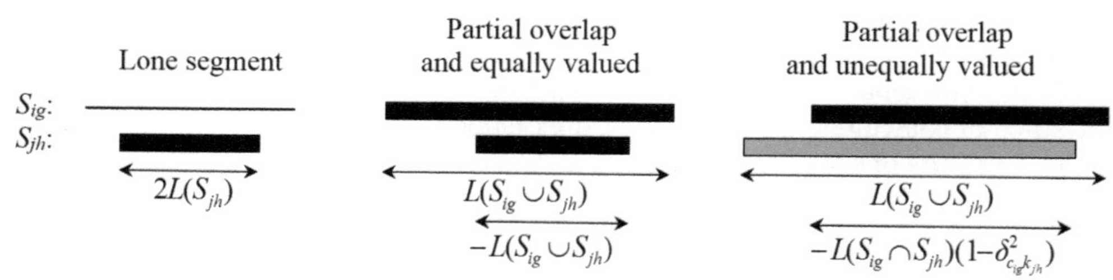


**FIGURE 9.1**
Three Configurations of Differences Among Identified Segments and ${}_U\delta_{igjh}$

is composed of two parts: the length of their union minus the length of their intersection, and if segments are valued differently, that intersection is weighted by their metric difference. If a segment intersects only with irrelevant matter, valued ϕ, the nominal metric is applied.

Formally, applied to any pair of intersecting segments, at least one of which is identified and considered relevant, whether valued or not, the difference function visualized in Figure 9.1 is defined by:

$$_{metric\,U}\delta_{igjh} = \begin{cases} L(S_{ig} \cup S_{jh}) - L(S_{ig} \cap S_{jh})(1 - {}_{\text{metric}}\delta^2_{c_{ig}c_{jh}}) & \text{iff } S_{ig} \cap S_{jh} \neq \{\} \text{ and } c_{jh} \neq \phi \neq c_{ig} \\ 2L(S_{ig}) & \text{iff } S_{ig} \subseteq S_{jh} \text{ and } c_{ig} \neq \phi = c_{jh} \\ 2L(S_{jh}) & \text{iff } S_{jh} \subseteq S_{ig} \text{ and } c_{ig} = \phi \neq c_{jh} \end{cases} \quad (9.1)$$

Note that (9.1) defines symmetrical differences: $_U\delta_{igjh} = {}_U\delta_{jhig}$

When two segments are identical in kind and location: $_U\delta_{igjh} = 0$

When two intersecting segments are of the same kind: $_U\delta_{igjh} = L(S_{ig} \cup S_{jh}) - L(S_{ig} \cap S_{jh})$

In the absence of any valuation: $_{\text{no metric}\ U}\delta^2_{c_{ig}k_{jh}} = 0$ and $_U\delta_{igjh} = L(S_{ig} \cup S_{jh}) - L(S_{ig} \cap S_{jh})$

For two unlike valued segments on the same location: $_U\delta_{igjh} = L(S_{ig} \cap S_{jh})\ _{metric}\delta^2_{c_{ig}c_{jh}}$

When two segments are of lengths one: $_U\delta_{igjh} = {}_{\text{metric}}\delta^2_{c_{ig}k_{jh}}$

The latter shows the difference function $_U\delta_{igjh}$ as a generalization of the difference functions $_{\text{metric}}\delta^2_{c_{ig}k_{jh}}$ defined in Chapter 5.

## 9.2 Its Observed Disagreement

The observed disagreement $_UD_o$ is the average difference $_Ud_{igjh}$ encountered in $m(m\text{-}1)/2$ pairs of continua unitized by $m$ observers:

$$_UD_o = \frac{\sum_i \sum_{j>i}^m \sum_{g,h}\ {}_{\text{metric}\ U}\delta_{c_{ig}c_{jh}}}{N_o} \tag{9.2}$$

$$\text{Where } N_o = \sum_{i=1} \sum_{j>i}^m \sum_{g,h} 1 \text{ iff} \begin{cases} S_{ig} \cap S_{jh} \neq \{\} \text{ and } & c_{ig} \neq \phi \neq c_{jh} \\ S_{ig} \subseteq S_{jh} \text{ and } & c_{ig} \neq \phi \text{ and } c_{jh} = \phi \\ S_{jh} \subseteq S_{ig} \text{ and } & c_{ig} = \phi \text{ and } c_{jh} \neq \phi \end{cases} \tag{9.3}$$

is the number of intersections whose differences are summed in the numerator of the observed disagreement $_UD_o$.

$c_a \neq k_b$ | $_UD_o \quad (\delta^2 = {}_{\text{metric}}\delta^2_{c_a k_b})$

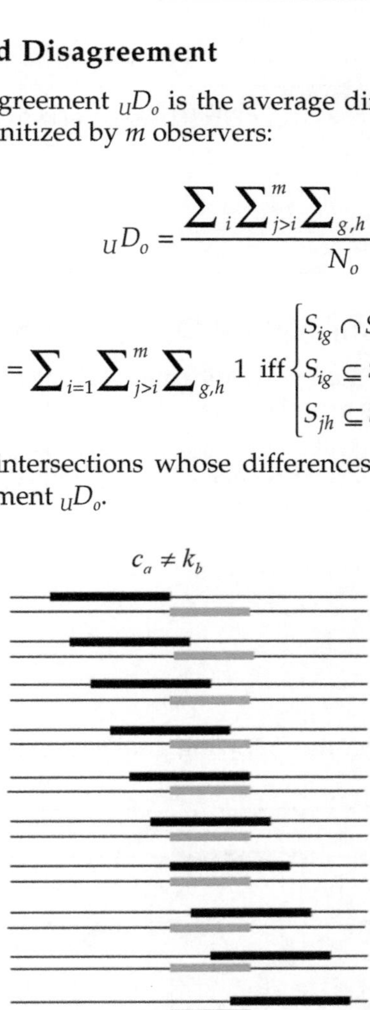

$a + b$

$a + b - 2 + 1 \cdot \delta^2$

$a + b - 4 + 2 \cdot \delta^2$

$a + b - 6 + 3 \cdot \delta^2$

$a + b - 2a + a \cdot \delta^2$

$a + b - 2a + a \cdot \delta^2$

$a + b - 2a + a \cdot \delta^2$

$a + b - 6 + 3 \cdot \delta^2$

$a + b - 4 + 2 \cdot \delta^2$

$a + b - 2 + 1 \cdot \delta^2$

Sum of differences: $a^2 + b^2 + ab \cdot \delta^2$

Number of differences: $a + b$

**FIGURE 9.2**
Permutations of the Intersections of Two Segments: Sums and Number of Differences.

## 9.3 Its Expected Disagreement

The expected disagreement pairs all segments identified as relevant by the $m$ observers, without reference to their original location on their continuum. To get a sense of the latter, consider the possible differences ${}_u\delta_{igjh}$ between two segments of lengths $a = 4$ and $b = 6$ in Figure 9.2.

Generalizing the proportion $\frac{a^2 + b^2 + ab\delta^2}{a+b}$ from Figure 9.2, to the pairing of all segments $S_{ig \text{ valued} \neq \phi}$ and $S_{jh \text{ valued} \neq \phi}$ with each other, except with themselves, gets us to the expected disagreement ${}_uD_e$ as the proportion of expected differences in proportion to the number of possible differences, analogue to the observed disagreement ${}_uD_o$ stated here without algebraic simplifications to make the derivation clear.

$$ {}_uD_e = \frac{\sum_i^m \sum_{j>i}^m \sum_{g,h} \left[ \left(L(S_{ig}\text{ valued} \neq \phi)\right)^2 + \left(L(S_{jh}\text{ valued} \neq \phi)\right)^2 + L(S_{ig}\text{ valued} \neq \phi)L(S_{jh}\text{ valued} \neq \phi)\,{}_{\text{metric}}\delta^2_{c_{ig}k_{jh}} \right]}{\sum_i^m \sum_{j>i}^m \sum_{g,h} \left[ L(S_{ig}\text{ valued} \neq \phi) + L(S_{jh}\text{ valued} \neq \phi) \right]} \tag{9.4}$$

## 9.4 *U*-Alpha for Freely Unitized and Valued Contiguous Segments

The ${}_u\alpha$-agreement for assessing the reliability of unitizing relevant matter, the segments identified by the observers, is defined as are all others, starting with (2.5):

$$ {}_u\alpha_{\text{metric}} = 1 - \frac{{}_uD_o}{{}_uD_e} \tag{9.5}$$

## 9.5 A Numerical Example

For a simple numerical example, consider the following unitizations by two observers:

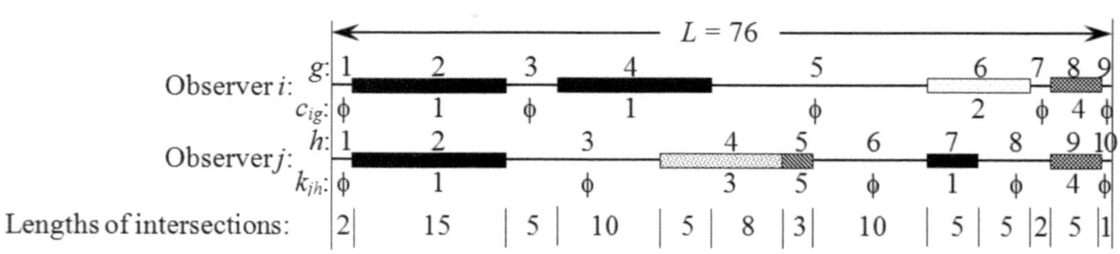

These reliability data depict the unitization of a continuum of length $L = 76$ by two observers who collectively identified nine segments, valued 1, 2, 3, 4, or 5 as indicated, and left ten gaps between segments labeled f. There are $N_o = 5$ intersections of segments that contribute to the observed disagreement: two perfectly matching pairs, two partially overlapping ones, and one lone segment. $_u a$ ignores all five intersections of irrelevant matter but embraces the one intersection of a relevant segment with a gap between two relevant sections.

When segments are considered categorized, unordered, and the nominal metric applies, according to (9.2), the observed disagreement for this example is:

$$_uD_o = \frac{(15-15(1-0))+(23-5(1-1))+2\cdot 3+(10-5(1-1))+(5-5(1-0))}{5} = \frac{0+23+6+10+0}{5} = 7.8$$

The calculation of the expected disagreement requires more computational steps than can be presented here. In fact, (9.3) calls for summing 1,548 differences from 9(9 – 1) = 72 possible pairs of segments, yielding:

$$_uD_e = \frac{23584}{1548} = 15.235$$

Whereupon $$_u\alpha_{\text{nominal}} = 1 - \frac{_uD_o}{_uD_e} = 1 - \frac{7.8}{15.235} = 0.488$$

The reliability of unitizing without their categorization or valuation is obtained by setting the difference function in $_u\delta_{igjh}$ of (9.1) to $_{\text{no metric}}\delta^2_{c_{ig}k_{jh}} = 0$. This would yield $_u\alpha_{\text{no metric}} = 0.515$. Naturally, by omitting the possible disagreements due to metric differences among valued segments, focusing only on their overlaps would yield agreements higher than when their metric differences are included.

## 9.6 Numerical Comparisons

To get a sense of how $_u\alpha$ behaves, consult the following Figure 9.3.

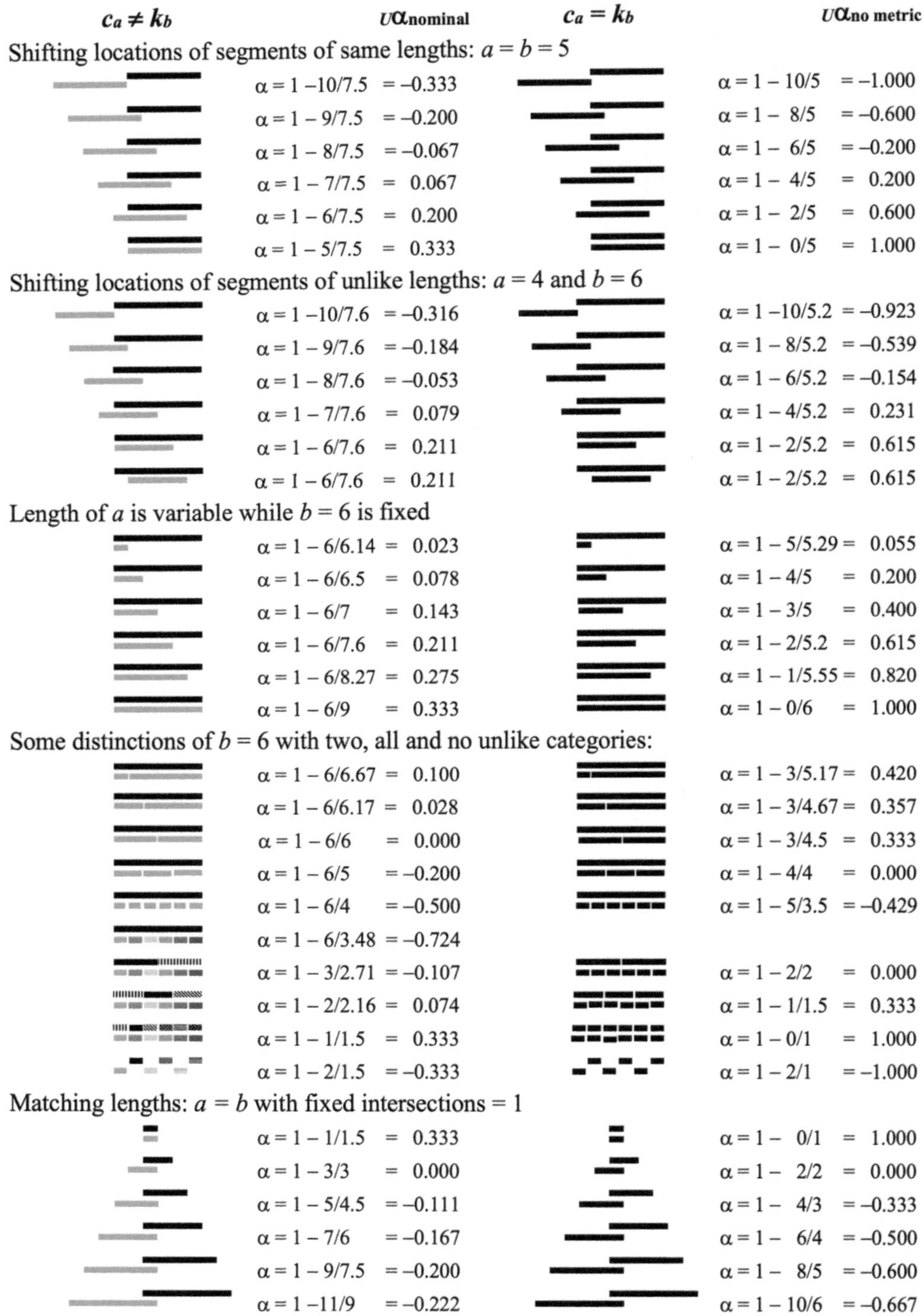


**FIGURE 9.3**
Several Instructive Comparisons of $_U\alpha$'s Behavior

# 10

# *Distinctions Drawn Within Continua*

Chapters 8 and 9 addressed several ways to obtain reliability measures for partitioning, unitizing, and segmenting continua. What underlies all of them are distinctions drawn within an otherwise undifferentiated continuum of known linear extension in texts, times, spaces, and other scaled entities.

A distinction separates parts without attributing qualities to them. Far too often, we are not aware that our observations of separate entities are not natural but the results of acts of our distinguishing them. So-called predefined units imply them to have been defined by linguistic means but thereafter they are considered unquestionable givens. Assuming units of analysis to be nonnegotiable facts renders concerns about the reliability of their unitizations moot. This is not to deny that some units are more natural or more easily distinguished than others. However, whether we count individuals, buy separate goods, or use words, they rarely exist independently of others. Individuals are born into families, learn the language of their communities, and work and live together with other individuals. All distinctions are made within contexts of larger systems and when the parts they distinguish are taken as independent, their embeddedness in the larger system is lost. Focusing on distinct units of analysis, taking them as unquestionable facts, has its advantages. Quantitative scholars whose declared starting point is the preexistence of countable units ought to acknowledge the embeddedness of such units in larger systems manifest in justifying causal connections, correlations, associations, and networks, and contingencies between them.

I would argue that distinctions are primary analytical acts. Taking what they distinguish as given may well be of practical utility but could be otherwise. Our use of language undoubtedly plays a role in distinguishing among the phenomena of this world. Communicating in this world amounts to coordinating the distinctions we make. There is a need to be able to measure the reliability of introducing distinctions in a continuum without reference to the resulting parts. While there are always reasons for drawing distinctions, this chapter does not care about the motivations of drawing distinctions nor about their results; only the extent to which independent agents, observers, analysts, conceptual systems, or practices deviate from each other regarding the distinctions they introduce in a finite continuum.

This chapter develops an agreement measure for assessing the reliability of drawing distinctions by identically instructed but independently working observers. Although distinctions are drawn everywhere, this chapter is limited to distinctions drawn within a finite continuum of finite lengths. It starts by presenting a measure of the replicability of drawing distinctions followed by a measure of the accuracy of distinctions.

## 10.1 The Replicability of Distinctions

### 10.1.1 Its Reliability Data

Following previous depictions of partitions of continua into identified segments and gaps between them, we conceive of the reliability data for distinctions as representing the replications of partitions of a continuum into undefined segments whose nature, substance, and relevance is beyond our concerns. Figure 10.1 gives an example of distinctions introduced by $m > 1$ observers into an unmarked continuum of finite length:

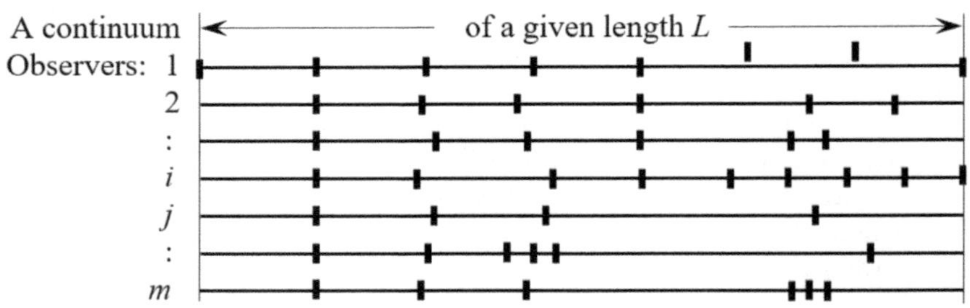


**FIGURE 10.1**
An Example of Distinctions Drawn Within a Continuum

In that figure, all $m$ observers perfectly agree on their first distinction. Their second exhibits minor deviations from a mean. In the third cluster, distinctions deviate more than in the second, even in their numbers, but appear intuitively related. Only the first four observers agree on their fourth distinction; the other three do not fall into line. The remaining distinctions seem quite chaotic. They barely concur.

In that figure other patterns may be worth mentioning. Observer 1 exhibits a regularity, distinguishing the continuum into segments of equal lengths. Maybe he or she is not a person but a mechanism, systematically applying a rule without deliberations. Observer 1's first and last mark on that continuum, including $j$'s last, cannot be considered distinctions as they coincide with the given beginning and end of the continuum. Not only do they make no difference in understanding a continuum; they also do not constitute a choice. Therefore we do not consider them distinctions proper.

In terms of a subset of the terminology introduced in Section 8.1.1 and Figure 8.2 in particular:

- The continuum begins at $End_{i0} = 0$ and ends at $End_{iN_i} = L$
- $N_i$ is the number of segments into which observer $i$ partitioned that continuum
- $N_i - 1$ is the number of distinctions observer $i$ made between 0 and $L$
- Distinctions are defined by the end $End_{ig}$ of segments $S_{ig}$ – excluding $S_{i0}$ and $S_{iNi}$
- For any two observers, $N_i \neq N_j$, but $End_{i0} = End_{j0} = 0$ and $End_{iN_i} = End_{jN_j} = L$

### 10.1.2 Its Difference Function

The difference between two distinctions, $End_{ig}$, made by observer *i*, and $End_{jh}$, made by observer *j* and nearest to $End_{ig}$, is defined as the square of the difference between the two. This difference is graphically illustrated by Figure 10.2 and defined by (10.1):

Graphically:

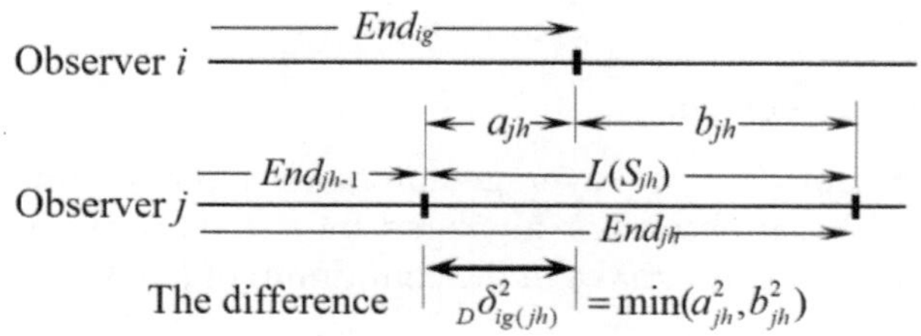


$$ {}_D\delta^2_{ig(jh)} = \min(a^2_{jh}, b^2_{jh}) \quad (10.1) $$

**FIGURE 10.2**
Definition of Differences between Any Two Distinctions

Note that the differences between two distinctions are not symmetrical: ${}_D\delta^2_{ig(jh)} \neq {}_D\delta^2_{jh(ig)}$. Their asymmetry is indicated by parentheses in their subscripts. "$(jh)$" denotes *j*'s segment $S_{jh}$ in which observer *i*'s distinction enters at $End_{ig}$ – not the other way around. It follows that:

When $End_{ig} = End_{jh}$: ${}_D\delta^2_{ig(jh)} = {}_D\delta^2_{jh(ig)} = 0$

When $End_{jh\text{-}1} < End_{ig} \leq End_{jh}$: $a_{jh} = End_{ig} - End_{jh\text{-}1}$; $b_{jh} = End_{jh} - End_{ig}$:

$$ {}_D\delta^2_{ig(jh)} = min(a^2_{jh}, b^2_{jh}) = min((End_{ig} - End_{jh\text{-}1})^2, (End_{jh} - End_{ig})^2). $$

$$ \text{Differences can range between: } 0 \leq {}_D\delta^2_{ig(jh)} \leq \left\| \frac{L(S_{jh})+1}{2} \right\|^2 \leq \left\| \frac{L+1}{2} \right\|^2 \quad (10.2) $$

where $\|x\|$ is the nearest lower integer of the decimal value of *x*.

For example:

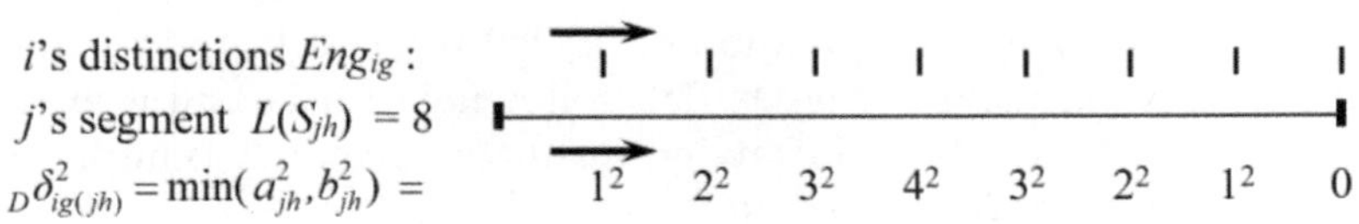


Here, the largest difference ${}_D\delta^2_{ig(jh)}$ for a segment of length $L(S_{jh}) = 8$ is $\left\| \frac{8+1}{2} \right\|^2 = 4^2$

### 10.1.3 Its Observed Disagreement

In these terms, the observed disagreement is defined by:

$$ {}_D D_o = \frac{\sum_{i=1}^{m} \sum_{g=1}^{N_i - 1} \sum_{j \neq i}^{m} \sum_{h=1}^{N_j} {}_D\delta^2_{ig(jh)}}{(m-1)\sum_{i=1}^{m}(N_i - 1)} \quad (10.3) $$

Where: $m$ = the number of replications by observers labeled 1, 2, …, $i$, …, $j$, …, $m$

$N_j$ = the number of segments distinguished by observer $i$

$N_i - 1$ = the number of distinctions made by observer $i$

$N_o = (m-1)\sum_{i=1}^{m}(N_i - 1)$ = the number of differences ${}_D\delta^2_{ig(jh)}$ summed in the numerator of ${}_DD_o$.

### 10.1.4 Its Expected Disagreement

The expected disagreement ${}_DD_e$ is the average among all possible differences within the segments distinguished by $m$ observers, averaged by $m(L-1)$ = the number of possible distinctions that $m$ observers could make in the continuum of length $L$.

$$
{}_DD_e = \frac{1}{m(L-1)}\sum_i^m \sum_g^{N_i}
\begin{cases}
\dfrac{2\left(\dfrac{L(S_{ig})-1}{2}\right)^3 + 3\left(\dfrac{L(S_{ig})-1}{2}\right)^2 + \dfrac{L(S_{ig})-1}{2}}{3} & \text{iff } L(S_{ig}) = \text{uneven} \\[2ex]
\left(\dfrac{L(S_{ig})}{2}\right)^2 + \dfrac{2\left(\dfrac{L(S_{ig})-2}{2}\right)^3 + 3\left(\dfrac{L(S_{ig})-2}{2}\right)^2 + \dfrac{L(S_{ig})-2}{2}}{3} & \text{iff } L(S_{ig}) = \text{even}
\end{cases}
\qquad (10.4)
$$

### 10.1.5 *D*-Alpha for the Replicability of Drawing Distinctions

$$
{}_D\alpha = 1 - \frac{{}_DD_o}{{}_DD_e} \qquad (10.5)
$$

### 10.1.6 A Numerical Example

The computation of ${}_D\alpha$ is illustrated by means of the numerical example introduced in Section 8.1.7 to exemplify the computation of ${}_u\alpha$, the replicability of partitioning a continuum into mutually exclusive segments. The following translates this example into the canonical form for reliability data of drawing distinctions, merely omitting references to units, their metric, and gaps between them:

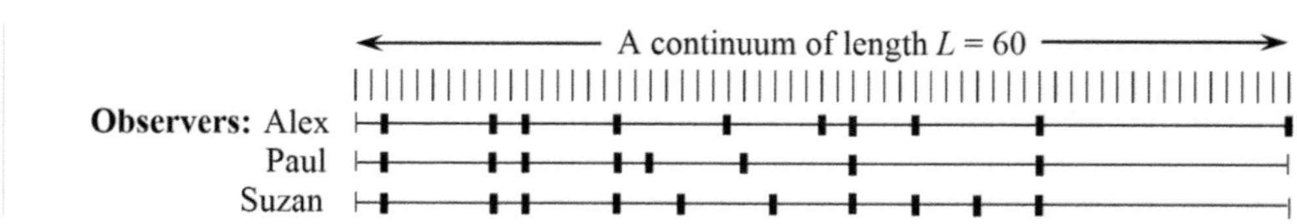

In these reliability data:

The number of segments distinguished: $N_{\text{Alex}} = 10$, $N_{\text{Paul}} = 9$, $N_{\text{Suzan}} = 11$
The number of distinctions made: $(N_{\text{Alex}} - 1) = 9$, $(N_{\text{Paul}} - 1) = 8$, $(N_{\text{Suzan}} - 1) = 10$
The number of differences to be summed: $(m-1)\sum_{i=1}^{m}(N_i - 1) = (3-1)(9+8+10) = 54$

Among these 54 differences, 38 are zero, which are omitted in calculating (10.3):

$$_D D_o = \frac{\sum_{i=1}^{m}\sum_{g=1}^{N_i-1}\sum_{j\neq i}^{m}\sum_{h=1}^{N_j} {}_D\delta^2_{ig(jh)}}{(m-1)\sum_{i=1}^{m}(N_i - 1)}$$
$$= \frac{1^2+2^2+4^2+3^2+2^2+2^2+1^2+2^2+2^2+3+2^2+3^2+2^2+4^2+4^2+4^2}{54} = 2.407$$

According to (10.4), $_D D_e = \frac{1}{m(L-1)}\sum_{i=1}^{m}\sum_{g=1}^{N_i}$ (The possible differences for each $S_{ig}$).
A few of these possible differences are calculated as follows:

$$\text{For } L(S_{\text{Alex1}}) = 2: \left(\frac{2}{2}\right)^2 + \frac{2\left(\frac{2-2}{2}\right)^3 + 3\left(\frac{2-2}{2}\right)^2 + \frac{2-2}{2}}{3} = 1 = 1^2 + 0$$

$$\text{For } L(S_{\text{Alex2}}) = 7: \frac{2\left(\frac{7-1}{2}\right)^3 + 3\left(\frac{7-1}{2}\right)^2 + \frac{7-1}{2}}{3} = 28 = 1^2+2^2+3^2+3^2+2^2+1^2+0$$

For $L(S_{\text{Paul8}}) = 12$: $= 146 = 1^2+2^2+3^2+4^2+5^2+6^2+5^2+4^2+3^2+2^2+1^2+0$
etc.

The sum of possible differences within all 30 segments is 1,524. This leads to the expected disagreement (10.4):

$$_D D_e = \frac{1524}{3(60-1)} = 8.610$$

and to the replicability of distinctions (10.5):

$$_D\alpha = 1 - \frac{_D D_o}{_D D_e} = 1 - \frac{2.402}{8.610} = 0.740$$

Naturally, by omitting all disagreements among the seven kinds of segments of the example from Section 8.1.3, their metric, and gaps between them, the reliability of distinctions $_D\alpha$ has good reasons to be larger than the $_u\alpha_{\text{nominal}} = 0.510$ which includes the disagreements between variously categorized segments.

## 10.2 The Accuracy of Distinctions

The general idea of accuracy was introduced in Section 1.4, expanded in Chapter 6, formulated for coding predefined units in Section 7.5 and for unitizing in Section 8.5. This section follows their leads and defines a measure of accuracy for drawing distinctions in terms defined in Section 8.1 and above. Notationally, it refers to the trusted standard by *s*, used to evaluate the accuracy of the distinctions made by $m > 0$ observers *i*.

### 10.2.1 Its Observed Inaccuracy

With the number $N_s$ of segments distinguished by the standard *s*, the observed inaccuracy compares all $\sum_{i=1}^{m}(N_i - 1)$ distinctions made by *m* observers *i* with that standard, applying the difference function ${}_D\delta^2_{ig(sh)}$ in (10.1) to each. The resulting observed inaccuracy becomes:

$$ {}_DD_o^s = \frac{\sum_{i=1}^{m}\sum_{g=1}^{N_i-1}\sum_{h=1}^{N_s} {}_D\delta^2_{ig(sh)}}{\sum_{i=1}^{m}(N_i - 1)} \tag{10.6} $$

### 10.2.2 Its Expected Inaccuracies

The expected inaccuracy equals the average of all differences from the standard *s*:

$$ {}_DD_e^s = \frac{1}{(L-1)}\sum_{h=1}^{N_s}\begin{cases} \dfrac{2\left(\frac{L(S_{sh})-1}{2}\right)^3 + 3\left(\frac{L(S_{sh})-1}{2}\right)^2 + \frac{L(S_{sh})-1}{2}}{3} & \text{iff } L(S_{sh}) = \text{uneven} \\[2ex] \left(\frac{L(S_{sh})}{2}\right)^2 + \dfrac{2\left(\frac{L(S_{sh})-2}{2}\right)^3 + 3\left(\frac{L(S_{sh})-2}{2}\right)^2 + \frac{L(S_{sh})-2}{2}}{3} & \text{iff } L(S_{sh}) = \text{even} \end{cases} \tag{10.7} $$

### 10.2.3 *D*-Alpha-*s* for the Accuracy of Drawing Distinctions

With the ${}_D\alpha$-accuracy for distinctions defined as usual:

$$ {}_D\alpha^s = 1 - \frac{{}_DD_o^s}{{}_DD_e^s} \tag{10.8} $$

### 10.2.4 A Numerical Example

The computation of accuracy is illustrated by adding a standard *s* (unlike in Section 8.5.1) to the reliability data used in Section 10.1.6:

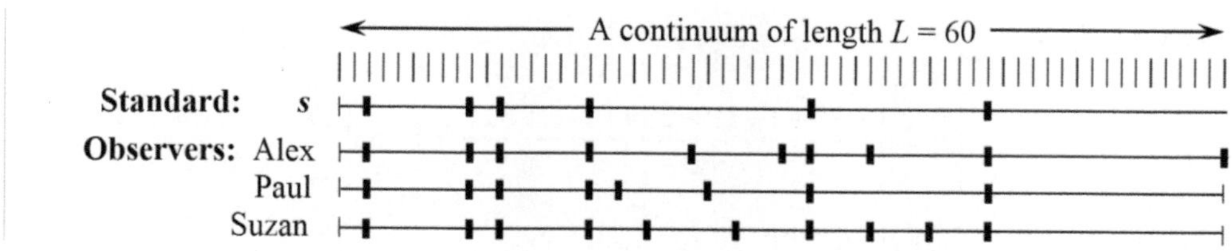


In these reliability data, 18 distinctions match with the standard. The three observers disagree within two large segments of the standard. In $L(S_{s5}) = End_{s5} - End_{s4} = 15$, Alex, Paul and Suzan draw two distinctions each, compared with none by the standard. Paul's $S_{Paul8}$ agrees with the standard's $S_{s6}$. But Alex draws one and Suzan two distinctions within the standard's undifferentiated segment, $L(S_{s6}) = End_{s6} - End_{s5} = 12$. These nine mismatching distinctions contribute to the observed inaccuracy. Numerically, by (10.6):

$$_{D}D_o^s = \frac{\sum_{i=1}^{m}\sum_{g=1}^{N_i-1}\sum_{h=1}^{N_s} {}_D\delta_{ig(sh)}^2}{\sum_{i=1}^{m}(N_i-1)} = \frac{7^2+2^2+4^2+2^2+7^2+4^2+5^2+4^2+4^2}{(10-1)+(9-1)+(11-1)} = 7.222$$

The expected inaccuracy of distinctions is actually the average number of differences possible within the standard, which shows 6 distinctions that leave $N_s = 7$ segments to calculate the possible inaccuracies one could encounter. Numerically, by (10.7):

For $L(S_{s1}) = L(S_{s3}) = 2$: $$\left(\frac{2}{2}\right)^2 + \frac{2\left(\frac{2-2}{2}\right)^3 + 3\left(\frac{2-2}{2}\right)^2 + \frac{2-2}{2}}{3} = 1$$

For $L(S_{s2}) = 7$: $$\frac{2\left(\frac{7-1}{2}\right)^3 + 3\left(\frac{7-1}{2}\right)^2 + \frac{7-1}{2}}{3} = 28$$

For $L(S_{s4}) = 6$: $$\left(\frac{6}{2}\right)^2 + \frac{2\left(\frac{6-2}{2}\right)^3 + 3\left(\frac{6-2}{2}\right)^2 + \frac{6-2}{2}}{3} = 19$$

For $L(S_{s5}) = 15$: $$\frac{2\left(\frac{15-1}{2}\right)^3 + 3\left(\frac{15-1}{2}\right)^2 + \frac{15-1}{2}}{3} = 280$$

For $L(S_{s6}) = 12$: $$\left(\frac{12}{2}\right)^2 + \frac{2\left(\frac{12-2}{2}\right)^3 + 3\left(\frac{12-2}{2}\right)^2 + \frac{12-2}{2}}{3} = 146$$

For $L(S_{s7}) = 16$: $$\left(\frac{16}{2}\right)^2 + \frac{2\left(\frac{16-2}{2}\right)^3 + 3\left(\frac{16-2}{2}\right)^2 + \frac{16-2}{2}}{3} = 344$$

Inserted into (10.7):

$$_{D}D_e^s = \frac{1+28+1+19+280+146+344}{60-1} = 13.881$$

and $_{D}D_o^s$ and $_{D}D_e^s$ into (10.8) yields the $_{D}\alpha^s$ for accuracy:

$$_{D}\alpha^s = 1 - \frac{_{D}D_o^s}{_{D}D_e^s} = 1 - \frac{7.222}{13.881} = 0.480$$

That $_{D}\alpha^s = 0.480$ for accuracy is significantly smaller than the $_{D}\alpha = 0.725$ for replicability that could be expected by visual inspection of the above reliability data. It turns out that all of the standard's distinctions happen to correspond to the distinctions on which the three observers agree perfectly, counting the eight remaining distinctions as inaccuracies. By contrast, replicability responds positively to the albeit smaller differences among these eight distinctions. For accuracy, most of them are substantial.

# 11

## *Text Mining and Information Retrieval*

### 11.1 Precision and Recall

The literature on information retrieval, text mining, pattern recognition, and binary classification of relevant versus irrelevant matter has promoted two measures to assess the qualities of search engines: *Precision* and *Recall* (van Rijsbergen, 1979; Manning et al., 2008). In that literature there is full consensus that an ideal search engine retrieves all and only relevant items and ignore all and only irrelevant ones.

In terms of the entries in a 2-by-2 matrix of items:

| | Relevant | Irrelevant | | |
|---|---|---|---|---|
| Retrieved | $a$ | $b$ | $a+b$ | |
| Not retrieved | $c$ | $d$ | $c+d$ | Unknown, at best estimated |
| Totals in data base | $a+c$ | $b+d$ | $a+b+c+d$ | |

**FIGURE 11.1**
Precision and Recall Visualized

$$Precision = \frac{a}{a+b} \tag{11.1}$$

$$Recall = \frac{a}{a+c} \tag{11.2}$$

Important for subsequent elaborations is to notice that both criteria aim at different explanations of the one cell in Figure 11.1, the amount of relevant information retrieved. This quantity is of course all that matters for research projects as well as designers of search engines.

While the two criteria for ideal search engines are logically plausible and easily stated, it is obvious that half of the denominator of the measure of *Recall* is virtually unknowable before actually searching the whole data base and observing what is truly relevant and what is not. However, the very motivation for employing search engines for finding and retrieving what matters is based on the acknowledgment that contemporary data bases tend to be too big for anyone to know what they contain. Indeed, if one were to know the content of a data base, there would be no point in searching it for what is relevant to a given project. This Catch-22 makes the measure of *Recall* an epistemological fiction, an assumption to know before proceeding to find out.

To salvage the measure of *Recall*, researchers have estimated the number $a+c$ of relevant items in a data base of known size from the proportion $a/b$ of the retrieved items. Such efforts project the property of the search engine to be evaluated onto the data base, including the very errors that *Recall* is meant to measure.

Experiments with applying search engines to data bases with known compositions have shown that *Precision* and *Recall* are inversely proportional: If a retrieval system returns more results, $a+b$, *Recall* is likely to increase but *Precision* is likely to fall (Voorhees and Harman, 2000: 14). This exemplifies empirical findings obtained under unrealistic conditions. There have been proposals to combine *Precision* and *Recall* into a single coefficient $F$ (van Rijsbergen, 1979), allowing users to weigh which of the two measures are more important. But none of these and many more efforts to modify the measure of *Recall* overcome the fundamental epistemological limitation of not knowing the composition of a data base before examining it by retrieving all items from it.

Unlike *Recall*, the measure of *Precision* is at least defined within actually retrieved items and therefore easily quantifiable. However, much of the information retrieval literature fails to recognize another epistemological obstacle. It pursues a naive conception of information as contained in the character strings of retrievable texts and unproblematically removable from their textual containers by their recipients. This use of the container metaphor leads to an objectivist conception of information as accessible to everyone (Krippendorff, 2019: 27–31), justifies ignoring human involvement in the use of search engines, and focuses instead on the algorithmic nature of search engines. The practice of searching data bases for what is useful is quite different. After all, it is researchers who anticipate what will be relevant to their project, formulate search terms which are assumed to be appropriate, utilize a search engine to apply them, and interpret the results. This empirical fact suggests that a good part of the measure of *Precision* reflects the ability of researchers to direct a search engine to relevant matter.

Surely, the nature of the search algorithm and its relation to how the data base it searches is organized limits researchers' ability to get to what they need to know. However, the expectation of what could inform a research project, which is reflected in the formulation of search terms and the distinction between items that inform a research project, $a$, and those that do not, $b$, is the product of interpretations by literary competent researchers with a mission and a need. The reliability of this distinction could be assessed by $_c\alpha$-agreement coefficients of the simplest kind, presented in Chapter 2. However, the unreliability of this distinction is only a part of what can get a research project into trouble.

Different search engines rarely perform alike, and all data bases adopt formats that make some items more accessible than others. Some search engines are designed to adapt to the search history of a user, Google's or Amazon's for example, and work differentially when used again. This makes their evaluation not only extraordinarily difficult but also limits their scientific use as a way to get a reliable picture of what is available. With all its flaws, *Precision* and *Recall* are not measures of ideal search algorithms but of the extent to which researchers are in control of their use.

Researchers learn from mistakes and software designers from the feedback they receive from the users of their systems. Software developers may hope for perfection; however, as research questions evolve, linguistic habits change, and data bases grow, I suggest we can expect only small improvements. Research that objectifies information and presumes that human interpretations have nothing to do with the successful use of search algorithms supports only one side of the equation. Agreement measures are needed that can separate the reliability and biases of search engines from the reliability of formulating appropriate search terms and judging the relevance of retrieved textual matter in support of scientific conclusions.

## 11.2 Reliability Data for Replications of Text Searches

When searching the same data base with different search engines, employing alternative search terms for getting to the same data, or employing several researchers to interpret search results, it may not be too difficult to measure the reliabilities of information retrieval regarding the relevance of retrieved items for a research objective.

Let the reliability data for text mining and information retrieval consist of

$1, 2, \ldots, i, j, \ldots, m$ = The searches of the same data base (alternative search terms, or independent researchers interpreting the results)

$R_i$ = The set of relevant items retrieved in search $i$

$I_i$ = The set of irrelevant items retrieved in search $i$

$\phi_i$ = The set of items not retrieved in search $i$ but by others

$R_i \cup I_i$ = The set of items retrieved in search $i$

$\bigcup_{i=1}^{m}(R_i \cup I_i) = T$ = The total set of items retrieved by $m$ searches (without duplications)

$\sum_i^m \sum_{j \neq i}^m |R_i \cap R_j|$ = The sum of all $m(m-1)$ pairwise intersections among all relevant items in $m$ search results

$|C|$ = The cardinality of the set $C$ = the number of elements $C$ contains

The sums of all intersections of the sets of items from reliability data of $m$ searches constitute a symmetrical coincidence matrix depicted in Table 11.1. To reduce the complexity of expressions in this table, its entries are not standardized as are the coincidence matrices in the preceding chapters. Also, because these sums will be used in the measures developed below, we use the shorter notations for clarity.

| | | | |
|---|---|---|---|
| $n_{RR} = \sum_i \sum_{j \neq i} \lvert R_i \cap R_j \rvert$ | $n_{RI} = \sum_i \sum_{j \neq i} \lvert R_i \cap I_j \rvert$ | $n_{R\phi} = \sum_i \sum_{j \neq i} \lvert R_i \cap \phi_j \rvert$ | $n_R = (m-1)\sum_i \lvert R_i \rvert$ |
| $n_{IR} = \sum_i \sum_{j \neq i} \lvert I_i \cap R_j \rvert$ | $n_{II} = \sum_i \sum_{j \neq i} \lvert I_i \cap I_j \rvert$ | $n_{I\phi} = \sum_i \sum_{j \neq i} \lvert I_i \cap \phi_j \rvert$ | $n_I = (m-1)\sum_i \lvert I_i \rvert$ |
| $n_{\phi R} = \sum_i \sum_{j \neq i} \lvert \phi_i \cap R_j \rvert$ | $n_{\phi I} = \sum_i \sum_{j \neq i} \lvert \phi_i \cap I_j \rvert$ | $n_{\phi\phi} = \sum_i \sum_{j \neq i} \lvert \phi_i \cap \phi_j \rvert$ | $n_\phi = (m-1)\sum_i \lvert \phi_i \rvert$ |
| $n_R = (m-1)\sum_j \lvert R_j \rvert$ | $n_I = (m-1)\sum_j \lvert I_j \rvert$ | $n_\phi = (m-1)\sum_j \lvert \phi_j \rvert$ | $n_T = m(m-1)\lvert T \rvert$ |

**TABLE 11.1**
Matrix of Coincidences in Reliability Data for $m$ Searches and Interpretations

Searches are perfect when all retrieved items are relevant. Under these ideal conditions, the sum of pairwise intersections of relevant items $n_{RR}$ would equal the total $n_T$. A first proposal for a coefficient that would assess the quality of information retrievals is the proportion $P^{IR}$ of relevant items among all those retrieved:

$$P^{IR} = \frac{n_{RR}}{n_T} = \frac{\sum_i \sum_{j \neq i} |R_i \cap R_j|}{m(m-1)|T|} \tag{11.3}$$

Table 11.1 highlights the two sums that (11.3) relates in the proportion $P^{IR}$. Evidently, if the $m$ searches retrieve only relevant items, $P^{IR} = 1$. In this case, its numerator, being one of the 3-by-3 = nine coincidences in Table 11.1 equals the matrix's total sum, leaving the other eight cells empty. As $P^{IR}$ deviates from 1, the eight cells account for the kind of disagreements that cause its demise. In effect, Table 11.1 amounts to a decomposition of the total number of pairs of items retrieved into nine possible coincidences, eight of which represent disagreements encountered in the reliability data.

(11.3) is a simple proportion of coincidences. It does not refer to any of the eight deviations from perfect reliability in Table 11.1, which quantify various errors. The expression (11.4) decomposes $P^{IR}$ into two parts, each being sensitive to different imperfections:

$$P^{IR} = \frac{n_{RR}}{n_T} = P^I P^R$$

$$P^I = \frac{n_{RR}}{n_{RR} + 2n_{IR} + n_{II}}$$

$$P^R = \frac{n_{RR} + 2n_{IR} + n_{II}}{n_T} \tag{11.4}$$

The first of the two proportions in (11.4), $P^I$ is the proportion of agreements on relevant items among those retrieved by all methods. While the denominator is the result of what all searches agree on, this proportion is determined by both the competencies of researchers in formulating search terms that are appropriate to the search engines employed, and more importantly by their ability to judge whether retrieved items bear the information relevant to answering the given research questions. This proportion generalizes the measure of $Precision = R_i / (R_i \cup I_i)$ from any one $i$ to the aggregate of $m$ searches.

The second proportion, $P^R$ is a measure of the replicability of the search engines; the extent to which they retrieve the same items when mining the same data base. It includes one part of the error that *Recall* was designed to assess – the failure to retrieve relevant items, here by reference not to what a data base contains but to what some searches retrieve but others fail to identify.

However, *Precision, Recall,* $P^{IR}$, $P^I$, and $P^R$, are all proportions. They suffer from the same problems that make percent agreement measures inadequate. Their zero point is far below what happens when chance governs the process, making their scale between 0 and 1 difficult to interpret in terms of whether – or the extent to which – data are not trustworthy. While $P^I$ and $P^R$ differentiate between human and machine errors, the other three proportions do not.

Evidently, the reliability data in the cells of Table 11.1 provide the basis of more detailed examinations of the quality of text mining and information retrieval efforts. We turn to this in the following sections.

## 11.3 *c*-Alpha-*IR* for Text and Information Retrieval

Because $P^{IR}$ is unrelated to what can be expected when chance governs the retrieval of relevant items, we overcome this deficiency by shifting attention to the observed disagreements that reduce $P^{IR}$ and subtract expected coincidences from those observed. This approach deviates from the ${}_c\alpha$-coefficients defined in (2.5) and in the preceding chapters by failing to acknowledge the extent to which the cells in Table 11.1 are due to chance, focusing only on the $n_{RR}$ which reflects both the quality of text mining and of information retrieval. It enumerates researchers' agreement on the relevant information among all $n_T$ items of Table 11.1. This focus on what matters as opposed to on how all categories (relevant, irrelevant, and not retrieved by all search engines) are distinguished is analogue to focusing on the numerator that *Precision* and *Recall* share. In terms of Table 11.1, the overall agreement coefficient thus becomes:

$$ {}_c\alpha^{IR} = 1 - \frac{{}_cD_o^{IR}}{{}_cD_e^{IR}} = 1 - \frac{n_T - n_{RR}}{n_T - \dfrac{n_R^2}{n_T}} \tag{11.5} $$

Observe that (11.3) behaviorally resembles the chance-corrected agreement coefficients discussed in Chapter 4 for nominal data. It measures 1 when all retrieved items are relevant, $n_{RR} = n_T$, it becomes 0 when agreement is chance, $n_{RR} = n_R^2 / n_T$, and it may be negative when information-bearing items are systematically below expectations.

Note that (11.5) is not limited to what *Precision* is aimed to assess. It is also affected by omissions that the conception of *Recall* had hoped to embrace.

Also, (11.5)'s exclusive focus on perfect agreements on relevant items $n_{RR}$, identified as providing relevant information to a research project, is only partly determined by what a search engine does. It is largely due to the purpose of searching a data base, its translation into the search queries and how retrieved documents might help answering the guiding research questions. This is one reason for ${}_c\alpha^{IR}$ to deviate from the common form of the definitions of ${}_c\alpha$-agreements in (2.5) and in subsequent chapters.

Finally, singular reliance on ${}_c\alpha^{IR}$ may not be advisable in all information retrieval efforts. On the one hand, researchers may want to preserve the possibilities of surreptitious discoveries of relevant information among retrieved items; on the other hand, it is advisable for researchers to consider the possibility that search engines may be biased regarding which texts they retrieve and deliberately search more broadly. Fundamentally, ${}_c\alpha^{IR}$ does not distinguish between the biases of search engines and the unreliabilities of the researchers deciding on the relevance of what they hoped to find. This speaks for a distinction between these two processes similar to the partition of $P^{IR}$ into the products $P^I$ and $P^R$ in (11.4).

$$ {}_cD_o^{IR} = {}_cD_o^{I} + {}_cD_o^{R} \tag{11.6a} $$

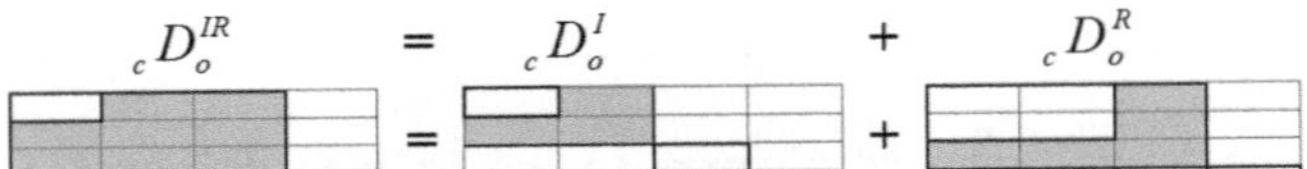

**FIGURE 11.2**
Three Accounts of Observed Disagreements in the Coincidences of Table 11.1

To start with, the observed disagreement in ${}_c\alpha^{IR}$ is decomposed into one part attributed to human uncertainties ${}_cD_o^I$ and another to the biases of the retrieval mechanisms ${}_cD_o^R$. This can be illustrated graphically by reference to the cells of Table 11.1.

The disagreements (11.6a) are quantitatively expressed by:

$$[n_T - n_{RR}] = [(n_{RR} + 2n_{RI} + n_{II}) - n_{RR}] + [n_T - (n_{RR} + 2n_{RI} + n_{II})] \tag{11.6b}$$

The corresponding expected disagreements do not add up as neatly because ${}_cD_o^I$ is a function of the numbers of items whereas ${}_cD_e^R$ is stated relative to $P^I/P^{IR}$ and $P^R/P^{IR}$.

These observed disagreements lead to the following separate measures.

## 11.4 c-Alpha-*R* for the Reliability of Search Engines

The more obvious of the two needed measures evaluates the reliability of the search engines that participate in generating the reliability data in Table 11.1. It relates to the measure of recall by focusing on what a data base evidently contains but some search engines recognize while others do not. In other words, it measures bias.

$${}_c\alpha^R = 1 - \frac{n_T - (n_{RR} + 2n_{IR} + n_{II})}{n_T - \dfrac{(n_I + n_R)^2}{n_T}} \tag{11.7}$$

When ${}_c\alpha^R = 1$, then all search engines agree on what they provide to the researcher and $n_\phi = 0$. When their retrievals vary, $n_\phi > 0$, ${}_c\alpha^R < 1$ indicates that search engines, while mining the same data base, retrieve different items. It provides evidence that the set of search engines is unreliable or biased, whether systematic or unexplained. Researchers cannot control these differences. They will have to acknowledge their uncertainties about whether a data base contains the texts they are looking to examine and cannot rely on the frequencies of the retrieved texts to infer the content of such data bases beyond what searching them yields.

However, it is also important to recognize that some search engines are designed to adapt to their users' search histories, habits, or associations with similar other users. Search engines that are intended to provide scientifically representative data should be replicable, not uncontrollably change their searches according to researcher past interest. Search engines are algorithms. Their scientific usability depends on searching data bases of unknown composition with perfect reliability. If search engines change in the process of surfing data bases, good intentions notwithstanding, comparing their results always implicates their users in unpredictable ways and becomes statistically problematic.

## 11.5 c-Alpha-*I* for Researchers' Competence of Handling Search Engines and Interpreting Their Results

Researchers' competencies involved in searching data bases for relevant information undoubtedly are complex. First, individual researchers may have only a scant understanding of how a chosen search engine works. It requires both linguistic knowledge and algorithmic understanding of how textual elements are identified. Second, they may have no clear understanding of how a data base is organized and the documents of interest are formatted. Third, texts do not simply contain information or are void of it; they need to be read and interpreted in view of existing research objectives. This requires complex knowledge of the potential implications of retrieved texts. As already mentioned, researchers may wish to preserve the possibilities of surprises. But they undoubtedly are bothered by too much irrelevant text.

Following the logic of (11.3) and (11.6a and b) leads to the following definition of the agreement coefficient ${}_c\alpha^I$:

$$ {}_c\alpha^I = 1 - \frac{(n_{RR} + 2n_{IR} + n_{II}) - n_{RR}}{n_{RR} + 2n_{RI} + n_{II} - \dfrac{(n_{RR} + n_{RI})^2}{n_{RR} + 2n_{RI} + n_{II}}} \tag{11.8} $$

Again, ${}_c\alpha^I = 1$ when the four cells in Table 11.1 containing $(n_{RR}+2n_{IR}+n_{II})$ items equal $n_{RR}$, i.e., all retrieved items are judged relevant. This condition corresponds precisely to that of perfect *Precision*. However, as irrelevant items become dominant, ${}_c\alpha^I$ converges to 0 at which point the identification of relevant items among those retrieved is indistinguishable from chance.

## 11.6 Their Relations

From the decompositions of the observed disagreements ${}_cD_o^{IR}$ in (11.6a and b) and the proportions of items that go in either coefficient, I suggest an approximation to the relationship between the three coefficients:

$$ {}_c\alpha^{IR} \approx \frac{p^I}{p^{IR}}\,{}_c\alpha^I + \frac{p^R}{p^{IR}}\,{}_c\alpha^R \tag{11.9} $$

## 11.7 Reliability Data for *m* Researchers Facing One Search Engine

When the results of only one search engine are available and judged for its relevance, then in the reliability data of Table 11.1 the rows and columns of intersections with omitted items are no longer occupied, and $n_\phi = 0$. This reduces this 3-by-3 coincidence matrix to a 2-by-2 matrix. Its total becomes $n_T \Rightarrow n_{RR}+2n_{RI}+n_{II}$, and ${}_c\alpha^{IR} \Rightarrow {}_c\alpha^I$ which proves the independence of (11.8):

$$ {}_c\alpha^{IR} = 1 - \frac{n_T - n_{RR}}{n_T - \dfrac{n_R^2}{n_T}} \Rightarrow {}_c\alpha^I = 1 - \frac{(n_{RR} + 2n_{IR} + n_{II}) - n_{RR}}{n_{RR} + 2n_{RI} + n_{II} - \dfrac{(n_{RR} + n_{RI})^2}{n_{RR} + 2n_{RI} + n_{II}}} \tag{11.10} $$

# 12

## *Diagnostic Devices and Remedial Actions*

The $\alpha$-agreements presented so far seek to give researchers the needed assurances for whether data – generated, found, or inherited – can be trusted in subsequent analyses. Chapter 15 will discuss criteria to decide whether to proceed or not. But besides this binary, $\alpha$-agreements may also inform efforts to improve the reliability of data. Such uses presuppose choices. I recognize three periods in most research effort during which reliability concerns result in different remedial actions:

1. In the process of developing a research project, one needs to:
   - Define the population of textual phenomena that could inform a research effort.
   - Develop coding instructions and measures that respond to various qualities of observations, readings of texts, and experimental designs yielding data meant to provide the information needed to answer the research questions about the phenomena of interest.
   - Sample of phenomena to test the reliability of the process of turning them into data.
   - Select qualified observers, coders, judges, and measuring instruments among those widely available in order to ensure replicability.
   - Locate sources of likely disagreements among replications of the data-making process
     - ❖ Among the kinds of phenomena chosen for analysis
     - ❖ In the definitions provided by the coding instructions regarding
       - Separate variables, Section 12.1.
       - Individual units of analysis regarding a variable, Section 12.2.
       - Default categories, Section 12.3.
       - Appropriate metrics for a variable, Section 12.4.
       - Ambiguous categories and values, Section 12.5.
     - ❖ Among replications, observers, or judges, selecting the most reliable ones, Section 12.6.
     - ❖ The recording medium used.

   All of which is in view of systematically testing and improving the reliability of the process of making data as long as it is feasible to restart the process until one is confident to proceed.
2. In the process of generating analyzable data with agreed-upon coding instructions and other computational aids, it is desirable that the reliability of the process be tested intermittently to make sure that the reliability is not decaying over time. During this phase of research, there are fewer options of correcting discovered unreliabilities, Section 12.7.

3. After data have been generated, remedial actions tend to be limited to either discarding unreliable variables or trying to extract reliable features from questionable data. Such efforts almost always amount to trading information thought to be relevant for reliable data, a trade-off that may well compromise answering the intended research questions. This concerns
   - Selecting the most reliable replications from multiple contributors, Section 12.8.
   - Deselecting variables that fall short of the required reliability, Section 12.9.
   - Separating default (no-information) categories from those that matter, Section 12.10.
   - Lumping ambiguities while retaining the most informative distinctions, Section 12.11.
   - Trying to correct explainable sources of systematic disagreements, Section 12.12.

Each of these three periods of a research project call for rather different actions which should not be confused or misused. As a bad example: After an international content analysis project had begun, the content analyst in one country decided to modify his coding instructions to counter a persistent source of unreliability. This was done with the best of intentions. However, instead of convincing the other participants to consider these corrections, he proceeded on his own, creating systematic cross-country incompatibilities that limited the intended international comparisons.

The following takes up the most promising methods of identifying problems and offers some remedial actions especially regarding the third phase. For simplicity, I will state these analytical devices in terms of coding of predefined units and trust the reader to apply the same logic to other data.

## 12.1 Concerning Separate Variables

Data languages tend to be composed of many logically independent variables, semantic domains, or groups of mutually exclusive categories. In the course of developing the data language for a research project, for example, in terms of the "actor – action – target" scheme, common in the English language, in terms of Harold Lasswell's (1948) formula for communication research: "who" "says what" "in which channel" "to whom" and "with what effect," or in rhetorical terms by "the argument in question," "evidence in support of it," "in opposition to it," and "consent on the winner or the conclusion," variables are contextually defined spaces that include all categories or values relevant for describing their alternatives.

In the definitions of variables, contexts matter greatly. For example, colors have standardized dictionary names, but a blue sky refers to something completely unlike feeling blue, i.e., depressed, down, and unhappy; having blue blood, i.e., being a member of the aristocracy; or wearing blue jeans. A variable that contains the category "blue" is meaningless without specification of the context of its use.

Most $\alpha$-coefficients measure the agreements among replications separately for each variable. They cannot respond to the contexts of variables except indirectly by the effect of the definitions of their variables. The following presents several ways to segment the reliability data of a variable so that weak features of variables are revealed and could be separately addressed.

| | | | | | | |
|---|---|---|---|---|---|---|
| **Standard** or **Surrogate**: | $k_1$ | . . . . | $k_u$ | . . . . . . . | $k_{N_u}$ | if available |
| **Units**: | 1 | 2 3 . . | $u$ | . . . . . . . | $N_u$ | |
| **Replications**: 1 | $c_{11}$ | . . . . | $c_{1u}$ | . . . . . . . | $c_{1N_u}$ | |
| ⋮ | ⋮ | | ⋮ | | ⋮ | |
| $i$ | $c_{i1}$ | . . . . | $c_{iu}$ | . . . . . . . | $c_{iN_u}$ | |
| $j$ | $c_{j1}$ | . . . . | $c_{ju}$ | . . . . . . . | $c_{jN_u}$ | |
| ⋮ | ⋮ | | ⋮ | | ⋮ | |
| $m$ | $c_{m1}$ | . . . . | $c_{mu}$ | . . . . . . . | $c_{mN_u}$ | |
| **Sums** $m_u \begin{cases} >0 \text{ for } {}_c\alpha^s \text{ and } k_u \\ >1 \text{ for } {}_c\alpha, {}_c\alpha^{sm}, {}_c\alpha^m \end{cases}$ | $m_1$ | . . . . | $m_u$ | . . . . . . . | $m_{N_u}$ | $n.. = \sum_{u=1}^{N_u} m_u$ |

**TABLE 12.1**
Canonical Form of Reliability Data Generalized to Accuracy, Surrogacy, and Decisiveness

## 12.2 Concerning Individual Units of Analysis of One Variable

The canonical form of reliability data in Table 6.7 contains the replications needed for obtaining the replicabilities of individual variables.

When replicability is the issue, standards are ignored, units $u$ have to be given $m_u > 1$ pairable values, and yield $m_u(m_u - 1)$ differences for each. Bypassing the construction of coincidence matrices with (5.1), for any one unit $u$, the observed disagreement is:

$$_{(u)c}D_o = \frac{1}{m_u - 1} \sum_{i=1}^{m_u} \sum_{j \neq i}^{m_u} {}_{\text{metric}}\delta^2_{c_{iu}c_{ju}} \leq m_u \tag{12.1}$$

$_{(u)c}D_o$ represents $m_u/n..$ of the contributions of $u$ to the observed disagreement $_cD_o$ to which all $N_u$ units contribute.

According to (6.12), the expected disagreement for all $N_u$ units $u$ is obtained from the marginal sums of its coincidences $n_{ck}$ or from Table 12.1:

$$_cD_e = \frac{1}{n.. - 1} \sum_{c=1}^{v} n_c \sum_{k=1}^{v} n_k \, {}_{\text{metric}}\delta^2_{ck} \leq n..$$

Adjusting for this numerical difference yields the measure of the replicability of valuing any one unit $u$:

$$_{(u)c}\alpha_{\text{metric}} = 1 - \frac{\frac{n..}{m_u} \, {}_{(u)c}D_o}{_cD_e} = 1 - \frac{n..(n.. - 1) \sum_{i=1}^{m_u} \sum_{j \neq i}^{m_u} {}_{\text{metric}}\delta^2_{c_{iu}c_{ju}}}{m_u(m_u - 1) \sum_{c=1}^{v} n_c \sum_{k=1}^{v} n_k \, {}_{\text{metric}}\delta^2_{ck}} \tag{12.2}$$

For accuracy, the observed disagreement for any one unit $u$ pairs $m_u > 0$ values with an existing standard $k_u$ and becomes:

$$_{(u)c}D_o^S = \sum_{i=1}^{m_u} {}_{\text{metric}}\delta^2_{c_{iu}k_u} \leq m_u \tag{12.3}$$

It has the same range as (12.1). According to (6.17), its expected disagreement ${}_cD_e^s$, obtained from the matrix of contingencies $x_{ck}$ for all $x_{..}$ pairable values:

$${}_cD_e^s = \frac{1}{x_{..}} \sum_{c=1}^{v} x_c \sum_{k=1}^{v} x_k \, {}_{\text{metric}}\delta_{ck}^2 \leq x_{..}$$

Making use of the same adjustment as for (12.2) yields the measure of the accuracy of valuing one unit $u$:

$${}_{(u)c}\alpha_{\text{metric}}^s = 1 - \frac{\frac{x_{..}}{m_u} \, {}_{(u)c}D_o^s}{{}_cD_e^s} = 1 - \frac{x_{..} \sum_{i=1}^{m_u} {}_{\text{metric}}\delta_{c_{iu}k_u}^2}{m_u \sum_{c=1}^{v} x_c \sum_{k=1}^{v} x_k \, {}_{\text{metric}}\delta_{ck}^2} \tag{12.4}$$

The reliabilities of individual units (12.2) and (12.4) for any one variable of questionable reliability may direct the designers of coding instructions to units that are worthy of examination as to why agreement on them is lacking and which definitions in the coding instructions need to be modified to accommodate unanticipated misconceptions.

It is also conceivable that the reliabilities of individual units may enable researchers to redefine the universe of texts they wish to analyze – not because their coding is unreliable but because they may be of an incompatible genre. For example, in a study of TV drama, we found that most units on which coders failed to agree were comics. Comics and TV shows with human actors are rather different genres. Therefore, we decided to explicitly ignore comics in our research into live TV shows.

However, it is important to warn against the misuse of these measures. It is tempting to drop unreliable units from the analyzed data and deceptively claim that the research results were based on reasonably if not perfectly reliable data. This practice would surreptitiously introduce the coders' difficulties or researchers' inability to anticipate the universe of the phenomena for which the values of a variable are defined into reliability considerations. This measure would undermine the fundamental conception of data as representing phenomena of analytical interest, not the competencies of the designers of coding instructions or of the coders hired to apply them.

## 12.3 Minimizing the Use of Default Categories

All researchers define their phenomena of interest in terms of the conceptions they have of them, the theories they consider applicable, the research questions they seek to answer, and they employ their own verbal and logical skills in devising suitable coding instructions. However, the world of phenomena is rarely organized in researcher's terms. Developing coding instructions from a sample of texts to be analyzed and testing their reliability may not cover all alternatives in the population of phenomena of analytical interest. What seems a plausible and exhaustively defined set of categories or values of a variable at the start of a research project may well be challenged by the emergence of unanticipated alternatives. For a far too obvious example, consider gender. The English language conceives gender as a binary attribute. This has not only made the life of transvestites, gays, and lesbians difficult but also presents problems when using such binaries when the gender of individuals is unknown or irrelevant. For example when individuals are referred to as occupying an office, as being of a certain age, or having committed a crime, it is not clear whether this is a male, female, or someone who wants not to be characterized as either. Also, referring to a child implies a range of ages but leaves open whether it is a boy or a girl. We encountered that problem when coding the gender of robots and comic characters. The customary strategy for coding unexpected kinds of phenomena is to add a default "catch all other" category to the set of conceptually motivated values of a variable, already discussed in Chapter 7, "unknown," or "true for all categories," "not applicable," or add an unlike number like "999" to a Likert scale for "cannot code."

The use of such default categories admits the inability of the designers of the coding instruments to anticipate all variations of phenomena of analytical interest. It preserves the requirement of variables to be exhaustively defined, does not bias the data by omissions, and it avoids forcing observers to apply categories that do not fit the given phenomena. Logically, this would preserve the distinctions among all other values; however, it invites a way for coders to proceed without much trying and is well known to be a likely source of unreliability.

A way to test whether a default category captures all alternatives that are clearly distinct from the defined categories or values of a variable is to construct a 2-by-2 coincidence matrix consisting of all information-bearing categories into one, say +, and the default category denoted by $\phi$:

| **Values**: | $\phi$ | $+$ | |
|---|---|---|---|
| $\phi$ | $n_{\phi\phi}$ | $n_{\phi+}$ | $n_{\phi}$ |
| $+$ | $n_{+\phi}$ | $n_{++}$ | $n_{+}$ |
| | $n_{\phi}$ | $n_{+}$ | $n_{..}$ |

for which (2.9) defines the binary agreement coefficient, here of the distinction between the default category and all otherwise meaningful categories or values in a variable by:

$$_{\phi|c}\alpha_{\text{binary}} = 1 - \frac{_cD_o}{_cD_e} = 1 - \frac{n_{+\phi}}{\frac{1}{n_{..}-1} n_+ n_\phi} \qquad (12.5)$$

There actually are quite interesting conclusions to be drawn from this simple agreement measure:

- A relatively high observed frequency $n_\phi$ in the default category reveals that the information-bearing categories of a variable cover only a small proportion of phenomena that the definition of a variable embraces. $n_\phi$ is the number for which the coding instructions fails to provide detailed definitions. It signals the need to examine whether the omission of such distinction imposes serious statistical limitations on the analysis of the resulting data. For example, if a variable is to record the race of say demonstrators with the stereotypical categories "Black," "White," and "Other," depending on where it is applied, in the U.S. this binary would not be applicable to very many people. The U.S. is the home of Asian, Indian, South American, Malaysian minorities, people with mixed backgrounds, and people that do not wish to be categorized either way. Statistically, the Other category is uninformative. If it turns out to be large, it may severely limit the use of this variable.

The ideal variables consist of a set of mutually exclusive and exhaustively defined values. While a default category solves the logical requirement for exhaustiveness, this logical benefit is countered by not knowing what ends up in the default category. There is the need for balancing reliability with the loss of analyzable details.

- If ${}_{\phi|c}\alpha_{\text{binary}}$ is perfect or relatively high, there is a good chance that the unanticipated phenomena that end up in the default category are clearly distinguishable alternatives. This condition leads to the recommendation of expanding the coding instructions by adding definitions of the missing categories to the variable and thereby minimizing the frequency $n_\phi$ of the no-information bearing default category. Another possibility is to examine just what categories of phenomena turn up in the reliable Other category and reanalyze the content of the default category and expand the set of defined values to include unanticipated but clearly distinct phenomena during the development phase of a research or after data were collected.
- If ${}_{\phi|c}\,\alpha_{\text{binary}}$ is low, this would indicate not only that there are phenomena that cannot be coded with the defined categories, but also that the supposedly informative categories are not too clearly defined and confusions among them spill over into the default category. This should signal the need to reconceptualize the information-bearing values.

Section 12.10 continues to discuss the options available once data have been collected.

## 12.4 Concerning the Use of Appropriate Metrics for a Variable

Categorizing units of analysis generates the most basic forms of data. They are often routed in nouns of natural language that merely distinguishes one category from what it is not. However, many phenomena possess more complicated qualities. For one, phenomena may occur in conceptually distinct dimensions, calling for data in terms of logically independent variables. Most α-agreements are defined primarily for individual and single-valued variables. Chapter 7 argued for the importance of allowing for the multi-valued coding of data.

This section addresses sources of unreliability due to mismatches between coding instructions' definitions of the ordering of single-valued variables and the nature of the phenomena to be recorded. Choices among such orderings need to satisfy two conditions.

On the one hand, an ordering of values needs to reflect the nature of the phenomena being turned into data. Attitudes regarding an issue may be strong, weak, or absent. Unemployment is discussed as a percentage of an employable population. Temperatures tend to be expressed in numerical degrees. Yet, listing codes alphabetically is an ordering that has nothing to do with their meanings, and using numerical codes may not imply a scale; for example, social security numbers or the numbers on the jerseys of football players.

On the other hand, all analytical techniques make specific demands on the nature of the data they are designed to handle. Computing correlations require interval scales. Network studies require bi-valued relationships between nodes. The availability of analytical techniques calls on researchers to formulate coding instructions that create data capable of being analyzed by the analytical techniques of their choice, which may but should not mismatch with the nature of the phenomena of interest. Emphasizing the metric of a chosen analytical technique over the metric underlying the phenomena of interest can create computable but unreliable data.

Since Stevens' (1946) groundbreaking distinctions among levels of measurement, the metrical properties of data are defined in terms of the mathematical operations applicable to them. We define them in terms of metric difference functions, (0.5), (5.7), and from (5.12) to (5.16). For identifying unreliabilities due to these mismatches, we are interested here only in the strength of their relationships with the nominal metric being the weakest and the ratio metric being the strongest.

$$\text{Nominal metric} < \text{Ordinal metric} < \underbrace{\text{Interval metric} < \text{Ratio metric}}_{\substack{\text{Polar metric} \\ \text{Circular metric}}} \qquad (12.6)$$

Stevens' work led to dual conclusions:

> *Analytical techniques that depend on the strength of one metric must not be applied to data whose metric strength is lower than theirs.*
>
> *The strength of the metric of an α-coefficient applied to given data should not exceed, ideally match the strengths of the metric that the intended analytical techniques require.* (12.7)

The first proposition acknowledges that whenever the metric strength of an analytical technique exceeds the metric strength of the data, it produces uninterpretable results – like averaging telephone numbers. Averaging assumes values to be on a scale. Telephone numbers are not. This applies also to the calculation of agreement. The metric strength of a ${}_c\alpha$-coefficient may be less than but must not exceed the metric strength of the data in question.

The second proposition is based on the conception of reliability as a measure of the extent to which data do not pollute their subsequent analysis. Therefore, the metric of the measured disagreements should conform to both; to the metric required by the chosen analytical techniques and the metric underlying the coded phenomena. It would be deceptive to publish agreement coefficients that are unrelated to what the data are used for.

The relationships among the ${}_c\alpha$-coefficients for different metrics are as follows. Whenever agreement is either perfect, ${}_c\alpha = 1$, i.e., pairs of matching values occur in the diagonals of coincidence or contingency matrices; or the mismatching values cells are randomly distributed in the off-diagonal cells of such matrices, they cannot reveal any ordering. In both cases the ${}_c\alpha_{metric}$ -coefficients for different metrics equal the ${}_c\alpha_{nominal}$ –coefficient. Graphically in Figure 12.1:

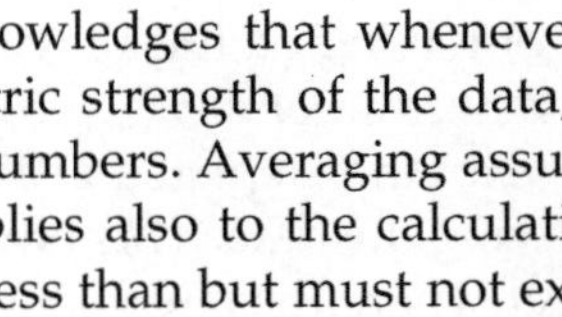

${}_c\alpha = 1.000$ ${}_c\alpha_{nominal} = {}_c\alpha_{ordinal} = {}_c\alpha_{interval} = {}_c\alpha_{ratio} = {}_c\alpha_{polar} = {}_c\alpha_{circular}$

**FIGURE 12.1**
Conditions for Which All α-Coefficients Are Equal

Any deviation from these two situations is registered in ${}_c\alpha_{metric}$-specific ways. ${}_c\alpha_{metric}$ increases when the distribution of disagreements in the off-diagonal cells of coincidence or contingency matrices conforms to its metric differences and it decreases when this distribution deviates or contradicts what its metric favors. While this is so for all metrics, Figure 12.2's graphic depiction concerns only the two most easily visualizable kinds:

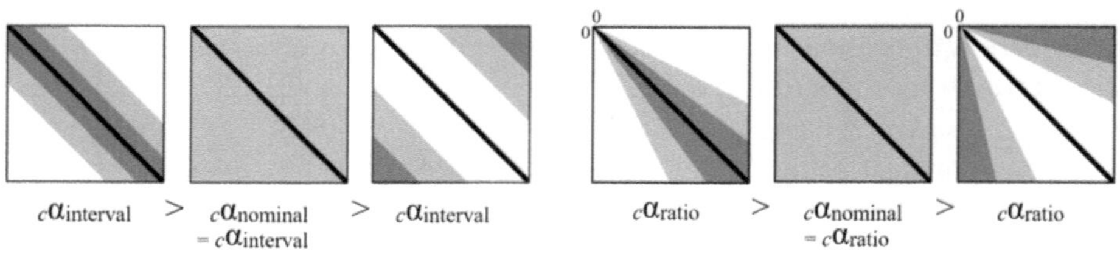


**FIGURE 12.2**
Conditions for Which α-Coefficients Differ Depending on Their Metric

This relationship offers clues as to how coders conceive of the metric nature of the phenomena they are describing irrespective of the metric that written coding instructions specify. Evidence of noncompliance with how coding instructions conceptualize the metric of a variable gives clues to what may have to be reformulated. Sometimes, there may be good reasons to trust competent observers' judgments and choose techniques for analyzing data that match the metric of the highest ${}_c\alpha_{metric}$-coefficient.

An important epistemological caveat of relying on the above considerations stems from the fact that the significance of the differences among ${}_c\alpha_{metric}$-coefficients disappears as the reliability of a variable increases. In the extreme, depicted in Table 12.1, perfect reliability reveals nothing about the metric underlying the data. This limits the above consideration to the earlier stages of a research project where disagreements are expected to be larger and remedial actions are more readily available.

## 12.5 Identifying Ambiguous Categories and Values

Variables consist of single- or multi-valued categories or scale values which, when unambiguously defined and applied alike by all observers, present no reliability problems for the variable in question. However, experiences have it that verbal definitions are often logically perfect but prove themselves unworkable in the face of actual situations for which they don't quite seem to fit. Designers of coding instructions can rarely imagine all, especially rare and unusual, phenomena that fall into the definition of a variable but find no clear place in its distinctions.

Recognizing this problem, designers of coding instructions tend to add a default category to catch situations for which the categories or values of a variable have no place. The possibility of expanding the definitions of values of a variable into the default category has already been discussed in Section 12.3. However, the problem of ambiguities among categories is another methodological challenge.

One response to ambiguities in natural language to bypass it by defining coding instructions with less linguistically dependent means of identification. For example, in a study of framing of cancer-related information in public discourse, Capella et al.

(Krippendorff and Bock, 2009: 253–265) developed extremely careful coding instruction of scaling sentences for whether they included:

(1) Probabilistic key words
(2) Slightly probabilistic key words
(3) Slightly deterministic key words
(4) Deterministic key words.

They realized that the wording of this probabilistic/deterministic dimension, a key variable in their research project, is uncommon in everyday language, quite difficult to handle, and would result in unreliable data had they not shifted the process of ranking sentences in these terms to another process. They developed four extensive lists of phrases and asked their coders to merely check in which lists the observed phrases occurred. While such lists rarely are exhaustive, unlisted phrases were more easily identified as belonging to one list or another.

This example focused on a solution to compensate for the occurrence of ambiguity. We now shift to two ways of identifying them quantitatively in reliability data.

Let a matrix of coincidences be as in the original Table 2.2:

| **Values:** | 1 | . | . | $k$ | . | $v$ | |
|---|---|---|---|---|---|---|---|
| 1 | $n_{11}$ | . | . | $n_{1k}$ | . | $n_{1v}$ | $n_1$ |
| . | . | . | . | . | . | . | . |
| $c$ | $n_{c1}$ | . | . | $n_{ck}$ | . | $n_{cv}$ | $n_c = \sum_{k=1}^{v} n_{ck}$ |
| $k$ | $n_{k1}$ | . | . | $n_{kk}$ | . | $n_{kv}$ | $n_k$ |
| . | . | . | . | . | . | . | . |
| $v$ | $n_{v1}$ | . | . | $n_{vk}$ | . | $n_{vv}$ | $n_v$ |
| | $n_1$ | . | . | $n_k$ | . | $n_v$ | $n_{..} = \sum_{c=1}^{v}\sum_{k=1}^{v} n_{ck}$ |

**TABLE 12.2**
Observed Coincidences, Column Highlighted for One Designated Value $k$

For replicability, the disagreements between all values $c$ and one target value $k$ are evident in the distribution of differences in the $k^{\text{th}}$ column in Table 12.2. It can be derived from the general form of (6.13) by omitting the sum of all columns $k$ from it:

$$_{(k)c}\alpha_{\text{metric}} = 1 - \frac{_{(k)c}D_o}{_{(k)c}D_e} = 1 - \frac{\sum_{c=1}^{v} n_{ck\ \text{metric}}\delta_{ck}^2}{\sum_{c=1}^{v} e_{ck\ \text{metric}}\delta_{ck}^2} = 1 - \frac{\sum_{c=1}^{v} n_{ck\ \text{metric}}\delta_{ck}^2}{\frac{1}{n_{..}-1} n_k \sum_{c=1}^{v} n_{c\ \text{metric}}\delta_{ck}^2} \tag{12.8}$$

(12.8) can be made more obvious in its nominal versions with its expected coincidence defined by (6.10): $e_{kk} = n_k\left(n_k\ -1\ \right)/\left(n_{..}\ -1\ \right)$:

$$_{(k)c}\alpha_{\text{nominal}} = 1 - \frac{_{(k)c}D_o}{_{(k)c}D_e} = 1 - \frac{\sum_{c=1}^{v} n_{ck\ \text{nominal}}\delta_{ck}^2}{\sum_{c=1}^{v} e_{ck\ \text{nominal}}\delta_{ck}^2} = 1 - \frac{n_k - n_{kk}}{n_k - e_{kk}} = 1 - \frac{n_k - n_{kk}}{\frac{n_k\left(n_{..} - n_k\right)}{n_{..}-1}} \tag{12.9}$$

(12.9) reveals the observed disagreement to be the difference between the ideal of perfect agreement $n_k$ and the actually observed coincidence $n_{kk}$. The expected disagreement takes the same form, but in reference to the expected coincidence $e_{kk}$. The fourth expression replaces $e_{kk}$ by its definition (6.10) quoted above.

The accuracy of individual values follows the same logic except that the values $k$ of the standard and $c$ of the replications belong to different sets and pose no problems of self-pairing. Whereas for replicability the choice of rows and columns is of no consequence, because contingencies $x_{ck}$ are asymmetrical, it is important to measure the accuracy of individual values with reference to $k$ as the value of the trusted standard:

$$_{(k)c}\alpha^{s}_{\text{metric}} = 1 - \frac{_{(k)c}D^{s}_{o}}{_{(k)c}D^{s}_{e}} = 1 - \frac{\sum_{c=1}^{v} x_{ck\ \text{metric}}\delta^{2}_{ck}}{\frac{1}{x_{\cdot\cdot}} x_{\cdot k} \sum_{c=1}^{v} x_{c\cdot\ \text{metric}}\delta^{2}_{ck}} \tag{12.10}$$

Applied to each value of a variable, (12.8) and (12.10) may reveal which categories or values are responsible for the $_c\alpha$-agreement of their variable and may guide efforts to clarify their definitions.

A drawback of this measure is that ambiguity, while conceptually associated as the quality of a value, is always manifest in confusions with particular other values, failing to distinguish among them. (12.8) and (12.10) measure the confusion of one with all other values, some of which may not be a problem to be solved.

Another approach that overcomes this drawback is to examine confusions among pairs of mismatching values. When data are unordered, one can calculate this reliability for all mismatching pairs. When values are ordered, only neighboring values should be considered in such examinations. In either case, the reliability of pairs of values becomes a binary reliability coefficient which does not respond to their metric differences. For replicability, the agreement coefficient of pairwise $c$–$k$ confusions is defined by:

$$_{c-k}\alpha_{\text{binary}} = 1 - \frac{_{c-k}D_{o}}{_{c-k}D_{e}} = 1 - \frac{n_{ck}}{\frac{(n_{cc}+n_{ck})(n_{kc}+n_{kk})}{n_{cc}+n_{ck}+n_{kc}+n_{kk}-1}} \tag{12.11}$$

The indicative ability of both measures may be demonstrated by means of the following 6-by-6 matrix of coincidences of unordered categories. For simplicity (12.11) is computed only for three neighboring values; between 4–5 and 5–6, which are perfect, and between 2–3 and 3–4, which together with 4–6 are the cause of considerable disagreement.

| **Values:** | 1 | 2 | 3 | 4 | 5 | 6 | |
|---|---|---|---|---|---|---|---|
| 1 | 32 | 5 | 7 | 6 | . | 4 | 54 |
| 2 | 5 | 28 | 18 | . | . | 4 | 55 |
| 3 | 7 | 18 | 20 | 24 | . | . | 69 |
| 4 | 6 | . | 24 | 30 | . | 28 | 88 |
| 5 | . | . | . | . | 58 | . | 58 |
| 6 | 4 | 4 | . | 28 | . | 40 | 76 |
| | 54 | 55 | 69 | 88 | 58 | 76 | 400 |

The overall replicability of these coincidences is ${}_c\alpha_{\text{nominal}} = 0.451$ against which its component measures can be judged:

$${}_{(1)c}\alpha_{\text{nominal}} = 1 - \frac{54-32}{\frac{54(400-54)}{400-1}} = 0.530 \qquad {}_{1-2}\alpha_{\text{binary}} = 1 - \frac{5}{\frac{(32+5)(5+28)}{32+5+5+28-1}} = 0.709$$

$${}_{(2)c}\alpha_{\text{nominal}} = 0.432 \qquad {}_{1-3}\alpha_{\text{binary}} = 0.568$$

$${}_{(3)c}\alpha_{\text{nominal}} = 0.144 \qquad {}_{1-4}\alpha_{\text{binary}} = 0.680$$

$${}_{(4)c}\alpha_{\text{nominal}} = 0.070 \qquad {}_{1-6}\alpha_{\text{binary}} = 0.801$$

$${}_{(5)c}\alpha_{\text{nominal}} = 1.000 \qquad {}_{2-3}\alpha_{\text{binary}} = 0.145$$

$${}_{(6)c}\alpha_{\text{nominal}} = 0.417 \qquad {}_{2-6}\alpha_{\text{binary}} = 0.787$$

$${}_{3-4}\alpha_{\text{binary}} = 0.020$$

$${}_{4-5}\alpha_{\text{binary}} = 1.000$$

$${}_{4-6}\alpha_{\text{binary}} = 0.113$$

$${}_{5-6}\alpha_{\text{binary}} = 1.000$$

Visual inspection of the matrix of coincidences easily confirms that value 5 is perfectly reliable. Values 2 and 6 are closest to ${}_c\alpha_{\text{nominal}} = 0.451$. The pair comparisons confirm the general picture but with more details. Value 5 is not confused with any other value. Confusions with value 1 and between 2 and 6 are above ${}_c\alpha_{\text{nominal}} = 0.451$ while all other confusions measure below ${}_c\alpha_{\text{nominal}}$.

The accuracy of *c–k* confusions in (12.12) is analogue to the above, responding to the asymmetry of contingencies and acknowledging that the pairs of values stem from different sets:

$${}_{c-k}\alpha^{S}_{\text{nominal}} = 1 - \frac{{}_{c-k}D^{S}_{o}}{{}_{c-k}D^{S}_{e}} = 1 - \frac{x_{ck} + x_{kc}}{\frac{(x_{cc}+x_{ck})(x_{kc}+x_{kk}) + (x_{kc}+x_{kk})(x_{cc}+x_{ck})}{x_{cc}+x_{ck}+x_{kc}+x_{kk}}} \qquad (12.12)$$

The two reliability measures for individual and pairwise confusions of values are mainly diagnostic. Section 12.9 will take up the problem of lumping categories or values, which is one of the few remedial actions possible after data have been generated and disagreements are unequally distributed.

## 12.6 Selecting the Most Reliable Replications, Observers, or Judges

If a research project involves generating data by human observers, measuring instruments or algorithms for identifying relevant qualities among textual and other phenomena, the most common problem is to assemble teams of human coders and software that can work together and are demonstrably able to replicate the data.

One problem that cannot be neglected is the need to hire observers and employ equipment that are both widely available, and train the observers with materials that can be communicated to those who might want to replicate the research or extend its findings into new dimensions. Unique experts, coders with unusual backgrounds, or intimate familiarity with the principal researcher's objective would not qualify as these qualities are unlikely to be found elsewhere and would make replications of the coding process elsewhere difficult. The criteria used for hiring human coders, the training sessions during which they learn what to do, as well as the computational aids used to record their judgments need to be published or at least be communicable to potential stakeholders.

Nevertheless, human beings bring implicit qualities to the task of generating data – some favorable to reliable work, others not – and it is advisable and in fact the general practice to hire more coders than eventually needed and let reliability tests decide who will do the job.

While developing the coding instructions, the main feedback is likely to come from groups of coders, preferably more than two, and the reliability data they generate should be treated as exploratory, at least to begin with.

However, measures of the replicability, pairing all independently working observers with each other, would not distinguish among their unequal qualities. One practical proposal is to compute the replicabilities of all $m-1$ out of a total of m observers and get an idea of whose omission increases the measured agreement among the others. Even if used repeatedly, converging to a group of which each member is better than those discontinued, it does not provide a measure related to how reliably one observer works in relation to what all have in common.

The surrogacy measures, developed in Section 6.6.4, are most suitable to evaluate the reliability of individual observers or methods of generating data relative to what all others have in common, their collective judgments, whether in the form of majorities, medians, or means. As discussed in Chapter 6, surrogacy inverts the customary direction of justifications, in accuracy from a trusted standard to inaccurate observers, in surrogacy from the trusted collective judgments of several observers to the candidate to be trusted as replacement.

The solution is found in Section 6.2.3 which introduced the idea of surrogacy with the help of binary data. Section 6.3.3 expanded the binary to more values among which computational conceptions of collectivity – majorities, modes, medians, and means – matter. Section 6.3.4 defined the measure of surrogacies ${}_{c}\alpha^{sm}_{\text{metric}}$ that could select among the most qualified candidates for generating reliable data, more generally the most trustworthy replication among many.

## 12.7 Maintaining the Reliability of Data While Research Is in Progress

Already mentioned was the need to ensure the reliability of ongoing coding practices, especially in larger projects. After a protocol for generating data was adopted and is being applied to a larger set of phenomena, actions to counter growing unreliabilities tend to be limited to corrections of unforeseeable decays in the quality of the generated data. We distinguish two cases.

The first focuses on the observers who are hired to face the phenomena of analytical interest, distinguish among them, code the resulting units, and record what they have seen, read, or judged. It is well established that long-term human coders develop habits, discover shortcuts, and enact expectations as they proceed. These much-cherished human qualities have the effect of lowering their reliability as time progresses. It is important, therefore, to test the process intermittently. If these changes in practices of coding are shared, they would not surface when the generated data are tested for their replicability. Therefore it is important to measure coders' performance against a standard of correctness. Such standards may be decided by panels of experts or previous coding before coders had become accustomed to the work and unknowingly deviated from their written instructions. Accuracy measures are developed in Section 6.2.2 for binary data, and in Section 6.3.2 for other kinds, found also in other sections of this book. Here, inaccuracies encountered serve mainly as warnings to coders turned unreliable – only in the extreme, replacing them.

The second example of using accuracy measures for controlling the quality of data comes from the application of crowd coding. Crowd-coded data are obtained from anonymous internet users of unknown qualifications, background knowledges, and commitment to do the best job they can. For these reasons it is essential to test their qualifications, not by asking questions they might not answer truthfully, but by introducing coding tasks similar to what researchers seek for their project but for which they know the correct answers. Accurately answering these test questions then serves as the criterion to accept or dismiss the crowd coder's contributions to the data to be analyzed. Accuracy measures mentioned above are appropriate for this purpose as well. Cut-off criteria will be discussed in Chapter 14.

## 12.8 Selecting the Most Reliable Data From Multiple Contributors

Sections 12.1 to 12.5 discuss testing the reliabilities of data-making processes in preparation for a full-blown research effort. At this early stage, replications rely on a sample of the population of data whose reliability is in question. If reliability tests on that sample are satisfactory, it is not unreasonable to generate the whole population of data by these means, proved to be reliable, except for intermittently checking whether the process is still stable, as discussed in the previous section.

One frequently raised question is whether the replications used to test the reliability of a sample could be included in an analysis, and which of them. If coding instructions are developed in conjunction with a sample of phenomena and are being modified where needed, especially when coders are aware of these modifications – may even have suggested them – then the inter-coder agreements among them can no longer be interpreted as indicators of the reliabilities of the subsequently analyzed data. These coders may refer to the communicable coding instructions but respond to what they had learned during the process of training, offering feedback, and thus participating in their development. This does not speak against including the phenomena on which reliability was tested, only against the data obtained during the development of reliable coding instructions. When coded by other and unbiased observers, passing reliability tests, there is no problem to include them in the final analysis.

When researchers are confident to proceed after initially satisfactory reliability tests and have good reasons to consider the reliability data to be representative of the population of data to be analyzed, being able to report satisfactory data reliabilities does not answer the question of which of the replications to include in the data to be analyzed. Similarly, if the coding of all phenomena is replicated, there remains the question of which of the replications should be analyzed.

When two observers, judges, raters, or measuring instruments are involved, their replicability measure provides no clue as to who the better coder would be. However, when three or more observers have contributed to the reliability data, the one with the highest surrogacy among them could arguably be considered to provide the most trusted replication. According to Section 6.3.4 surrogacy measures the extent to which a possible substitute agrees with what characterizes all replications.

However, while the highest surrogacy of any one replication would arguably be the best choice, surrogacy measures should not be confused with publishable measures of data reliability. As discussed in Section 6.3.4, surrogacy is not a reliability measure of data. It indicates the reliability of a potential surrogate to fit what the remaining the replications collectively share, their majorities, modes, medians, or means, not the actual coding by all. It follows that, while surrogacy measures are enormously helpful in selecting the best replication, it must not be confused with publishable replicability measures.

Replicating all data will always give the best assurances of knowing reliability of data – avoiding the problem of representative sampling of reliability data – but when data are large and the phenomena to be turned into data are complex, replicating all coding gives the best assurance that the data are reliable as measured.

With several replications available, it has been tempting and become a practice to reconcile disagreements among coders by mutual consultation after completing coding of all data. This involves all coders jointly examining their independently obtained disagreements, discussing each case and coming up with what would be the collective judgment of all observers involved. Intuitively, resolving disagreements by mutual consent is arguably closer to perfect agreement than by independent judgments, and this may well be so as observers may recognize where they failed for whatever reason and become convinced as to how some phenomena should have been interpreted.

Creating one jointly agreed-upon set of coded data out of several slightly diverging ones certainly gives the impression of being the perfect solution; however, it may not be so in fact. I carried out experiments with a large number of observers coding a set of documents. Afterwards, several groups of coders worked separately to reconcile their disagreements. Then the replicability of the reconciled data was measured. It turned out to be not that much better than without reconciliation. When looking at the taped reconciliation effort, I noticed that theses reconciliations were guided not only by logical arguments about the correct interpretations of the coding instructions, but also by the emerging social relationships among observers. Being made to realize mistakes made introduces inequalities among observers. Some were more articulate than others, and observers with more prestige were more likely yielded to. Even where inequalities were not played out, disagreements were often resolved by saying the equivalent of "You let me have it last time, I yield to you in this case." Such social processes can direct reconciliation efforts in ways that have less to do with the phenomena to be represented in the data. Without providing agreement measures, the claim that reconciled data are more reliable has to be rejected. In the absence of replicability measures of independent reconciliation efforts, the replicability measured before reconciliation is the only publishable measure – subjective feelings to have done the best one can do notwithstanding.

## 12.9 Deselecting Variables That Fall Short of the Required Reliability

As discussed in Section 12.1, most reliability measures concern independent variables. Their independence is logical. Their use to describe phenomena that may not naturally comply with logically independent accounts has been identified as a hindrance to generalizing their reliabilities across different semantic domains. This important point presents a problem for analyzing the relationship among variables – associations, correlations, factor analyses, network analyses, and more.

After data have been generated, finding some variables to be below acceptable levels and contemplating discarding them altogether, leaves researchers very few remedial options. I will sketch four that concern whole variables.

The first is pursuing less ambitious research questions, answerable with fewer but reliable variables. Based on the practice of developing efficient coding instructions that would represent the chosen phenomena of analytical interest in all relevant dimensions needed to answer adopted research questions, losing variables that were expected to provide the needed information most likely has the effect of not reaching the stated goal of the research. Short of giving up – although this may well be an option and starting all over – the more common response is to reformulate the research question, reducing its scope to what the reliable data might be able to answer. This is often disappointing not only to their principal investigators but also to funding agencies who may have posed larger questions and expected corresponding answers.

The second is to downgrade the validity of the outcome. Reducing the reliability standards to include variables that would otherwise have to be rejected may mean that the analytical results can no longer be considered as making contributions of hoped-for significance. The option to lower the reliability standards and include variables otherwise rejected has also the effect of downgrading the potential validity of proposed research and reducing the social relevance of the proposed research. In academia this option is often justified by declaring a research merely exploratory in the hope of stimulating discussion of the ideas and generating more reliable support in future efforts. However, the unfortunate reality is that researchers in such situations may be embarrassed to report the unreliability of their data, omit mentioning it altogether, and mislead readers into believing findings to be valid, not merely exploratory. Section 15.4 will discuss an alternative to relying on the reliability of data, but this path requires another kind of rigor.

The third is to anticipate possible unreliabilities and build enough redundancy into the data so that the loss of some unreliable variables can be compensated by more reliable others providing the same information perhaps in slightly different forms. Redundancies may support alternative paths to what could bring a research project to valid conclusion. This is by far the most productive strategy but by no means a foolproof one. In practice, it is difficult to anticipate which variables are likely to be in trouble and proposed alternatives may not be perfect either, leading back to the first option of downgrading the research questions. However, building redundancies into coding instructions is the best way to prevent unreliable variables from ruining a project altogether.

The fourth is to examine the disagreements among the values of variables and try to trade the information they were expected to provide for increasing the reliability of the remaining values. This will be the topic of the following two sections.

## 12.10 Separating Default (No-Information) Categories From Those That Matter

In scholarly work, reliable data need to inform conclusions about phenomena that researches claim to be their concerns. The most commonly discussed obstacle of data being informative of the desired conclusions of a research project stems from the lack of agreement among observers about what they observed, read, judged, or measured and recorded in code.

However, there is another obstacle to data being informative: the appearance of unanticipated and often rare phenomena; unanticipated because replicability requires communicable coding instructions that are predefined by definition. Unanticipated phenomena may well fall into the semantic domain of a well-defined variable but have no place among its values and therefore cause disagreements.

The most common solution for coping with such phenomena is to add a default category to the values that constitute a variable. Default categories are meant to preserve the logical requirement of variables to be exhaustive and their values to be mutually exclusive. However, default categories inform the conclusions of a research only by negating the applicability of any of the other values of a variable: "other," "not applicable," "absent," "undecidable," or "unknown," in public opinion surveys "no answer," and when rating with numerical scales "999." Section 12.3 discussed the use of such categories during the development period of research. Here we address remedial actions on three undesirable constraints on the analysis of variables that contain such default categories.

First, unless a variable is a binary one, defined by the presence or absence of a single quality, the frequencies of default categories become increasingly meaningless the more values they deny. The category "not applicable" may occur in contrast with any number of well-defined values. It assumes different meanings in the contexts of different values and prevents the use of most analytical techniques; computing associations or correlations, as one example. The frequencies of default categories are simply uninformative about phenomena possessing no metric. For example, computing averages, in variance analyses; network conceptions effectively destroy a variable that may have been conceived of as providing information toward a conclusion of a proposed analysis.

A small anecdote comes to mind. At the beginning of a large content analysis of TV violence, we defined the variables of our concerns. The very first reliability test of these variables turned out to be disastrous. We examined their coincidence matrices in the hope of being able to identify ambiguous categories, lump them into one, and saving the remainder for further refinement. After finding such possibilities in the frequency tables, we looked into their definitions, and were dismayed to find several variables reduced to the distinction between "applicable" and "not applicable." Such a distinction said more about our own conceptual inabilities than about TV violence. Reporting on the frequencies of "other," "cannot code," "undecidable," or "unknown" would not tell anyone anything about television violence but certify our inability to operationalize our research questions.

Somewhat idealized, we had hoped our coincidence matrices would be of the first kind, did not expect data of the second kind, but could have lived with data of the third kind. Obviously these cases can be considered entirely independent of each other.

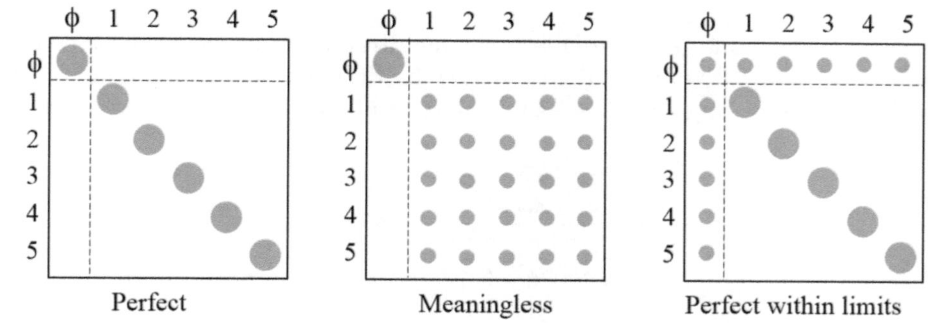


Second, as soon as a default category is introduced into a variable with ordered values, ordered metrics are no longer analytically available. Consider the variable age. A person's age is a positive number that enables computing averages or encourages examinations of their distributions on a scale. But if researchers find a word like "unknown" or "999," the analytical possibilities of analyzing age shrink drastically. The inclusion of a default category forces the analysts to ignore all intended metrics of its values, having to analyze such data as unordered nominal categories.

Third, it is well known that observers vary considerably regarding the thoroughness with which they approach their coding tasks. When the identification of a textual unit with one of the defined categories is obvious, observers are not tempted to consider the default category as an option. However, as soon as the phenomena to be coded require more than the usual attention, the availability of a default category offers coders the possibility of avoiding difficult decisions. It is simply easier to check "cannot code" than to carefully weigh where a phenomenon would belong. It follows that the use of default categories is heavily influenced by the attentiveness and dedication of observers. Such unpredictable human qualities are difficult to measure, and certainly should not enter the data where they introduce unreliabilities that can disqualify a variable from being taken seriously.

Once data have been generated with variables that include a default category, the only real actions researchers can take are to analytically separate the information bearing values from the category that has little analytical value. This brings us to how to measure their reliabilities in such situations. One can presume that a trusted standard is unlikely to make use of default categories. Therefore we start with the coincidence matrix of such variables in the center of Figure 12.3, with the default category labeled $\phi$:

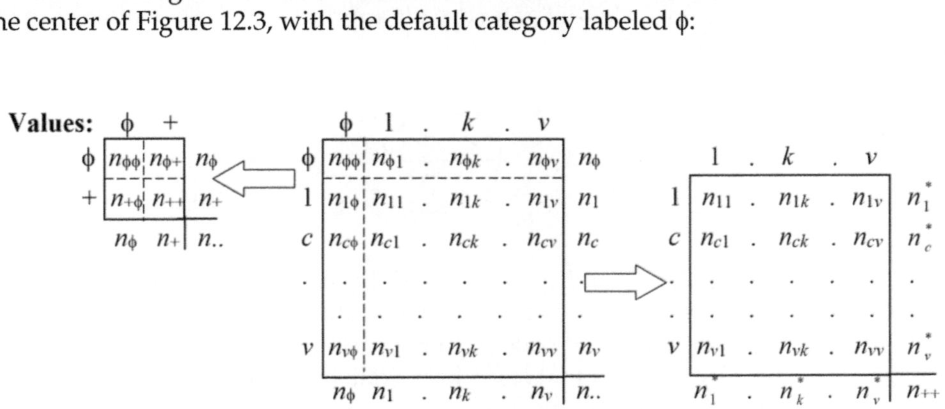


**FIGURE 12.3**
Coincidence Matrices With and Without a Default on Uninformative Category $\phi$

Inasmuch as default categories possess no ordinal qualities, as discussed in Chapter 5, and the metric applicable to a variable is always the metric of its weakest value, the agreement coefficient for evaluating coincidence matrices that include a default category is always limited to treating all values as unordered categories. According to (2.7):

$$_c\alpha_{\text{nominal}} = 1 - \frac{_cD_o}{_cD_e} = 1 - \frac{\sum_{c=\phi}^{v}\sum_{k\neq c}^{v} n_{ck}}{\frac{1}{n_{..}-1}\sum_{c=\phi}^{v} n_c \sum_{k\neq c}^{v} n_k} = 1 - \frac{n_{..} - \sum_{c=\phi}^{v} n_{cc}}{n_{..} - \sum_{c=\phi}^{v} \frac{n_c(n_c-1)}{n_{..}-1}} \tag{12.13}$$

Note that its double sums cover all rows and columns from $\phi$ to $v$ of the matrix in the center of Figure 12.3.

The 2-by-2 coincidence matrix on the left of Figure 12.3, already discussed in Section 2.2 and used in Section 12.3, can now be seen as representing the sum of cell entries in the four quadrants of the coincidence matrix involving all values, indicated by dotted lines:

$n_{\phi\phi}$ = The frequency of the default category $\phi$ coinciding with itself in the larger matrix

$n_{\phi+} = n_{+\phi} = \sum_{c-1}^{v} n_{c\phi}$ = all coincidences of informative values with $\phi$

$n_{++} = \sum_{c-1}^{v}\sum_{k=1}^{v} n_{ck}$ = the sum of coincidences among all informative values

Introduced as (2.9) and discussed in Section 12.3, the binary agreement on the distinction between all informative values + and the default category $\phi$ becomes:

$$_{\phi|c}\alpha_{\text{binary}} = 1 - \frac{_cD_o}{_cD_e} = 1 - \frac{n_{+\phi}}{\frac{1}{n_{..}-1} n_+ n_\phi}$$

It measures the reliability of the distinction between the default category and all information-bearing values summed into one category. Its implication has been discussed above and has now been defined as an $\alpha$-agreement.

Finally, the agreement among all information-bearing values, the values that are worth analyzing, and only these, is a submatrix of the coincidence matrix in the center of Figure 12.3. In the absence of a default category all appropriate single-valued difference functions are applicable. By (6.11):

$$_c\alpha_{\text{metric}} = 1 - \frac{_cD_o}{_cD_e} = 1 - \frac{\sum_{c=1}^{v}\sum_{k=1}^{v} n_{ck}\ {}_{\text{metric}}\delta_{ck}^2}{\frac{1}{n_{++}-1}\sum_{c=1}^{v} n_c^* \sum_{k=1}^{v} n_k^*\ {}_{\text{metric}}\delta_{ck}^2}$$

Evidently, (6.11) provides the reliability measure of the substantially defined values of a variable, the very values that could have an impact on the research findings and need to be reliable to lead to potentially valid conclusions. Decomposing the matrix in the center of Figure 12.3 into the 2-by-2 matrix that leads to the reliability of the default category and into the $v$-by-$v$ matrix of properly defined values has the advantage of now applying any appropriate metric to these values, otherwise restricted to the nominal metric.

The above referred to replicability data. However, the same logic of decomposing the matrix of all coincidences applies accuracy, surrogacy and decisiveness as well, relying on (6.14), (6.20), and (6.23). Section 8.2 spells out the special provisions needed when applying the above logic to reliability data for coding segments within partitions of textual continua. It also includes a numerical example for this kind of data.

The point of this section was to salvage the metric reliability of a variable with default categories by separating the reliability associated with the default category from the reliability of the information bearing units so valued. This was accomplished by considering the information-bearing data conditional on units not being judged uninformative. This makes the reliability of information-bearing data also conditional on the same pairs of judgments, effectively excluding the frequencies $n_{\phi+}+n_{+\phi}$ of disagreements between informative values and the default category shown in Figure 12.3. Now consider a research project with several such variables. It is empirically unlikely that in each variable the same units are pairably judged information bearing. The relationship between any two or more information-bearing parts of variables numerically reduces with the intersections of units judged informative – as more variables are to be linked the number of analyzable units supporting their analysis shrinks.

There is not that much one can do about these mathematical facts except to limit the use of default categories, and where this not possible, limit at least the number of such variables that an analysis needs to relate for its conclusions.

## 12.11 Lumping Ambiguous Values and Preserving Informative Distinctions

It is important to note that all reliability coefficients are proportions, in the case of α, of information in one variable about the phenomena to be analyzed. Proportions are averages, and averages do not reflect the variance of the values averaged. In reliable variables, all of their values contribute to the α-agreement measure 1.000, providing only information to an analysis. In imperfect variables reliability measures can hide unreliable values behind reliable ones. It follows that whenever a variable consists of more than two categories, it is advisable at least to check whether each value can be trusted and if not to take precautionary actions.

This section is concerned with salvaging variables that turn out to be of questionable reliability as a whole but contain values that could still make reliable contributions. This calls for measures that do not merely identify sources of correctable unreliabilities in the coding instructions, which was the aim of Section 12.5, but locate disagreements that could inform remedial actions to eliminate their effects on the analyzable data.

Most observed disagreements can be explained as the result of ambiguities, not necessarily evident in the logically distinct definition of values but in how well they fit the nature of the phenomena they are meant to represent. Linguistically, ambiguity tends to be attributed to particular definitions. However, in the process of generating data, it is always evident in the inability to distinguish one category or value from others, in empirical confusions among alternatives meant to be distinct. The reliability measures that would aid the analysis have to focus on the distinctions that are still viable within otherwise unreliable variables. Distinctions are also fundamental to concepts of information, by Shannon (Shannon & Weaver, 1949) measured in bits, here merely looked for in the hope of providing an analysis with the needed information about the phenomena attended to.

For example, in the following 7-by-7 matrix of observed coincidences measures ${}_c\alpha_{nominal} = 0.589$, which is significantly below the usual cut-off criteria for acceptable reliabilities.

| | *a* | *b* | *c* | *d* | *e* | *f* | *g* |
|---|---|---|---|---|---|---|---|
| *a* | 20 | . | . | . | . | . | . |
| *b* | . | 6 | 6 | . | 6 | . | 1 |
| *c* | . | 6 | 6 | . | 6 | . | 1 |
| *d* | . | . | . | 32 | . | 8 | 1 |
| *e* | . | 6 | 6 | . | 6 | . | 1 |
| *f* | . | . | . | 8 | . | 16 | . |
| *g* | . | 1 | 1 | 1 | 1 | . | 26 |

${}_c\alpha_{nominal} = 0.589$

Category *a* is evidently unambiguously distinct from all other categories, hence perfectly reliable individually, unlike all others. Category *g* is less so, disagreeing with *b, c, d*, and *e* but only once each. Among the other categories confusions are larger and not so obvious. They certainly are responsible for the less than acceptable reliability of the whole variable.

Categories labeled *b, c*, and *e* show a discernable pattern of total confusion between them. The most obvious response to such a finding is to search for explanations of why *b, c*, and *e* are virtually indistinguishable in these reliability data. Confusions of this nature may have syntactical, conceptual, or clerical origins which may be explored by looking into how they are phrased, exemplified, and encoded. However, after having generated all the data in anticipation of proceeding to their analysis, the issue is no longer what would improve their definitions in the coding instruction, but what if anything can be done to salvage the reliable parts of an overall unreliable variable. The questions we cannot answer here are: How would a project be affected if one were to lump them into one category? Without reference to a project, what we could ask and answer is whether the variable could be brought within an acceptable range of reliability by lumping ambiguous categories.

To identify unreliable distinctions among values, here among unordered categories that could possibly be lumped into one, thus creating a simpler but more reliable variable, the following suggests an algorithm that makes use of previously defined reliability coefficients and converges to a redefined variable with optimal reliability. It involves the following four steps:

1. Compute ${}_c\alpha_{nominal}$ on the whole coincidence matrix as a baseline.
2. Compute ${}_c\alpha_{binary}$ for all possible distinctions among the $v$ unordered categories. Stop if none of them is significantly unlike ${}_c\alpha_{nominal}$.
3. Select the distinction with the largest ${}_c\alpha_{binary}$
4. For each set of values distinguished by 3, one at a time, compute ${}_c\alpha_{binary}$ for all possible distinctions and go back to 3.

(12.14)

This algorithm generates a hierarchy of conditional binary distinctions of decreasing reliability. It stops when all individual categories have their places, giving the researcher the ability to choose a cut-off point or another alternative that balances the increasing distinctiveness (amounts of information possible) with the decreasing reliability of the resulting variable.

In the above numerical example, the largest ${}_c\alpha_{binary} = 1.000$ is measured for *a* | *b,c,d.e,f,g*. Among the remaining six categories, the next largest distinction measuring ${}_c\alpha_{binary} = 0.973$ is between *b,c,e,g* | *d,f*. The distinction *b,c e* | *g* measures larger than between *d* | *f*, i.e.,

${}_{c}\alpha_{binary}$ = 0.846 > ${}_{c}\alpha_{binary}$ = 0.475. The latter amounts to a significant drop. All binary distinctions between *b,c,e* measure ${}_{c}\alpha_{binary}$ = 0.060 and between any two, ${}_{c}\alpha_{binary}$ = 0.042. Graphically, the conditional ${}_{c}\alpha_{binary}$ of this procedure and the resulting ${}_{c}\alpha_{nominal}$ would give the researcher some choices.

| Hierarchy of max(${}_{c}\alpha_{binary}$) distinctions | | All distinctions | ${}_{c}\alpha_{nominal}$ |
|---|---|---|---|
| *a,b,c,d,e f,g* | | *a,b,c,d,e f,g* | |
| *a* \| *b,c,d,e,f,g* | ${}_{c}\alpha_{binary}$ = 1.000 | *a* \| *b,c,d,e,f,g* | ${}_{c}\alpha_{nominal}$ = 1.000 |
| *b,c,e,g* \| *d,f* | ${}_{c}\alpha_{binary}$ = 0.973 | *a* \| *b,c,e,g* \| *d,f* | ${}_{c}\alpha_{nominal}$ = 0.980 |
| *b.c,e* \| *g* | $\alpha_{binary}$ = 0.846 | *a* \| *b,c,e* \| *d,f* \| *g* | ${}_{c}\alpha_{nominal}$ = 0.912 |
| *d* \| *f* | ${}_{c}\alpha_{binary}$ = 0.475 | *a* \| *b,c,e* \| *d* \| *f* \| *g* | ${}_{c}\alpha_{nominal}$ = 0.820 |
| *b* \| *c,e*<br>*c* \| *b,e*<br>*e* \| *b,c* | ${}_{c}\alpha_{binary}$ = 0.060 | *a* \| *b* \| *c,e* \| *d* \| *f* \| *g*<br>*a* \| *c* \| *b,e* \| *d* \| *f* \| *g*<br>*a* \| *b,c* \| *d* \| *e* \| *f* \| *g* | ${}_{c}\alpha_{nominal}$ = 0.661 |
| *b* \| *c*<br>*b* \| *e*<br>*c* \| *e* | ${}_{c}\alpha_{binary}$ = 0.042 | *a* \| *b* \| *c* \| *d* \| *e* \| *f* \| *g* | ${}_{c}\alpha_{nominal}$ = 0.589 |

Interestingly, although entirely expected, any distinction between the three categories *b*, *c*, and *e* carries virtually no information and is in fact most detrimental to the agreement of this variable. Partitioning the set of values into four lumped categories: *a* | *b,c,e* | *d,f* | *g* would yield the following 4-by-4 coincidence matrix and measure a respectable ${}_{c}\alpha$-agreement:

| | *a* | *b c e* | *d f* | *g* |
|---|---|---|---|---|
| *a* | 20 | . | . | . |
| *b c e* | . | 54 | . | 3 |
| *d f* | . | . | 64 | 1 |
| *g* | . | 3 | 1 | 26 |

${}_{c}\alpha_{nominal}$ = 0.912

The decision of whether the loss of unreliable information in the 7-by-7 matrix due to lumping categories to a 4-by-4 matrix is worth the gain in reliability from ${}_{c}\alpha_{nominal}$ = 0.589 to 0.912 depends on whether the two lumped categories *b,c,e* and *d,f* can be made sense of in subsequent analyses. If lumping the categories *d* and *f* does not make sense conceptually, one could leave them distinct. Lumping only *b,c,e* would reduce the reliability of this variable further 0.912 to 0.820 which is still better than the original 0.589.

The procedure described in (12.14) was applied to unordered categories. In nominal data all distinctions among a set of categories are possible and worth trying. When a variable is ordered, the same logic applies except that the distinction to be explored must not violate the ordering of its values. For example, with $v$ = 7 ordered values, the algorithm would require testing only $v$–1 = 6 binary distinctions:

1|2,3,4,5,6,7; 1,2|3,4,5,6,7; 1,2,3|4,5,6,7; 1,2,3,4|5,6,7; 1,2,3,4,5|6,7; 1,2,3,4,5,6|7

For ordered data, (12.14) would reach the optimal partition faster that for unordered data. However, the result could only be regarded as ranked and analyzed by methods appropriate for ordinal data.

## 12.12 Correcting Sources of Systematic Disagreements

Systematic disagreements may well force ${}_{c}\alpha$ below what can be expected by chance, ${}_{c}\alpha < 0$, but this is not necessary so. When observers understand and apply the same coding instructions to the same set of phenomena, systematic disagreements are unlikely.

Systematic disagreement becomes evident whenever measures of correlations or associations yield values larger than compatible agreement measures do. The reason for this difference lies in the fact that correlations and associations measure the statistical dependency between two variables whereas agreement measures respond to how similar two variables are. In the following two contingency matrices, the left one shows a perfect regression line between two interval scales which visibly deviates from the diagonal. In the other matrix, associations are perfect except for the systematic confusion of the 3rd and the 5th category of two nominal variables.

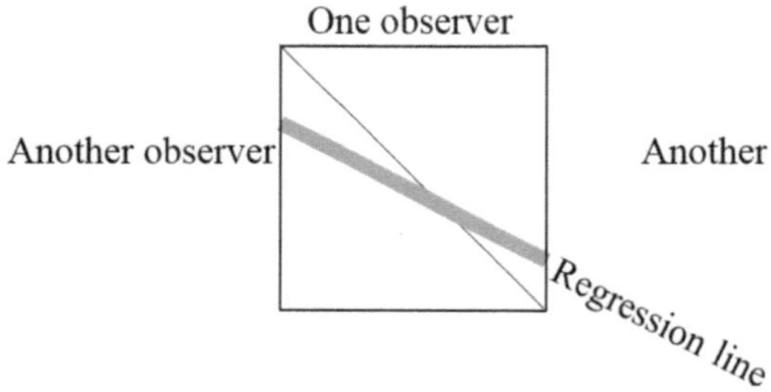


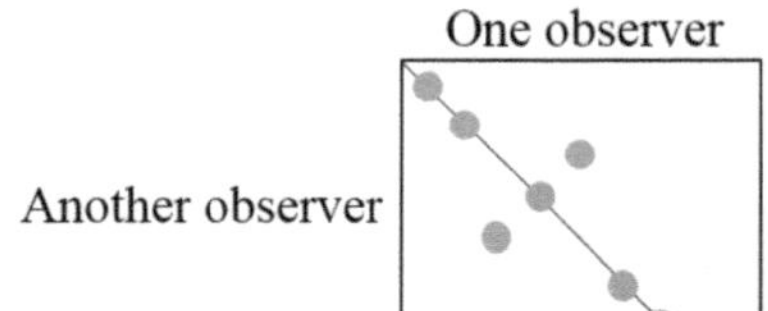


Disagreements are systematic to the extent they cannot be explained as chance events. Also, they escape being identifiable as such in the coincidence matrices that omit everything that is irrelevant for obtaining the agreements for estimating replicability of data. Coincidence matrices omit the identities of the observers who generate the data for being irrelevant in assessing replicabilities. This is not so where contingency matrices are used for accuracy and surrogacy but they can also be obtained by pairing observers when replicability is the issue. Therefore, to visualize the nature of systematic disagreements, we have to rely on their contingency matrix representations, exemplified in the two examples above.

Before proceeding, recall Karl Pearson's (1926) ingenious proposal, discussed in Chapter 4, for his "coefficient of racial likeness" $r_{xx}$ – racial because it was initially developed for measuring the capacity of human skulls in Burma with the aim of finding racial differences. Tildesley (1921) reported having been advised by Pearson to apply his product-moment correlation coefficient for cross-tabulations of interval data that consist of the X–Y contingencies by combining them with their inverses, the Y–X contingencies into a single table. Accordingly, the above regression equation and its inverse would create the following coincidence matrix whose joint regression line is the 45-degree diagonal from which the regression equations visibly deviate.

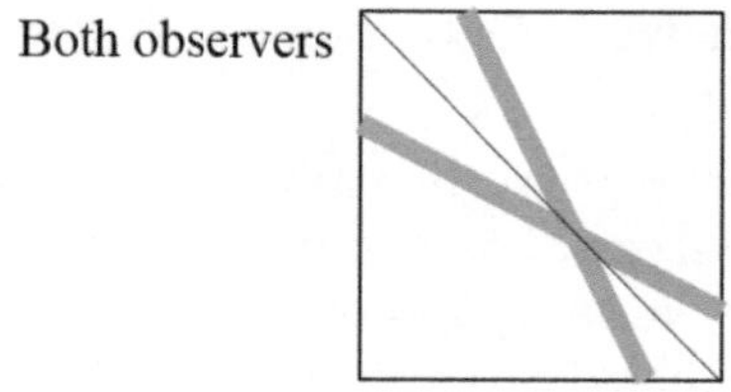


Incidentally, applying Pearson's method to the second example would not make a difference.

Pearson's method of transforming any asymmetrical contingency matrix into a symmetrical coincidence matrix gave rise to agreement coefficients for interval data and ultimately to so-called intraclass correlation coefficients, which fundamentally measure agreements between variables – see also Krippendorff (1970a).

The fact that systematic disagreement is ignored in measures of correlation and association while responded to in measures of agreement to estimate the replicability of data renders their difference an important indicator of the presence of systematic disagreements in data.

What are some of the sources of systematic disagreements and what can be done to take advantage of their systematicity?

- By far the most common source of systematic disagreements is the unknowing application of different coding instructions, an earlier and a recently revised one. This may cause any two coders' valuations to correlate highly while disagreeing between each other. If and only if it can be established which coding instruction was correct, then there is a possibility of recoding the systematically deviating data and render them analyzable reliably.
- Another source of systematic disagreement occurs when the opposites of a polar scale are unclearly defined. For example, when asked to rate a product from 1 to 9, someone may consider 0 to have no flaws while others consider 0 to be the most undesirable product. Suppose each observer's coding is perfectly determinate, with the endpoints of the nine-point interpreted in opposition to each other, the following contingencies are perfectly related, but the agreement measure is close to zero.

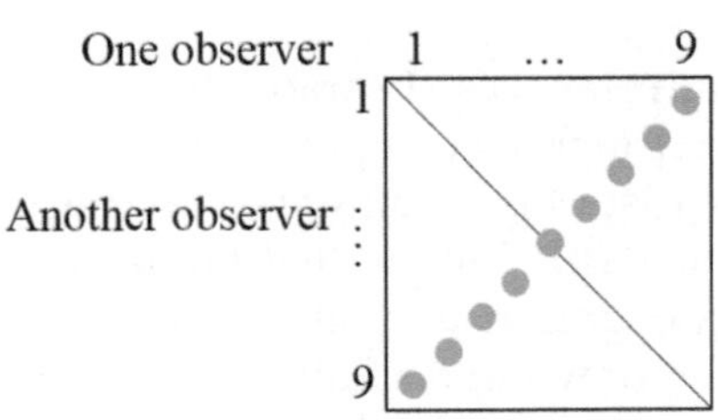


The systematicity of these disagreements is easily detected by the difference between measures of correlation or association, here perfect, and of agreement, here virtually zero. In reality, differences are rarely as radical and obviously correctable as in the above example. However, to recode any one observer's records requires an additional inquiry about which of the two interpretations of the scale was intended in subsequent analyses. Only then can systematic disagreement be eliminated from unreliable data.

- A different but not uncommon source of systematic disagreements is due to consistent but unlike predilections by observers. For example, some shy away from extreme judgments while others see the world in oppositions. This is the bias visualized in the first of the two examples presented above. If consistently pursued, these different judgment styles yield a regression line that deviates from perfect agreement. These unequal conceptual conceptions reduce the reliability of the data, which may, however, be corrected by applying a linear transformation of the contingencies that brings the regression line closer to the diagonal of perfect agreement – as suggested in the left of the two matrices depicted earlier and below.

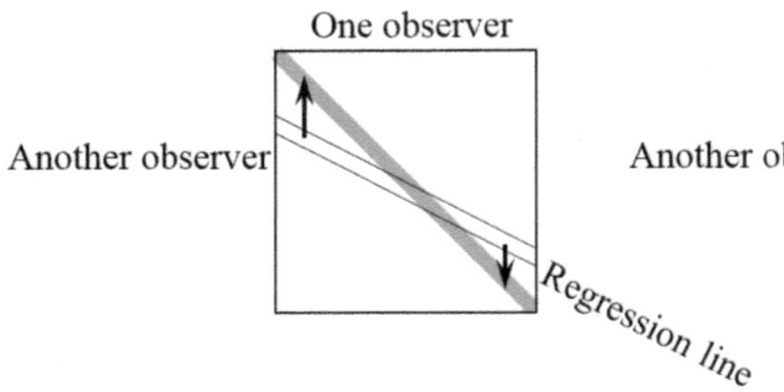


Analogously, the matrix on the right of the above exemplifies a common systematic disagreement among nominal data: the confusion of two or more categories, here between the 3rd and the 5th category. The path to correct the error may be visually obvious but not quite so in practice. Whether the one observer's 3rd and 5th category should remain unaltered while the other observer's category 5 should become 3 and vice versa (as suggested by the arrows), this has profound consequences for what the resulting data on these two categories mean when analyzed. Obviously, whether systematic disagreements are present and correctable is one thing. As there always are at least two ways to correct this error, only interviews with the observers and the designers of the coding instructions can determine which direction possible corrections may take.

- Many sources of systematic disagreement can be traced to the medium used to record coded data. There is much progress in using computer-aided recording devices which are not entirely free of systemic distortions. Long lists of categories to choose from are known to be unreliable. Miller's (1956) law that humans make mistakes when there are more than seven plus or minus two alternatives to choose from applies here. Even if long lists are broken down into smaller chunks, those listed first often are more likely chosen. This exemplifies that the psychology of conceptualization may override even carefully crafted coding instructions. There are many systematic disagreements associated with nearly all technologies. They are not easily analyzable for lack of replications by other means.
- There are number of less tractable but avoidable systematic disagreements known to occur as coders develop individual habits. As one example, when coding instructions use numbers to designate variables – their values, the units of analysis to which they

are to be assigned, the rows and columns of data sheets, and more – they are easily confused. Observers make sense of such and other complexities by creating convenient shortcuts and stick to them throughout a coding task. These are individual-specific coding habits that because they may not be shared among independently coding observers, turn up in systematic disagreements. They are measurable but tricky to correct.

In other words, measured systematic disagreement is not always correctable, especially when their correlations or associations are low to begin with. The condition of the latter raises the question of whether the gain in the reliability of the recoded data is worth the effort.

Perfect correlations or association always indicate either that data already are in perfect agreement or can be recoded in order to become so. The simplest possible example for both cases is:

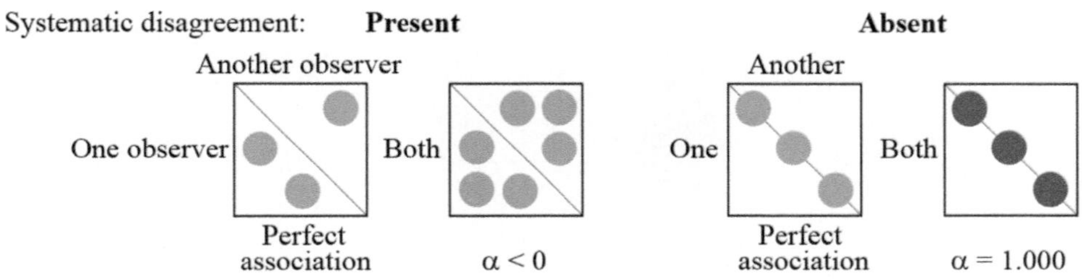

# 13

# *Some Special Applications*

Previous chapters explored the reliability of data that are quite common in social research, product development, text analysis, educational evaluations, and computational linguistics, even medical research. However, reliability issues arise in numerous areas concerned with data that are application-specific, posing problems not necessarily shared among many. This chapter reports on a few instances that were not considered worth a whole chapter but whose problematic nature may well resonate with some researchers' reliability concerns.

## 13.1 Coping With Violations of the Mutual Exclusiveness of Categories

ATLAS.ti is a Qualitative Data Analysis (QDA) software for coding written text (Friese, 2019). It grew out of the tradition of grounded theory, a philosophy that encourages analytical categories to emerge in processes of careful reading of texts. It opposes the application of predefined systems of descriptive categories which, in the course of working through a text, may turn out to be unable to capture how it is read by the analysts, yielding distorted representations. This represents a challenge for qualitative scholars to plausibly claim that other scholars would come to the same conclusions when facing the same phenomena.

Conceptual grounding resembles ethnographic approaches to textual analysis (Garfinkel, 1967; Altheide, 1987). It is an effort to stay as close as possible to the meanings of textual matter to those who generate and live with it. According to Strauss and Corbin (1994) the process starts with asking questions, and continues to the iterative sampling of texts that could answer or revise these questions. Detailed reading of texts is expected to lead to segmentations of texts and groupings into conceptually similar kinds. One of the basic concepts of grounded textual analysis is that of a semantic domain.

A semantic domain names a space of distinct concepts with shared meanings, e.g., "colors," "mental illnesses," "emotions," "gender issues," "personalities," or "sentiments." Semantic domains may be named abstractly but are always context dependent. For example, the concept of color is different when applied to the sky, a dress, a national flag, an ethnic group, a traffic light, or the state of being drunk. The gender of German nouns is not correlated with the gender of living organisms.

The contexts of semantic domains are often implied when themes, propositions or quotes form paradigmatic patterns. For example:

"_____ says _____ to _____ with the intention to _____ but causing _____ "

defines the semantic domains of speakers, utterances, addressees, and intended and unintended consequences. Or:

" _____ diagnoses _____ to have _____ ."

The first semantic domain is that of experts in an empirical domain, the second of objects in that domain, and the third the problems in that domain. That is, say, when the first is a medical professional, the second is a patient and the third an illness. Many semantic domains concern attributes of objects, actions, people, or categorizations of abstract ideas.

This admirable philosophy has come into conflict with several realities that qualitative researchers are facing. One is the nature of QDA-software. While vastly expanding the analytical capabilities of their users, it also limits them to working within interfaces that entail rules for identifying and recording their users' interpretations of text segments. The other is the increased demand of the academic community in general and its publishers in particular, to provide evidence that qualitative text analyses are reliable, trustworthy, after which its conclusions can be taken seriously, especially when they have social consequences and incur costs.

Although no software can accommodate all QDA ideals, ATLAS.ti preserves the fundamental openness of coding, and facilitates segmenting texts by highlighting character sequences, assigning them to multiple semantic domains and the categories that each embrace. It also provides aids for searching within texts and offers various summaries of the data users have created. At the time of this writing, ATLAS.ti enables the computation of the first three α-agreement coefficients in Figure 13.1, which offer a reasonably complete picture of the extent to which the resulting data can be trusted. The fourth has a dual purpose. It is intended to provide reliability measures for coding within each semantic domain and focus individual coders' attention on violations of the mutual exclusiveness requirement of a semantic domain. The four agreement measures are defined on the following data:

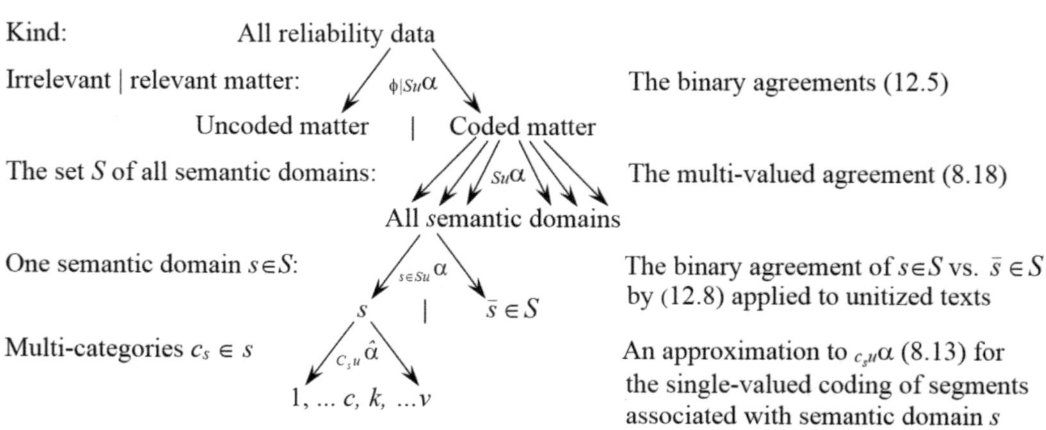


**FIGURE 13.1**
Agreement Coefficients Supported by ATLAS.ti

The distinction between irrelevant and relevant segments or uncoded and coded matter is primary and common, see Table 12.3. It calls for a binary agreement measure, analogue to (12.5), applied to different coders' partitions of the volume of text into these two categories. It indicates the extent to which coders can reliably distinguish between what matters to a research project and what does not and establishes whether the data ultimately analyzed have a clear focus of attention.

When evaluating the reliability of diverse sets of semantic domains – someone's age, party affiliation, educational level achieved, hobbies, or profession – the categories within these semantic domains are ignored in favor of the extent to which the combination of semantic domains were reliably identified. Multi-valued conceptions of agreement, developed in Chapters 7 and 8, are the obvious candidates and (8.18) applies.

The reliability of any one semantic domain in contrast to all others was discussed in Section 12.5. (12.8) provides one agreement coefficient for each semantic domain and evaluates the extent to which it stands on its own in the context of all other semantic domains.

A fundamental requirement of all reliability considerations is the ability to decide whether two codes are the same or different. However, ATLAS.ti could not prevent their users from violating the mutual exclusiveness requirement of codes within any one semantic domain. It allowed for uninterpretable uncertainties and contradictions to be part of the generated data. For example, people cannot be simultaneously present at different locations, be of two ages at the same time, or claim two opposing political affiliations. However, they can travel from one place to another, speak about their childhood experiences, or lie about their political beliefs. Finding two mutually contradictory codes in one semantic domain signals a lack of distinguishing several semantic domains. It turns out that single-valued reliability coefficients are not applicable to such confusions and are unable to give a hint of how reliable a semantic domain otherwise is.

Violations of the requirement of mutual exclusivity may not be the result of carelessness but are more likely rooted in inadequate definition of a semantic domain. A car cannot be blue and black, but it may have a blue exterior and black seats, which are two separate semantic domains. Facts and lies about facts belong to two semantic domains, not one. To enable measuring reliabilities despite such confusions, I devised an alternative difference function, one that applies to any one named semantic domain, regardless of whether its categories are used as mutually exclusive or conflicting. It enables ATLAS.ti to compute inter-coder agreements and aid resolving such problems.

Recall the multi-valued difference function for nominal data (7.2), discussed in Chapter 7, and visualized in Figure 7.1. It discounts all differences between categories of intersecting sets:

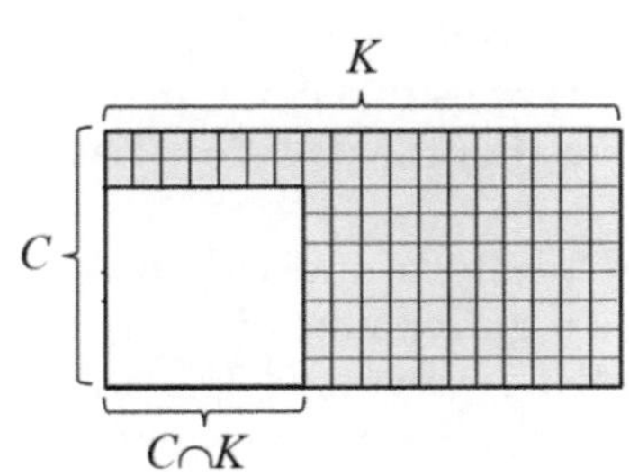


$$_{\text{nominal}}\Delta_{CK} = |C||K| - |C \cap K|^2$$

A difference function that would apply to multi-valued sets without regarding the values shared by two semantic domains as agreeing and treating them instead as single-valued is comparatively visualized in Figure 13.2 and defined by (13.1):

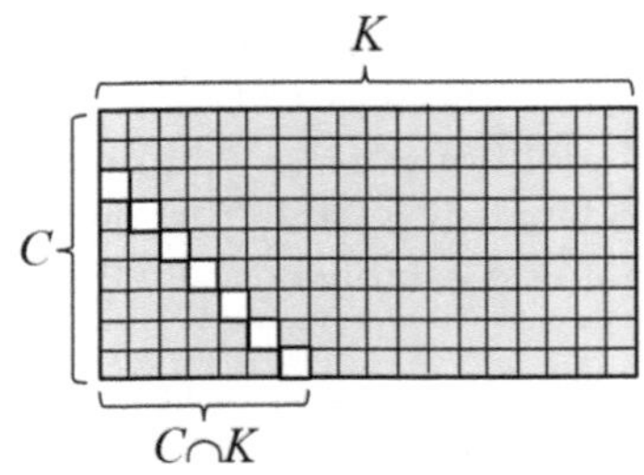


$$\nabla_{CK} = |C||K| - |C \cap K| \quad (13.1)$$

**FIGURE 13.2**
Multi-Valued Difference Function That Ignores Memberships in Sets

Merely replacing ${}_{\text{metric}}\Delta_{CK}$ in the observed disagreement (8.15) by $\nabla_{CK}$:

$$_{C_s u}\hat{D}_o = \frac{\sum_C \sum_K \ell^*_{CK} \nabla_{CK}}{\sum_C \sum_K \ell^*_{CK} |C||K|} \quad (13.2)$$

As well as in the expected disagreement (8.17):

$$_{C_s u}\hat{D}_e = \frac{\sum_C \sum_K \varepsilon^*_{CK} \nabla_{CK}}{\sum_C \sum_K \varepsilon^*_{CK} |C||K|} \quad (13.3)$$

results in an agreement coefficient, notationally distinguished by adding a circumflex:

$$_{C_s u}\hat{\alpha} = 1 - \frac{_{C_s u}\hat{D}_o}{_{C_s u}\hat{D}_e} \quad (13.4)$$

It has these properties:

- When a semantic domain is single-valued, and all categories are mutually exclusive, (13.1) to (13.4) is numerically equal to $_{c_s u}\alpha_{\text{nominal}}$ (12.9) for the single-valued coding of data.
- When a semantic domain is coded without complying with the mutually exclusiveness requirement, (13.1) to (13.4) accepts that possibility. But unlike (8.18), it treats the shared values as single-valued, as if their joint membership in a set did not matter. Thus, $_{C_s u}\hat{\alpha}$ can be interpreted as an estimate of $_{c_s u}\alpha_{\text{nominal}}$, which is the agreement that one could conceivably achieve after the violations have been corrected.

One function of this measure is to inform individual ATLAS.ti coders of where they violated the mutual exclusiveness requirements of categories within a semantic domain, enabling them to make the needed corrections. The analyst of the agreement among several coders' work is informed of whether the measure is the correct reliability measure $_{c_s u}\alpha_{\text{nominal}}$ or its approximation $_{C_s u}\hat{\alpha}$, whether the results of a reliability test is publishable or an invitation to ask coders to make the needed corrections. The reliability of coding within semantic domains is probably most important for subsequent analyses of coded segments of text.

The use of the difference function (13.1) is not limited to the valued segments of textual continua, ATLAS.ti's domain. It can serve the same purpose when assessing the reliability of predefined units, as discussed in Chapter 7, where similar violations could conceivably occur.

## 13.2 Predefined Units With Unique Categories to Be Identified

Coding instructions commonly define the same set of variables and its categories to code all units of analysis, whether predefined or the result of unitizing textual continua. A thermometer has a scale with a fixed range, and social psychologists may consider individual behavior to be social or asocial. One researcher, Steven Corman[1], approached me with an interesting reliability problem that did not fit this conception of data making.

Corman studied the choices that readers make when interpreting words or phrases in texts. More specifically, he presented readers with concepts for which a dictionary listed a flexible number of interpretations. Word dictionary entries vary not only numerically from one term to another; the set of alternative interpretations is also specific to each word. Although a dictionary lists distinct words, they may be related, whether by common prefixes or different suffixes varying a stem like "help," "helper," "helpful," or "helpless." *Webster's Collegiate Dictionary* lists six meanings for "help." Literary competencies become evident when readers choose the same meaning as a result of the context in which a word appears.

The situation is not too different when judging the answers to multiple choice questions – ignoring for a moment that teachers have decided which answers count as the correct one.

Corman was not interested in correctness but in measuring the agreement among two readers. α-coefficients too respond to agreements or disagreements of selections without adjudicating which one is correct. (Chapter 11, addressing issues of text mining and information retrieval, was an exception as the sole purpose of surfing a data base is to find relevant information.) The solution to Corman's problem turned out to be simple:

- Corman had 1, 2, …, $u$, …, $N_u$ textual units
- Each unit $u$ had $n_u$ unique interpretations
- Two readers were asked to select one out of $n_u$ possible interpretations for each unit $u$
- He observed $A \le N_u$ matching interpretations
- His problem was to find a correction of the observed disagreements $N_u - A$ by what could be expected by chance, when readers failed to understand the context in which $u$ occurred or picked one out of the $n_u$ interpretations without apparent reason

For each $u$, the probability of two readers to agree by chance is $n_u^{-1}$.

$$\begin{aligned}&\text{The observed disagreement is: } D_o = N_u - A\\ &\text{The expected disagreement is: } D_e = N_u - \sum\nolimits_{u=1}^{N_u} n_u^{-1}\end{aligned} \qquad (13.5)$$

[1] Steven R. Corman, personal communication 2012.2.13.

And the $\alpha$-agreement becomes: $_{(u\text{-specific})c}\alpha = 1 - \frac{D_o}{D_e} = \frac{N_u - A}{N_u - \sum_{u=1}^{N_u} n_u^{-1}}$ (13.6)

Although Corman did not intend to employ more than two readers for his project, had he increased that number from 2 to $m$: 1, 2, …, $i$, $j$, …, $m$; and made sure that everyone judged all $N_u$ units of texts, the agreement coefficient for this situation would be:

$$_{(u\text{-specific})c}\alpha = 1 - \frac{D_o}{D_e} = \frac{\frac{m(m-1)}{2}N_u - \sum_{u=1}^{N_u}\sum_{i=1}^{m}\sum_{j>i}^{m} 1 \text{ iff observer } i \text{ and } j \text{ agree}}{\frac{m(m-1)}{2}N_u - \sum_{u=1}^{N_u}\sum_{i=1}^{m}\sum_{j>i}^{m} n_u^{-1}} \quad (13.7)$$

From the fact that the two sums $\sum_{i=1}^{m}\sum_{j>i}^{m}$ count the agreements among $m(m$-1)/2 pairs of judgments regarding each textual unit, whether observed or by chance, it should not be difficult to verify that (13.7) generalizes (13.6). Whenever the number of readers shrinks from $m$ to 2, (13.7) becomes (13.6).

## 13.3 Alpha Extended to Conditions Where Units Are Ranked

A researcher named Mehta[2] approached me with reliability data that are simple and easily described but for which agreement measures were unknown.

He had asked

- $m$ judges, 1, 2, …, $i$, $j$, …, $m$
- to rank 1, 2, …, $u$, …, $N$ objects

and was in search of a way to measure their reliability.

Usually, predefined units of analysis are considered unordered and independent of each other and hence enumerable. However, Mehta's objects were no longer merely distinct. They were assumed to possess qualities relative to each other that would enable judges to rank them. Perfect reliability would be indicated if all $m$ ranks agree. A suitable agreement coefficient would have to numerically decline as observed disagreements among diverse rankings approach being chance events, i.e., their ranking can no longer be related to any ordering of the objects that observers had faced. It would have to be negative if disagreements were systematic, below what independent judges could achieve without looking at what they were asked to rank.

Mehta assured me that each judge ranked all $N$ objects. There would be no ties and no omissions.

Terminologically:

- Let $r_{iu}$ be the rank between 1 and $N$ that the $i^{\text{th}}$ judge assigned to object $u$ and $r_{ju}$ assigned by the $j^{\text{th}}$ judge to $u$.

[2] I could no longer retrieve Mehta's communications.

- The observed disagreement can be defined quite obviously as the sum of all squared differences among the $m(m-1)$ pairs of ranks assigned to $N$ objects:

$$_\rho D_o = \sum_{u=1}^{N} \sum_{i}^{m} \sum_{j \neq i}^{m} \left(r_{iu} - r_{ju}\right)^2 \tag{13.8}$$

Mehta's problem was to find an expression for what could be expected when the ranking was without merits.

Over a century ago, Charles Spearman (1904) dealt with the situation of two judges ranking $N$ objects, yielding $\sum_{u=1}^{N}\left(r_{iu} - r_{ju}\right)^2$ squared differences.

He conceptualized what I now call the expected disagreement as the absence of any pattern among the squared difference among possible ranks. He proved that when $N$ ranks showed absolutely no pattern, as in the following 7-by-7 difference matrix:

| | 1st | 2nd | 3rd | 4th | 5th | 6th | 7th |
|---|---|---|---|---|---|---|---|
| 1st | **0** | $1^2$ | $2^2$ | $3^2$ | $4^2$ | $5^2$ | $6^2$ |
| 2nd | $1^2$ | **0** | $1^2$ | $2^2$ | $3^2$ | $4^2$ | $5^2$ |
| 3rd | $2^2$ | $1^2$ | **0** | $1^2$ | $2^2$ | $3^2$ | $4^2$ |
| 4th | $3^2$ | $2^2$ | $1^2$ | **0** | $1^2$ | $2^2$ | $3^2$ |
| 5th | $4^2$ | $3^2$ | $2^2$ | $1^2$ | **0** | $1^2$ | $2^2$ |
| 6th | $5^2$ | $4^2$ | $3^2$ | $2^2$ | $1^2$ | **0** | $1^2$ |
| 7th | $6^2$ | $5^2$ | $4^2$ | $3^2$ | $2^2$ | $1^2$ | **0** |

the average sum of their squared differences would be $\frac{1}{N}\sum_{c=1}^{N}\sum_{k=1}^{N}\left(c^{th} - k^{th}\right)^2 = \frac{N\left(N^2-1\right)}{6}$. With reference to the above matrix of all squared differences among $N = 7$ out of $N^2 = 7^2 = 49$ pairs of ranks, this sum would be:

$$\frac{1}{7}\sum_{c=1}^{7}\sum_{k=1}^{7}(c-k)^2 = \frac{1}{7}392 = 56 = \frac{7\left(7^2-1\right)}{6}$$

From this, Spearman proposed his famous rank order coefficient $\rho$ (rho) – here acknowledged by adding $\rho$ as a subscript to $D_o$, $D_e$, and $\alpha$.

Because for $m$ judges, $_\rho D_o$ sums all differences between $m(m-1)$ pairs of rank orderings, the corresponding expected disagreement would have to be:

$$_\rho D_e = m(m-1)\frac{N\left(N^2-1\right)}{6} \tag{13.9}$$

And the $_\rho\alpha$-agreement for ranking $N$ objects by $m$ judges would become:

$$_\rho\alpha = 1 - \frac{_\rho D_o}{_\rho D_e} = 1 - \frac{\sum_{u=1}^{N}\sum_{i}\sum_{j \neq i}^{m}\left(r_{iu} - r_{ju}\right)^2}{m(m-1)\frac{N\left(N^2-1\right)}{6}} \tag{13.10}$$

It is worth pointing out that in the literature, Spearman's ρ is almost universally referred to as a rank order 'correlation' coefficient. However, the way ρ counts the differences among two ranks and estimates the expected ranks, illustrated by the above symmetrical 7-by-7 difference matrix, conforms to what replicability coefficients do. His rank order differences equal the non-standardized metric difference (5.13) exemplified with Table 5.6, equal to the ordinal difference (5.12) in the special case of ranks occurring with the same frequency. I would conclude that Spearman's $\rho$ does not measure correlations but agreements proper.

## 13.4 Agreements Among Groups

Garcia Nobles,[3] a sociologist from Brazil, was determined to measure the agreement between four groups of diverse number of members, rating several social events on an interval scale. As the number of scores each group provided varied, it seemed obvious to rely on their easily comparable average score. This conception brought him to the canonical form of reliability data in Table 13.1, just as in Table 5.1:

| **Units:** | 1 | 2 | ... | $u$ | ... | $N$ |
|---|---|---|---|---|---|---|
| Group A | $\bar{a}_1$ | $\bar{a}_2$ | ... | $\bar{a}_u$ | ... | $\bar{a}_N$ |
| Group B | $\bar{b}_1$ | $\bar{b}_2$ | ... | $\bar{b}_u$ | ... | $\bar{b}_N$ |
| Group C | $\bar{c}_1$ | $\bar{c}_2$ | ... | $\bar{c}_u$ | ... | $\bar{c}_N$ |
| Group D | $\bar{d}_1$ | $\bar{d}_2$ | ... | $\bar{d}_u$ | ... | $\bar{d}_N$ |

**TABLE 13.1**
Reliability Data of Group Averages

Nobles had no problem calculating the overall α-agreement (5.8), for interval metric differences (5.13), but became uncertain about what ${}_c\alpha_{\text{interval}}$ would tell him. When he looked more closely into the distributions of scores within individual groups, he observed several cases where averages were virtually identical but the variances that gave rise to them differed greatly.

First, he lost confidence in the conception of α. But then he realized the culprit was not its conception of agreement but applying it to averages. He concluded, correctly, that his use of average scores hides the very differences among the scores that were most important to his project. This motivated him to explore the other extreme, comparing the actual scores of each group with one another. This turned out to be too complex, at which point he approached me.

I am using Figure 6.1 as a template for the following Figure 13.3, which shows the decomposition of all pairs of values within any one unit of a larger set into two intra-group sets of pairs plus the inter-group pairs between the two groups, which was how I understood his second attempt at analyzing agreements among groups.

[3] I lost the date of his request and ultimately his connection.

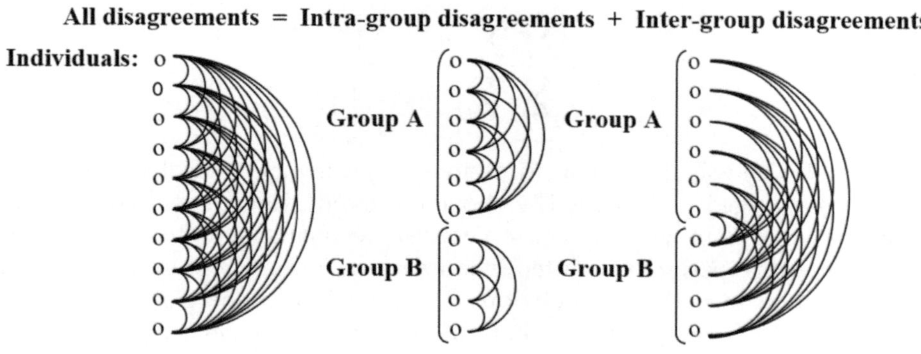


**FIGURE 13.3**
All Coincidences Decomposed into Two Intra-Plus Inter-Group Coincidences

Nobles had already observed and expressed frustration about the fact that the observable inter-group disagreements were heavily influenced by the number of intra-group disagreements, a condition he considered a distraction from his analytical mission.

Luckily, I had worked out the multi-valued difference function between pairs of sets of values, described in Chapter 7, (7.2) for nominal data and (7.3) generalized to all metrics. This difference function pairs all values in paired sets of values but considers the values in the intersections of these sets as in agreement, graphically depicted in Figure 7.1, here reproduced as Figure 13.4 together with the two difference functions, (13.11) being the special case of (13.12).

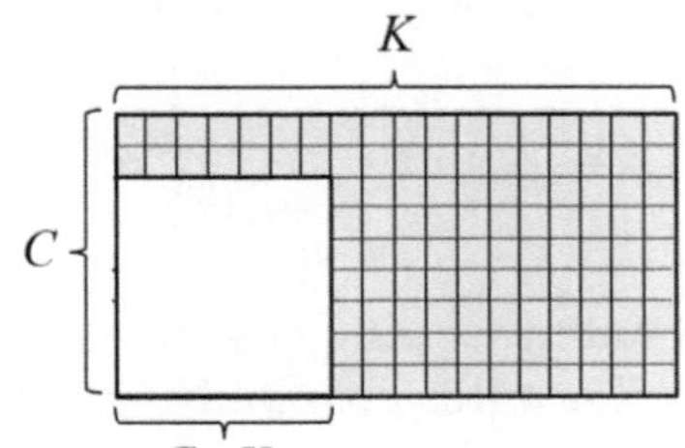


$$_{\text{nominal}}\Delta_{CK} = |C||K| - |C \cap K|^2 \quad (13.11)$$

$$_{\text{metric}}\Delta_{CK} = \sum_{c \in C} \sum_{k \in K} {}_{\text{metric}}\delta^2_{ck} - \sum_{c \in C \cap K} \sum_{k \in K \cap C} {}_{\text{metric}}\delta^2_{ck} \quad (13.12)$$

**FIGURE 13.4**
The Multi-Valued Difference Function Defined in (7.2) and (7.3)

This function would not consider the pairs of values in intersecting sets as different. It enumerates only the differences between values not shared between two sets. But its use would solve only one of three analytical problems Garcia Nobles was hoping to solve. A first was to give due credit for what two groups shares. A second concerned the unequal sizes of his groups, which he didn't want to dominate his analysis. This called for expressing the summed disagreements relative to the number of scores paired. This was precisely what the multi-valued disagreement function was designed to accomplish. In terms of the coincidences $n_{CK}$ between two sets of values $C$ and $K$ within all units, summed as in (7.1), this observed disagreement (7.5) is:

$$ {}_{C}D_{o} = \frac{\sum_{C}\sum_{K} n_{CK} \,{}_{\text{metric}}\Delta_{CK}}{\sum_{C}\sum_{K} n_{CK} \,|C||K|} \tag{13.13}$$

Third, Garcia Nobles was interested primarily in agreements associated with each unit, the social events judged and rated. The above observed disagreement could serve him merely as a baseline from which the observed disagreements for individual units would be deviating. For this reason I suggested reliance on an α-coefficient for individual units, discussed in Section 12.2. This is an obvious derivative of (7.5), here of (13.13), wherein $i$ and $j$ signify Garcia Nobles' groups A, B, C, or D, whose members provide one set $C_{iu}$ or $K_{ju}$ of ratings for each event $u$:

$$ {}_{(u)C}D_{o} = \frac{\sum_{i}\sum_{j \neq i} \,{}_{\text{metric}}\Delta_{C_{iu}K_{ju}}}{\sum_{i}\sum_{j \neq i} |C_{iu}||K_{ju}|} \tag{13.14}$$

For both, ${}_{C}D_{o}$ and ${}_{(u)C}D_{o}$, the expected disagreement (7.6) is:

$$ {}_{C}D_{e} = \frac{\sum_{C} n_{C\cdot}\sum_{K} n_{\cdot K} \,{}_{\text{metric}}{}'' {}_{CK}}{\sum_{C} n_{C\cdot}\sum_{K} n_{\cdot K} \,|C||K|} \tag{13.15}$$

which leads to the two α-agreements of interest to Nobles:

$$ {}_{C}\alpha_{\text{metric}} = 1 - \frac{{}_{C}D_{o}}{{}_{C}D_{e}} \text{ and } {}_{(u)C}\alpha_{\text{metric}} = 1 - \frac{{}_{(u)C}D_{o}}{{}_{C}D_{e}} \tag{13.16}$$

Garcia Nobles did not inquire about measuring how far each group deviated from what all groups have in common. When I saw this possibility, he was no longer responding, and I suppose he had solved his problems and moved on.

Such a measure would be one of the surrogacies for each group. It could have provided Garcia Nobles with additional comparisons. Here it is, developed in Section 6.6.4, discussed in Section 12.6, but now stated in his data's terms:

Let $\bar{c}_u$ = the mean of all scores assigned by members of several groups to unit $u$

$y_{\bar{c}\cdot}$ = the frequency of all $N$ means $\bar{c}_u$

$K_u$ = the set of scores assigned by members of the targeted group to unit $u$

$y_{\cdot K}$ = the frequency of sets $K$ used by the targeted group.

The observed disagreement is: $$ {}_{(K)C}D_{o}^{sm} = \sum_{u=1}^{N} {}_{\text{metric}}\Delta_{\bar{c}_u K_u} \tag{13.17}$$

The expected disagreement is: $$ {}_{(K)C}D_{e}^{sm} = \frac{1}{N}\sum_{\bar{c}} y_{\bar{c}\cdot}\sum_{K} y_{\cdot K} \,{}_{\text{metric}}\Delta_{\bar{c}K} \tag{13.18}$$

And group $K'$ surrogacy becomes: $$ {}_{(K)C}\alpha_{\text{metric}}^{sm} = 1 - \frac{{}_{(K)C}D_{o}^{sm}}{{}_{(K)C}D_{e}^{sm}} \tag{13.19}$$

## 13.5 Circumscribing and Coding Two-Dimensional Images

Todor Todorov,[4] a freelance computer programmer working in Berlin, Germany, challenged me with the problem of assessing the reliability of identifying rectangularly framed images in larger areas, illustrations, photographs, and texts (not moving images).

Let each image $C$ have a horizontal extension $C_x$, a vertical extension $C_y$, a location relative to a reference point that is common to all circumscribed images, and a single value $c$:

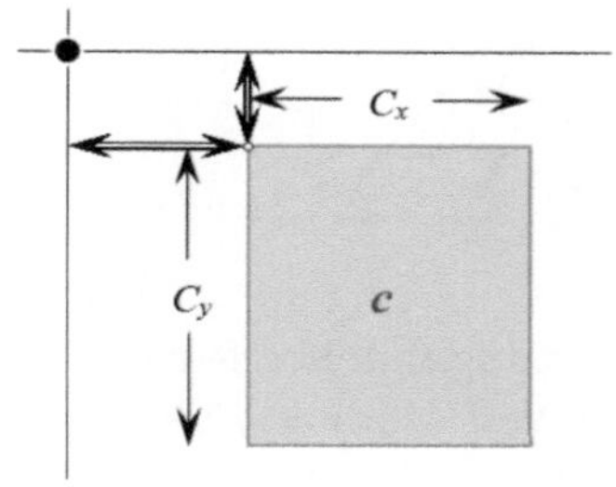


The intersection between two rectangularly framed images $C$ and $K$ is:

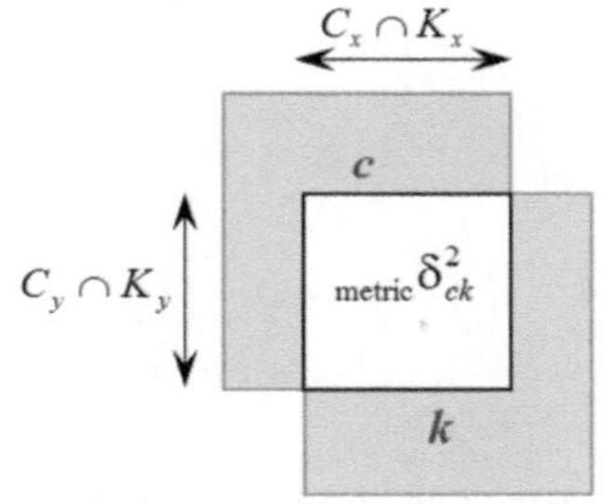


To define the difference between two individually circumscribed images, we draw on the difference function (9.1) between two segments of a continuum. It counts as a difference everything that is outside their intersection but weighs the intersection by a standardized metric difference function. Following this conception, the difference between two separately identified images $C$ and $K$, valued $c$ and $k$ respectively becomes:

$$_{\text{metric}}[\cdot\cdot]_{CK} = C_x \cdot C_y + K_x \cdot K_y - 2(C_x \cap K_x)(C_y \cap K_y)(1 - {}_{\text{metric}}\delta^2_{ck}) \quad (13.20)$$

A brief comment on what $_{\text{metric}}[\cdot\cdot]_{CK}$ responds to. Regardless of in which units the sizes of images and their locations are measured, the gray areas in the above figures are simply summed. When the two values $c = k$, their intersection is considered in agreement and is excluded from the sum. However, when the two images are valued differently, the appropriate metric difference function determines the extent to which the intersection contributes to the difference between $C$ and $K$. For example, when the images concern problematic areas within x-rays and the associated values are medical diagnoses, which are unordered and call for the use of the nominal difference function (5.10), when examiners agree on their diagnosis the intersection is subtracted. When they disagree on

[4] Todor Todorov, personal communication 2018.6.20.

their diagnosis, the intersection is not subtracted. When one observer's circumscribed image is paired with the absence of another, the whole area of the one observer's image is considered disagreement.

The canonical form of reliability data, constructed analogue to Table 5.2, will have to show $N_u$ units of analysis $u$ and up to $m$ recognition devices or observers $i$ or $j$, who mark an area considered relevant, and assigns it a code. The entries of such an $m$-by-$N_u$ matrix contain the specifications of an image $C_{ju}$, including its location, extensions and a code.

Applying (12.20) to all $N_u$ units $u$, each of which contains $m_u$ observers' judgments, leads to the observed disagreement:

$$_{[\text{metric}]}D_o = \sum_{u-1}^{N_u} \frac{1}{m_u - 1} \sum_{i=1}^{m} \sum_{j \neq i}^{m} {}_{\text{metric}}[\cdot\cdot]_{C_{iu}K_{ju}} \quad \text{iff } m_u > 1 \tag{13.21}$$

Constructing a matrix of coincidences might not be feasible as the areas circumscribed are unlikely to occur with frequencies above 1. Without reference to coincidences, the expected disagreement must pair all $C_{iu}$s with each other but not with themselves.

$$_{[\text{metric}]}D_e = \frac{1}{(N_u \sum_{u=1}^{N_u} m_u) - 1} \sum_{u_1=1}^{N_u} \sum_{i=1}^{m_{u1}} \sum_{u_2=1}^{N_u} \sum_{j=1}^{m_{u2}} {}_{\text{metric}}[\cdot\cdot]_{C_{iu_1}C_{ju_2}} \tag{13.22}$$

The α-agreements become as usual:

$$_{[\text{metric}]}\alpha = 1 - \frac{_{[\text{metric}]}D_o}{_{[\text{metric}]}D_e} \tag{13.23}$$

While this agreement measure is applicable to the replicability of recognizing faces of people cleared to enter a room, identifying brands in photographic images, diagnosing medical problems in x-rays, or reading the license plate numbers of traffic-violating cars, its expected disagreement is so far limited to one kind of object. Coding multiple faces, diverse medical problems, for example, requires further developments.

## 13.6 A Variance-Independent Agreement Coefficient

To infer the reliability of data from agreement measures, variance is important. Its absence does not allow an analyst to make distinctions, report differences, apply customary statistical analyses, and answer complex research questions. This why I object to criticism in Feinstein and Cicchetti (1990) and Cicchetti & Feinstein (1990) of κ, by implication also of α, for converging to zero when variance disappears – see Sections 3.8, 14.1.2 and 14.1.3. A few analysts have echoed Feinstein and Cicchetti's concerns, some have excavated Bennet, et al. (1954) *S*, Gwet (2002, 2008) proposed his infamous $AC_1$, and Zhao, et al. (2012) argued for redefining reliability as coder difficulties. Some of these efforts have been critically examined in Section 4.2.3.

However, agreement measures can be used for purposes other than assessing the reliability of data and the possibility of an agreement measure that is not affected by lopsided frequency distributions could be useful.

Christian Waloszek[5] wanted to assess the degree of homogeneity among about 250 raters regarding seven educational tasks, each evaluated by eight test questions, all of which used the same 7-point semantic differential scale whose polar opposites were well defined. Therefore, he had a matrix of 250 independent raters by $7 \times 8 = 56$ units of analysis but was interested only in the agreements among raters regarding the seven educational tasks and the eight questions applied to each. He initially assumed that $\alpha$ would not be applicable to seven or eight units but observed correctly that $\alpha$ responded to the uneven frequency distributions of categories or values, in his case of the seven values ranging from –3 to +3. He became discouraged when discovering a case where raters almost unanimously selected one extreme on this scale. The absence of variations in data yielded near zero $\alpha$-measures.

In particular, he calculated the means and standard deviations for each task and each question. This assumed interval data, while his ratings clearly had a scale with polar opposites. The first suggestion I made was to apply the polar difference function to his data. Recall from Chapter 5.

(5.13) defines the metric differences as: $_{\text{interval}}\delta^2_{ck} = \left(\frac{c-k}{c_{\text{largest}} - c_{\text{smallest}}}\right)^2$ and

(5.15) defines the polar differences as: $_{\text{polar}}\delta^2_{ck} = \frac{(c-k)^2}{(c+k-2c_{min})(2c_{max}-c-k)}$

To illustrate their differences for his 7-point scale, the upper triangle of Table 13.2 shows the differences for the interval scale and its lower triangle shows the differences for the polar scale.

| **Values:** | -3 | -2 | -1 | 0 | 1 | 2 | 3 |
|---|---|---|---|---|---|---|---|
| -3 | **0** | .028 | .111 | .250 | .444 | .694 | 1 |
| -2 | .091 | **0** | .028 | .111 | .250 | .444 | .694 |
| -1 | .222 | .056 | **0** | .028 | .111 | .250 | .444 |
| 0 | .333 | .125 | .029 | **0** | .028 | .111 | .250 |
| 1 | .500 | .257 | .111 | .029 | **0** | .028 | .111 |
| 2 | .714 | .444 | .257 | 125 | .056 | **0** | .028 |
| 3 | 1 | .714 | .500 | .333 | .222 | .091 | **0** |

**TABLE 13.2**
Interval differences above and polar differences below, standardized for 7 values

[5] Christian Waloszek, personal communication 2019.12.19–2020.1.13.

As already observed in Section 5.3.5, the polar differences weight more heavily near their poles, like ratio metrics do, rather than around their midpoint where they respond like interval differences do. For measuring the agreement among Waloszek's data, the polar difference function makes intuitive sense and is methodologically correct. However this was not his main concern.

To obtain agreements that are independent of the frequencies with which the seven values of his scale are used, one can draw on the α-coefficient for individual categories, each of which is an agreement coefficient proper and has the same range. They are defined on a coincidence matrix, which can easily be constructed from Waloszek's data, whether for the seven educational tasks, the eight questions asked for each, or 56 units of his analysis. According to Table 12.2:

| **Values:** | 1 | . | . | $k$ | . | $v$ | |
|---|---|---|---|---|---|---|---|
| 1 | $n_{11}$ | . | . | $n_{1k}$ | . | $n_{1v}$ | $n_1$ |
| . | . | . | . | . | . | . | . |
| $c$ | $n_{c1}$ | . | . | $n_{ck}$ | . | $n_{cv}$ | $n_c = \sum_{k=1}^{v} n_{ck}$ |
| . | . | . | . | . | . | . | . |
| . | . | . | . | . | . | . | . |
| $v$ | $n_{v1}$ | . | . | $n_{vk}$ | . | $n_{vv}$ | $n_v$ |
| | $n_1$ | . | . | $n_k$ | . | $n_v$ | $n_{..} = \sum_{c=1}^{v} \sum_{k=1}^{v} n_{ck}$ |

The disagreements between all values $c$ and one target value $k$ occur in the $k^{\text{th}}$ column highlighted above. According to (12.8), the agreement for just this one value $k$ is:

$$_{(k)c}\alpha_{\text{metric}} = 1 - \frac{_{(k)c}D_o}{_{(k)c}D_e} = 1 - \frac{\frac{1}{n_k}\sum_{c=1}^{v} n_{ck\ \text{metric}}\delta_{ck}^2}{\frac{1}{n_{..}-1}\sum_{c=1}^{v} n_{c\ \text{metric}}\delta_{ck}^2} \quad \text{iff} \begin{cases} n_k > 1 \\ n_{..} = \sum_{k=1}^{v} n_k > 1 \end{cases} \tag{13.24}$$

My suggestion to was to compute $_{(k)c}\alpha_{\text{polar}}$ for each value and report them individually, together with their frequencies $n_k$. However, Waloszek was looking for a single variance-independent coefficient for all seven values. This could be obtained by averaging all $v$ agreement measures $_{(k)c}\alpha_{\text{polar}}$ and reporting it with their standard deviations.

This average provides the variance-independent α-agreement measure $_{(\bar{k})c}\alpha^{=}_{\text{metric}}$:

$$_{(\bar{k})c}\alpha^{=}_{\text{metric}} = \sum_{k=1}^{v} \frac{n_k}{n}\, {_{(k)c}\alpha_{\text{metric}}} \tag{13.25}$$

Its superscript " = " was to indicate that variance-independence.

To be clear, averages of the agreements on individual categories should not be interpreted as agreement measures for inferring the reliability of data. Waloszek was not interested in reliability but finding measures of the homogeneity among respondents to educational tasks and to questions about them. For his purpose (13.24) and (13.25) sufficed – at least to start.

## 13.7 A Generalization of Percent Agreement to More than Two Coders of Ordered Data

After calculation these 7+8 agreement measures, Waloszek recognized, correctly, that the expected disagreements, summed either over all questions, all tasks, or both, are not entirely justifiable as a baseline for every one of its ${}_{(k)c}\alpha$-measures as the distributions over the tasks were systematically different and the responses to questions exhibited justifiable differences among their distributions. Six out of eight questions were about emotions; the other two covered types of efforts. Moreover, (13.25) did not give him a measure of the homogeneity for any one of the 56 units of analysis.[6]

Recall that the expected disagreement of $\alpha$ serves as its baseline for the absence of any relationship between the data and the nature of the units coded, indicated by $\alpha$'s zero value. He thought that this criterion was not important when assessing the homogeneities among raters. I agreed. Having 250 raters for any one task and any one question allows the construction of 7-by-7 coincidence matrices, one for each of the 56 units to which 250 subjects contributed in effect 250(250–1) pairs of values.

The $v$-by-$v$ coincidences $n_{ck}$ among $n_{..}$ subjects for one unit is implied by (5.1):

$$n_{ck} = \frac{1}{n_{..}-1}\sum_{i=1}^{n_{..}}\sum_{j\neq i}^{n_{..}} 1 \text{ iff} \begin{cases}\text{subject } i \text{ rates the unit } c\\ \text{and subject } j \text{ rates it } k\end{cases}$$

The resulting coincidences invite the calculation of disagreements that do not distinguish between observed and expected coincidences; hence $\alpha$ would not make sense, but measures of disagreements could.

My first suggestion to Waloszek was to consider a gestalt switch from interpreting his aim not one of measuring the homogeneity among raters but the lack of agreement among raters, according to (5.6):

$$0 \leq \text{Lack of Agreement} = {}_{\text{metric}}D_o = \sum_{c=1}^{v}\sum_{k=1}^{v} n_{ck}\,{}_{\text{metric}}\delta_{ck}^2 \leq max({}_{\text{metric}}D_o) \qquad (13.26)$$

(13.26) gave Waloszek a measure with an unambiguous ideal of ${}_{\text{metric}}D_o = 0$, meaning perfect agreement from which ${}_{\text{metric}}D_o$ deviated. I was concerned that the upper limit of ${}_{\text{metric}}D_o$ varied with the size of the data which rendered (13.26) no longer comparable across different data. This was not of concern for Waloszek as all tasks and questions asked were answered on 7- or 8-point scales.

I decided to explore that maximum. It could be had if the number $n_{..}$ of values were equally distributed over the $v^2$ cells of a $v$-by-$v$ coincidence or contingency matrix:

$$max({}_{\text{metric}}D_o) = \frac{n_{..}}{v^2}\sum_{c=1}^{v}\sum_{k=1}^{v} {}_{\text{metric}}\delta_{ck}^2 \text{ iff} \begin{cases}\text{there are } v > 2 \text{ values}\\ \text{for none of which } n_c = 0\end{cases} \qquad (13.27)$$

The complement of the observed disagreement, expressed relative to its maximum, becomes an agreement measure $A_{\text{metric}}$ that varies between 0 and 1:

[6] Christian Waloszek, personal communication 2020.2.17.

$$\text{Total absence of agreement } = 0 \le A_{\text{metric}} = 1 - \frac{{}_{\text{metric}}D_o}{max({}_{\text{metric}}D_o)} \le 1 = \text{Perfect agreement} \quad (13.28)$$

I wanted to alert Waloszek to the fact that $max({}_{\text{metric}}D_o)$, the determinant of $A_{\text{metric}}$'s zero point, is empirically unlikely to be rendering the interpretability of its values closer to its zero point not too clear, when I recognized this very feature of $A_{\text{metric}}$ to be similar to the percent agreement $P_o$ as discussed in Section 4.1. Perfect agreement, $A_{\text{metric}} = P_o = 1$, is unquestionable. To observe $P_o = 0$, requires a considerable amount of coordination among two raters, each knowingly choosing a category the other does not. Perhaps a little less unrealistic, $A_{\text{metric}} = 0$ would be observable when all raters would happen to distribute their ratings evenly over all possible ones. These two conditions are different but sufficiently similar to explore their difference.

How are they related?

For nominal data and $v$ categories:

$$\sum_{c=1}^{v}\sum_{k=1}^{v} {}_{\text{nominal}}\delta_{ck}^2 = v(v-1) \text{ and} \sum_{c=1}^{v}\sum_{k=1}^{v} n_{ck}\, {}_{\text{nominal}}\delta_{ck}^2 = n.. - \sum_{c=1}^{v} n_{cc}$$

from this and (13.28), it follows that:

$$0 \le A_{\text{nominal}} = 1 - \frac{n.. + \sum_{c=1}^{v} n_{cc}}{\frac{n..}{v^2}v(v-1)} = \frac{n..\frac{v-1}{v} - n.. + \sum_{c=1}^{v} n_{cc}}{n..\frac{v-1}{v}} = \frac{-\frac{n..}{v} + \sum_{c=1}^{v} n_{cc}}{-\frac{n..}{v} + n..} \approx \frac{\sum_{c=1}^{v} n_{cc}}{n..} = P_o \le 1$$

The noticeable difference $n../v$, subtracted from both the numerator and the denominator of the %-agreement $P_o$, does not affect either measure when agreement is perfect but leads to a slightly less extreme zero point for $A_{\text{metric}}$ (all values are equally likely). This encouraged me to consider $A_{\text{metric}}$ to be an improved generalization of the %-agreement $P_o$ to various metrics, any number of raters, and no requirement of communication to disagree at all costs.

$$\text{For ordinal and interval: } max({}_{\text{ordinal \& interval}}D_o) = \frac{n..}{v^2}\frac{v^2(v^2-1)}{6(v-1)^2} < max({}_{\text{nominal}}D_o) < n.. \quad (13.29)$$

Although the zero points of $A_{\text{metric}}$ are more plausible than for $P_o$, I have the same objections to using $A_{\text{metric}}$ in reliability assessments as discussed in Section 4.1 for %-agreement. However, Waloszek was only interested in comparing the homogeneity of various answers to questions for which (13.28) was perfectly suitable.

$A_{\text{metric}}$'s proportion (13.28) can be considered a generalization of the common but fairly limited %-agreement $P_o$ to more than two observers, judges, or coders and agreeing or disagreeing on variously ordered phenomena. It may increase the use of its proportions in areas other than reliability.

# 14

# *Statistical Considerations*

Reliability coefficients are statistical coefficients of agreements observed among independent replications of processes of generating data according to communicable coding instructions. As such they have several properties in common with all quantitative indices. This section reviews those applicable to $\alpha$; some of them were already mentioned in previous sections but are here put into a shared context. This chapter reviews three broad statistical properties. Section 14.1 concerns the statistical meanings of $\alpha$ and what they indicate. Section 14.2 addresses sampling issues, and Section 14.3 deals with $\alpha$'s distribution.

## 14.1 Statistical Meanings of $\alpha$

Preceding chapters discussed, adopted, and formalized numerous agreement measures in conformity to several prerequisites for being interpretable as indices of the reliability of data. This section raises issues of interpretations.

### 14.1.1 The Range of $\alpha$'s Reliability Interpretations and Its Scale

The simplest statistical feature of $\alpha$, already discussed in Chapters 1 and 2, is the range of its interpretability as a reliability coefficient:

$$-1 \le \left\{ \begin{array}{l} -\text{Systematic disagreement} \\ \pm\text{The granularity of data} \end{array} \right\} 0 \le \alpha = 1 - \frac{D_o}{D_e} \le +1 \tag{14.1}$$

- Under conditions of perfect agreement, the observed disagreement $D_o = 0$ and $\alpha = 1.000$.
- Absent any statistical relationship between the variance in the data and diversity of phenomena to be represented by them, the expected disagreement $D_e = D_o$ and $\alpha = 0.000$.

  Section 4.2 already discussed that not all available agreement coefficients conform to $\alpha$'s definition of zero agreement. Several commonly described as chance-corrected agreement coefficients have adopted conceptions of chance unlike $\alpha$'s.

- $\alpha$ may assume negative values. While negative $\alpha$-values are outside the range of $\alpha$'s meaningful reliability interpretation, their occurrence signals conditions that could conceivably be addressed.

The range (14.1) names two causes of negative $\alpha$-values. Systematic disagreement among replications can result from at least three conditions. First, observers literally agree to disagree on judging particular units of analysis. This would violate the condition that replications are to be independent of each other. It blatantly undermines the possibility of interpreting the resulting data in terms of reliability. Second, observers may unknowingly apply different coding instructions to the same phenomena or analysts compare the coding of different phenomena. Third, observers may be trapped in stable but mismatching habits, independent of what they are expected to code, for example falling asleep. Section 12.12 discusses some conditions under which the systematicity of disagreement may be corrected to restore the reliability of such data.

Incidentally, percent agreement $P_o = 0$, not to be confused with the absence of agreement, is an indication of the maximum possible systematic disagreement observable among two observers who must communicate with each other so that one does not chose the category chosen by the other. It follows that $P_o$'s scale between 0 and 1, or 0 and 100% has little to do with the reliability of data (except when agreement is 100% and reliability data exhibit sufficient variance). In assessing the reliability of generating data, the use of %-agreement should be discouraged.

The other condition affecting deviations from $\alpha$'s zero point is the granularity of the reliability data. This is a byproduct of small reliability sample sizes. $\alpha$ is a proportion. For $\alpha$ to measure 0 exactly, the observed and expected disagreements would have to be identical: $D_o/D_e = 1$. Because $D_o$ is a count and $D_e$ is an average, it is mathematically rare that they are identical; hence, $\alpha = 0$ is nearly always approximated. Granularity has no effect on $\alpha = 1$. It tends to be very small and becomes insignificant as sample sizes become large.

The relationship between the granularity of data and the degrees of freedom in tabulating these data has been discussed in Section 3.4. That section also demonstrated $\alpha$'s equal interval scale. For nominal data $\alpha$'s intervals are $1/D_e$ generally – except $2/D_e$ for binary data – equally distributed over the whole range between $\alpha = 1$ and the smallest $\alpha$ possible for given data. For ordered data, the size of the intervals is smaller and not as easily stated. The point applicable to all metrics is that the scale of $\alpha$'s deviations from perfect agreement is just like the deviations from its absence. There is no bias punishing one part and elevating another part of $\alpha$'s scale.

This equal interval scale is important for interpreting $\alpha$. However, because reliability tests are conducted in the hope of observing perfect agreement or only minimal deviations from that ideal, it makes sense to some to use a different scale, one that values deviations from perfect agreement unlike the deviations from zero. The psychological effect of notable deviations from perfect agreement when the actual number of mismatching categories seem few have lead Perreault and Leigh (1989) to propose a coefficient $I_r$ which is the square root of the positive range of Bennett et al.'s (1954) $S$. The square root of any number between 0 and 1 is larger than the square-rooted number and gives the impression, and only that, of data having a higher reliability. Such scale transformations could be applicable to $\alpha$ as well. However, square-rooting a coefficient is a mathematical transformation of its scale and applied to $\alpha$ would not change what it indicates and cannot be recommended.

There is another source for α's sensitivity; the variance in the data, as already discussed in Chapters 0 and 4. This will be elaborated in the following section.

### 14.1.2 Variance and Alpha's Interpretability as a Reliability Coefficient

The role of variance in inferring the reliability of data from the measured agreements is often overlooked. α recognizes variance to be essential to assessing the reliability of data. As discussed in Chapter 1, all data need to provide enough information to compute correlations, define indices, select among competing hypotheses, and ultimately answer given research questions or support decisions on courses of action. When discussing the conditions under which observed agreement can serve as indication of the reliability of data worthy of analytical attention, Section 1.1 mentioned the needed relationship between the diversity of phenomena of analytical interest and the number of options provided in coding instructions to capture that diversity in order to provide the information needed to answer given research questions.

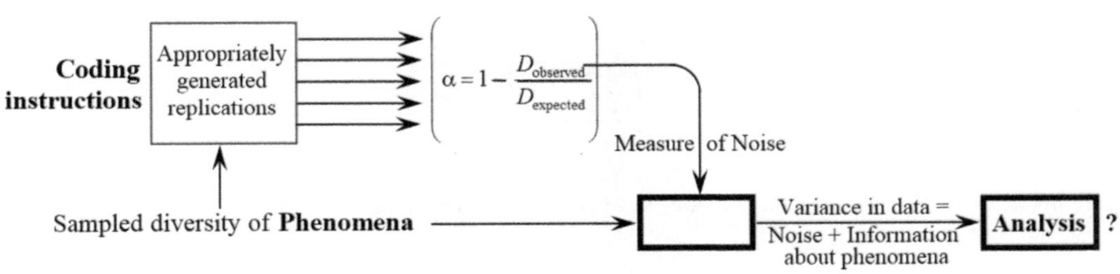


**FIGURE 14.1**
The Variance of Data Consisting of Noise and Information About Phenomena (2)

Figure 14.1 depicts the process of generating data as a channel of communication from the diversity of sampled phenomena to their analysis. When data contain imperfections, α becomes the proportion of the information in data about the diversity of phenomena as opposed to unrelated variance or noise. The extent to which data are not polluted by noise or disagreements is the extent to which the generated data can be trusted in subsequent analyses.

In this larger picture I am taking, variance in the data is a defining feature of the quality of any data. After all, "data" is the plural of "datum." α is a proportion of the extent to which the variance in data relates to the diversity of phenomena of analytical interest as opposed to irrelevant variations. Variance is necessary to make that distinction and determine whether data are worth analyzing or could lead researchers to potentially invalid conclusions for failing to provide the needed information to select among the answers to the questions they are pursuing.

To get a feel of various patterns of reliability data, Figure 14.2 depicts several examples in their canonical form, labeled A through P, representing the categorizations of seven units by three observers without missing data. The 21 cells in these 3-by-7 matrices contain categories in the form of dots whose differences are indicated by their sizes and shades. The figure locates these reliability data in two dimensions: Vertically, by their calculated α-coefficients and horizontally, by their variances.

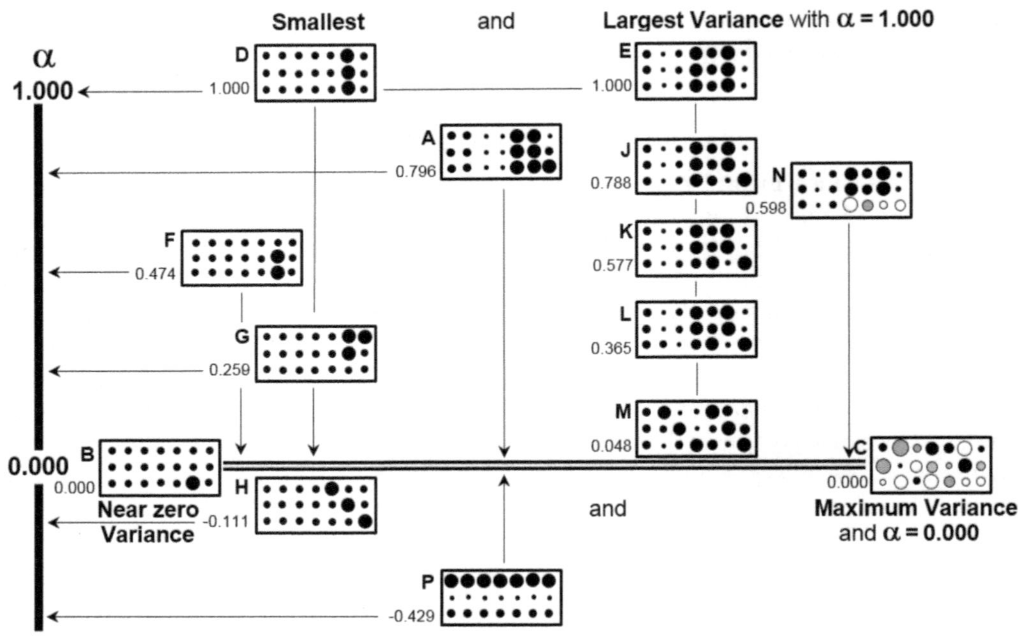


**FIGURE 14.2**
Reliability Data Within the Dimensions of Observed Agreement and Variance

The most typical reliability data exhibit matching categorizations for most units and a few mismatches. The example A in Figure 14.2 may serve this purpose. It consists of three categories with frequencies of 7 each. Six of its units are identically categorized. There are six units with matching values, and one with mismatching values. ${}_c\alpha = 0.796$ is slightly less than the often-considered lowest acceptable data reliability of 0.800. Reliability data can be located in two dimensions; a horizontal scale of the amount of variance between zero and its maximum whose numerical values are not of concern here, and the vertical scale measures α-agreements from its less interesting negative values to perfect agreement, ${}_c\alpha = 1.000$. Reliability data A are quite common. Only one of its seven units are systematically confused. On the scale between 0 and 1, it is closer to 1.

Near zero variance is exemplified by reliability data in B. It consists of 20 identical categories and one deviant. Observers do not exhibit any evidence of being able to distinguish among the two qualities assigned to the seven units. The lone deviant categorization does not make the data more informative about the phenomena of interest. This is reflected in the α-agreement of $\alpha = 0.000$. This example demonstrates also the relation between α and the information that data have to provide. Without some evidence of the ability to draw concurrent distinctions among the valuations of units of analysis, data do not provide information.

The most extreme case of zero variance consists of 21 identical categories contained in our 3-by-7 tables. Such data might be obtainable when observers were asleep, responded mindlessly, or their measuring instruments stopped working, in either case being unresponsive to possible differences among phenomena of interest. Absent any noticeable pattern within such data, researchers could not make comparisons, explain or find correlations, and would not learn anything about the diversity of the phenomena of possible interests. In this extreme case, observed and expected disagreements are equal and absent, and hence zero, and the α-agreement measures $\alpha = 1-0/0 = 0$ by definition. Just as in B, this example also demonstrates the meaninglessness of percent agreements for reliability assessments. Without reference to unequal distributions, without variance in reliability data, 100% agreement or 90% in B may sound impressive but says nothing about the analyzability of data.

The other extreme on the variance scale is exemplified by reliability data in C. Here, each of the 21 cells of the 3-by-7 matrix is occupied by a different category. Its variance is maximum; in communication theoretical terms, it contains only white noise. Such data cannot possibly communicate anything about the phenomena coded to researchers but for reasons opposite to B and the case of zero variance. Whereas the latter contains only agreements, C contains only disagreements. Their α-agreement is zero, as it should be, but for opposite reasons. In C, observed and expected disagreements are equal and 21 and there is no agreement on any unit, hence $\alpha = 1-21/21 = 0.000$.

What do the α-agreements of the other examples in Figure 14.2 reveal? On the opposite end of the α-agreement dimension we find the two reliability data D and E whose reliability is perfect: $\alpha = 1.000$. However, they differ radically regarding their variances. D contains only two categories, E seven. Within the constraints of our tables, D exemplifies perfect agreement with the smallest variance possible – two categories with frequencies of 18 and 3 respectively – E exemplifies perfect agreement with the largest variance possible – seven categories, three of each kind. D and E are the two extremes of seven possible perfect agreement coefficients whose tables are not shown. They demonstrate the proposition that the number of data measuring perfect agreement is always smaller than those measuring zero.

What are the implications of the unequal variances in D and E, both measuring $\alpha = 1000$? One is the amount of information they provide to a research project. D represents one reliable distinction between one unit and the remaining six. E represents reliable distinctions among seven units of analysis, and hence providing more information than D does.

Another implication is their unequal vulnerabilities to disagreements. When reducing the variance in D by one, from consisting of 18 and 3 categories to 19 and 2 in F, the agreement coefficient drops from $\alpha = 1.000$ by about half to $\alpha = 0.474$. Further loss of variance to 20 and 1 of B yields $\alpha = 0.000$. By contrast, adding confusions of categories to E, two at a time, and keeping their variance constant, leads from E to J, to K, to L and adding a few more to become M. Evidently, when the variance is large, α-agreements drop far more slowly than when the variance is small. This leads to the conclusion that perfectly reliable data with the smaller variances are more vulnerable to coding errors than data with larger variances. This is true whether their variance is constant as in the path from E to J to K to L and to M, or from D to G and to H, or when their variance is depleted as in the path from D to F to B and to the maximum variance.

Another path of decaying α-agreements can be explored by adding novel categories to reliability data. For example, adding one new category at a time to E's seven perfectly agreeing categories, reduces its $\alpha = 1.000$ to 0.896, then to 0.794, to 0.695, and to 0.598 of N. Adding ten more categories to N ends up in C of $\alpha = 0.000$.

Finally, data in P exemplify the presence of the largest possible systematic disagreement. Its three observers are perfectly correlated but never agree. P and A have the same variance. They demonstrate two properties of α. The first is that variance alone does not explain anything about the reliability of data. Their distribution matters. Second, as illustrated elsewhere, for 2-valued or binary data, α can be –1.000. However, as the number of categories increase, the maximum negative α moves away from that numerical extreme. With three categories, P's $\alpha = -0.429$. As the number of categories increases the minimum α converges to zero.

In the range of α's reliability interpretations, between $\alpha = 0$ and $\alpha = 1$, α is a measure of the proportion of information about phenomena of interest in data. This is the proportion on which researchers can rely on these data. The examples suggest that perfectly reliable data can lose their trustworthiness in two ways: by confusing categories across units or adding random variances to the data. Both ways reduce α and can end up with negative values when disagreements are systematic. Systematic disagreement is explainable by observers' predilections that have no meaningful relations to what reliable data are to stand for.

The importance of variance for establishing whether data provide information about phenomena to be analyzed needs to be kept in mind, especially in the following section.

### 14.1.3 The "Paradox" of Inadequate Variation Revisited

In the history of assessing the reliability of data, percent-like agreement measures, used by Osgood (1959) and Holsti (1969), revamped by Aickin (1990), to name but three, had already been replaced by chance-corrected agreement measures starting with Benini's (1901) β and in contemporary literature by Scott's (1955) π, Cohen's (1960) κ, Fleiss' (1971) *K*, and most recently Gwet's (2008) $AC_1$. Section 4.2 compared them to ${}_{c}\alpha_{nominal}$. This section sharpens the issues subsequently raised by Feinstein and Cicchetti (1990) with the help of the previous section.

Oddities of the popular kappa-coefficient have been discussed for some time (Krippendorff, 1978, 1980; Kraemer, 1979; Brennan & Prediger, 1981; Grove et al., 1981; Zwick, 1988). Feinstein and Cicchetti restated them in the form of two "paradoxes" of Cohen's κ (kappa), already mentioned in Chapter 3. The publication of their paradoxes found numerous echoes (Berry, 1992; Goldman, 1992; Sargeant & Martin, 1998; Gwet 2002, 2008; Viera & Garrett, 2005; Zhao, Liu, and Deng, 2012). Zhao et al. outdid Feinstein and Cicchetti by extending the list of paradoxes to thirteen and adding nine abnormalities to it – see my critical comments to their efforts (Krippendorff, 2012). The first of Feinstein and Cicchetti's paradoxes, described as "high agreement low kappa" is due to skewed distributions of the frequencies of categories in the reliability data. As in any move from examples E to B of Figure 14.2, it is mathematically inevitable that percent agreement converges to 100% as the distribution becomes more skewed. What these authors find paradoxical is that when variance declines, percent agreement increases but κ declines as well, eventually converging to zero. By focusing on the necessity of reliable data to possess variance, Section 4.2 and the previous section offered explanations of this property.

Feinstein and Cicchetti's second paradox consists of finding that unequal proclivities of observers for particular categories cause κ to increase. In other words, κ registers disagreements in the two observers' uses of categories as agreements (Krippendorff, 1978). Section 4.2.2 noted that κ's chance agreement has little to do with the absence of a relationship between the variance in data and the diversity of phenomena they are expected to represent. Instead, κ corrects its observed agreement by what is expected when two observers' use of categories are statistically independent of each other. This second paradox and all of its well-known oddities is limited to Cohen's κ and other coefficients that define chance as in correlation and association measures. It has been discussed in Section 4.2.2, does not apply to π, α, and *K*, and is not of concern here.

By contrast, Feinstein and Cicchetti's first paradox pertains to all chance-corrected agreement measures, including α. Section 3.6 already mentioned that the proposition "high agreement low kappa" privileges the measure of percent agreement over its chance-correction. Chance-corrected agreement coefficients were developed to overcome the inadequacies of percent agreement. Now privileging percent agreement is a step back into the history of inadequate reliability measurements. It is fueled by the misconception that percent agreement should positively correlate with coefficients capable of indicating the reliability of data. It blatantly ignores the purpose of reliability tests, which is to establish whether data can be regarded as reliable proxies for diverse phenomena and ultimately provide the evidence needed to select among valid answers to the guiding research questions. As argued in the preceding section, for generated data to inform conclusions about what data are meant to represent, variance is essential and information is what drives empirical research. As Claude E. Shannon (1949) taught us a long time ago, channels of communication without variance cannot convey information. Section 3.6 already dissolved Feinstein and Cicchetti's first paradox by placing it in the larger context of empirical research:

*As information in data (about the phenomena of interest) declines so does their reliability.* (14.2)

This reinterpretation acknowledges the central importance of phenomena-related diversity and conceptual variables that are capable of representing this diversity in data. In the absence of variance, data are mute and percent agreement is always 100%. This is the implication of Figure 14.1 which is convincingly illustrated in Figure 14.2. For example, E has both, D is reliable as well but with only a minimum of variance. As already discussed, D, having the smallest possible variance of perfect agreement is vulnerable to losing its reliability when minor errors enter. The paradoxical nature of the statement "high agreement and low kappa" is the result of a misconception that focuses on something other than what makes data reliable.

In reality, there are ample examples of phenomena of analytical interest that exhibit very low variances; for example, extremely rare diseases, global warming, minimal variations of the gravity on earth, and universally agreed-upon cultural traits. Generating analyzable data when variance is low is a methodological challenge. The literature suggests several ways to cope with such situations empirically. Testing for the occurrence of a rare disease calls for sample sizes large enough to include statistically significant numbers of cases or sampling within a smaller population known to be more likely susceptible to the disease. Evidence of global warming calls for longitudinal data and multiple sources of evidence for otherwise barely noticeable changes. Deniers of global warming tend to argue from analyses of present observations. For Galileo Galilei, gravity was a constant, manifest in the weight of objects. Isaac Newton, in his effort to predict the movement of planets, defined gravity as a function of unequal masses attracting each other. He analyzed observable phenomena outside of the constancy of gravity on earth. Cultural anthropologists allow themselves to be challenged by interacting with members of cultures other than their own whose behaviors do not conform to their expectations. Only by making a concerted effort to find variation can one start explaining phenomena that are normally stable, common, and undistinguished.

Searching for and finding conditions under which phenomena are variable and hence analyzable is one thing, generating data to establish the reliability of such data is another. Both analytical tasks require variability, however. Following are three examples that demonstrate the epistemological importance of creating variances when the reliability of data is at stake.

Fire extinguishers tend to have pressure gauges. In everyday life their pointers never change their position. The owner of a fire extinguisher has no clue as to whether its dial indicates the hydrostatic pressure inside it or is stuck and dysfunctional. Therefore, fire extinguishers need to be regularly checked by professionals who are able to de- and re-pressurize the extinguisher and certify their reliability. Without the ability to actively vary the pressure and observe changes, there is no way to be sure whether the dial measures what it is meant to indicate.

Communication is impossible to experience without observing its consequences. For example, getting no response to repeated email messages to someone does not tell their sender whether these messages actually left their computer, were received and ignored, went to a non-working address, or the addressee was not in attendance. It takes active interventions by another means of communication to explain the absence of response.

In content analyses, when coders describe all units identically, as in B of Figure 14.2, achieving 100% agreement, statistically naive researchers may be pleased but should be suspicious and examine why this is happening. Maybe the coding instrument was too insensitive to distinguish among the nuances of unlike units of analysis. Maybe the coders were tired and did not look hard enough to notice otherwise evident differences. Maybe they shared preconceptions and saw everything from a singular perspective. Knowing that agreement matters, coders could have agreed to assign the same category to every unit. They could have coded what they expected to see, not what they would have seen with an open mind. Researchers hoping to analyze data that were generated by employed coders would not know the difference unless variations are evident.

As these examples suggest, for reliability tests to establish whether data with very low variance can be trusted, researchers may have to artificially create the needed diversity of phenomena. This is justified by distinguishing two analytical tasks. Testing the reliability of data and testing the reliability of the coding instructions. To test the latter, researchers may well create phenomena that coders should be able to distinguish. Once the reliability of the coding process is established, the artificial phenomena can be removed from the data to provide information about real phenomena, not the ones created to test an instrument.

Incidentally, the difficulty of establishing reliability when errors are rare is also found in engineering. A U.S. Government handbook of engineering statistics states: "The Paradox of Reliability Analysis: The more reliable a product is, the harder it is to get the failure data needed to 'prove' it is reliable!" (NIST, 2012, Section 8.1.3). I would call this a serious inconvenience, not a paradox. Unable to test the reliability of large technological systems, engineers are known to estimate the reliability from their weakest components, by exposing them to a variety of conceivably extreme circumstances. The statistical reality of testing generally stable technological systems is not unlike the testing of the reliability of data with relatively low variance.

As suggested earlier, Feinstein and Cicchetti's first paradox seems to be the result of having adopted a conception of reliability that fails to acknowledge the role of information in the larger context of empirical research.

### 14.1.4 The Effects of Unreliability on the Reportable Frequencies

Unaccountable variance or noise in the data can impede reaching valid conclusions. Analytical techniques vary in sidestepping or amplifying the effects of error variances in data on the final results. Therefore, it is not easy to generalize the effects of unreliable data on the conclusions supported by them. Section 15.4 will discuss a methodology for studying the effects of unreliability. However, for very simple analytical steps, like reporting frequencies, such effects may be explored.

Section 3.3 demonstrated how observed coincidences can reconstruct as a function of the product of ${}_{c}\alpha_{\text{nominal}}$ and perfectly reliable coincidences plus the product of $(1-{}_{c}\alpha_{\text{nominal}})$ and the expected coincidences. So, if all preconditions for generating reliability data are met, ${}_{c}\alpha$ equals the proportion of information in the variance of the generated data, i.e., the proportion of values whose meanings the researchers can be sure of among the total that includes errors and misjudgments.

Reporting observed frequencies together with the reliability of the variable may not tell the whole story. Even when data measure $_c\alpha = 0.800$, which is generally taken to be an acceptable level of reliability, the observed frequencies may not look as good as they seem. For example, when a research project aims at reporting the observed frequencies $n_a$ of category *a* among several categories out of a total of $n_{..}$, only 80% of this frequency is unquestionable, the remaining 20% remain uncertain.

For once, the α-agreement of a variable is the average of the agreements of its values. The frequencies of one of its values may be more reliable or less than the others. Section 12.5 suggests how to evaluate the reliability of individual categories. Assuming all values are equally affected by noisy data, let me merely explore which frequencies for one observed category *a* one may want to report. Referring to (12.9), and assuming $_c\alpha_{\text{nominal}} = 0.800$, Figure 14.3 decomposes the observed frequency $n_a$ of a category *a* which is what a researcher is likely to report as finding.

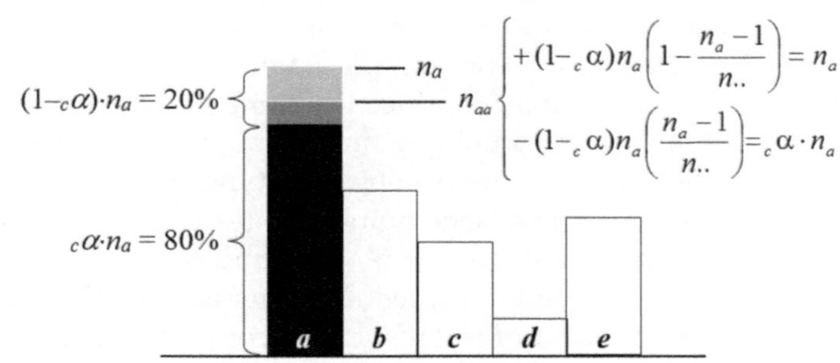


**FIGURE 14.3**
Unreliability Affecting Reported Frequencies When $_c\alpha = 0.800$

As already mentioned, with a measured reliability of $_c\alpha = 0.800$, only 80% of the observed frequency $n_a$ is unquestionable. $n_a$ overestimates the frequency known to be reliable by up to 20%. By reference to a coincidence matrix for this variable, one may be inclined to report the frequency $n_{aa}$ on which all observers agree. However, $n_{aa}$ is not certain either as it includes agreements by chance which may vary between the sum of all uses of *a*, $n_a$, and the smallest frequency $_c\alpha \times n_a$ one can trust as indicated in Figure 14.3. The difference $n_{a.} - n_{aa}$ is what should probably be reported as error.

Whether to report $n_a$, $n_{aa}$, or $_c\alpha \times n_a$ as findings may have to be made dependent on the costs of using polluted frequencies (see Chapter 15). A 20% uncertainty regarding what can be reported with some certainty may well make a big difference when it comes to predicting close elections, deciding on risky investments, making medical diagnoses, or rendering decisions in court that could make a difference between life and death.

## 14.2 Sampling Issues

Reliability data can be considered adequate in size when its calculated α-agreement is generalizable to the reliability of the population of data to be analyzed at a chosen level of statistical significance. Smaller than adequate sample sizes reduce the certainty by which the reliability of the population of data can be inferred from the calculated $\alpha$-coefficient.

It is important to realize that agreement measures focus on any one variable to begin with. Research projects tend to generate data in terms of many variables. Some have numerous, others have few categories. Some represent common and easy distinctions, others rare and difficult ones. Such dissimilarities may lead to unequal demands on the size of adequate sample sizes. When the sample size of data in numerous variables is the issue, the variable that calls for the largest sample size is the one to be chosen for determining the sample size for all.

For any one variable, the size of its reliability data equals the number of pairable values that contribute to the agreements measured. When predefined units are single-valued, this number of pairable values depends not only on the number of units judged and the number of observers judging them, but also on how many categories or values are available for recording these judgments, and the rarity of the phenomena to be attended to. Moreover, the statistically adequate number of pairable values depends also on the desirable agreement to be achieved and the level of significance of the anticipated findings.

Within such limits, to be discussed in the following subsections, the number of pairable values needed for an agreement measure to be an unquestionable indicator of the reliability of data can be accomplished variously. For example, coding a large number of phenomena by the minimum of two observers may produce the same number of pairable values as having a large number of observers coding the minimum of two phenomena. Recording the answers to a few questions by very many subjects is typical of public opinion research and psychological tests. Categorizing a large number of textual units by only two coders is the common practice of many content analysts. The tendency to use a small number of human coders tends to be economically justified. Training and employing human beings for the task of recording phenomena of interest can become expensive, especially when this task is not easily definable. However, either extreme may well satisfy the statistical certainty with which an $\alpha$ can be reported but runs into inadequacies of a different kind. The optimum is likely between these two extremes, considering other criteria.

Methodologically, the choice of the proportion of observers × units has to do with whether the data-generating process is generalizable in all of its dimensions. The need of observers to be representative of the stakeholders of research will be addressed in Section 14.2.1, of recording units to be representative of a population of such units in Section 14.2.2, and of categories or values of variables in Section 14.2.3. Before getting to these dimensions, several clarifications are in order.

The two previous sections drew their examples from the single-valued coding of predefined units, mainly because the role of variance in ${}_c\alpha$-coefficients was easier to explain and pair comparisons are relatively obvious. Finding adequate sample sizes is a problem also in assuring the reliability of the unitizations of continua. ${}_u\alpha$-coefficients pair segments of a continuum and the number of pairable segments too must be sufficiently large for ${}_u\alpha$-coefficients to be interpretable as indicators of the reliability of data. If that number is found lacking, one may increase either the number of their replications by adding observers, unitizers, or annotators to the reliability tests or by asking them to unitize longer texts or additional documents. It may be noted that ${}_u\alpha$ and ${}_{\phi|u}\alpha$ developed in Chapter 8 concern partitions of continua and ${}_U\alpha$ in Chapter 9 concerns all possible pairs of coded segments. However, pairability is limited to the intersections of valued segments when ${}_{cu}\alpha$ and ${}_{cU}\alpha$ are evaluated.

When data are multi-valued, two cases need to be distinguished. Most multi-valued agreement coefficients rely on difference functions (7.2). (7.4), and (13.1) discussed in Chapters 7, 8, and 13. Such difference functions are defined for sets of values and treat the values as recoverable member of sets. (7.15) defines a standardized multi-valued difference function that uses the values in sets merely to establish the degrees of resemblance between sets. Their sizes are determined without reference to the size of sets and their sample sizes are a function of the number of pairable sets.

Finally, when the reliability of distinctions $_D\alpha$ within continua is at stake, as developed in Chapter 10, the number of distinctions equals one minus the number of intervals an observer distinguishes. Although $_D\alpha$ pairs distinctions, for estimating the required number of distinctions, the lengths of the intervals between them remain critical, with the smallest interval being decisive.

The lesson learned from the discussion of Zhao et al.'s example of reliability data about rare phenomena in Section 3.6 applies to all of these cases. Adequate sample sizes are a function of the most unlike kind of phenomena that observers need to distinguish from all other phenomena. If a rare phenomenon occurs only once, as exemplified in the reliability data D of Figure 14.2, even if the agreement is perfect, data are extremely vulnerable to decay by minor mismatches. For reliability data to be adequate in size, they need to be reasonably robust.

To obtain adequate sample sizes for $\alpha$ to be statistically significant, Bloch and Kraemer's (1989, p. 276) provide an expression, their Formula 3.7, to estimate the number of units required for 2-by-2 tables to show statistically significant relations. In the context of evaluating reliability, and with the possibility of employing more than two observers, we are concerned with pairs of values, which are twice the number of units in Bloch and Kraemer's formula. Accordingly, the minimum *number of pairable categories* (not of the recording units a la Block & Kraemer) required to interpret $\alpha$ as a measure of the reliability of data is a function of the value or set of values that occurs least often or the smallest interval distinguished:

$$T(P_{min}, \alpha_{ok}, p) = 2z_p^2\left(\frac{(1+\alpha_{ok})(3-\alpha_{ok})}{4(1-\alpha_{ok})P_{min}(1-P_{min})} - \alpha_{ok}\right) \tag{14.3}$$

Where:

$T(P_{min}, \alpha_{ok}, p)$ = the required number of pairable values, kinds of sets, or distinctions

$P_{min}$ = the smallest probability among values $c$ of any one variable when data are single-valued or computed as members of multi-valued sets $C$

= the smallest probability among sets $C$ of values reliability is computed with standardized multi-valued difference functions

= the proportionally smallest interval $^{L(Sig)}/_L$ between any two distinctions when the reliability of distinctions is computed with $_D\alpha$

$\alpha_{ok}$ = the minimum $\alpha$ below which data would have to be rejected as unreliable

$p$ = the desired level of statistical significance, represented in the $z_p$-values for one-tailed tests

I am not taking the restatement (14.3) of Bloch and Kraemer's formula as the last word on statistically adequate sample sizes for $\alpha$. There might well be other ways of estimating the required number of pairable values, sets of values, or distinctions. But this expression provides a convincing approximation.

Because calculating the required number of pairable values, sets, or distinctions in Replications × Units reliability data matrices is somewhat involved, Table 14.1 lists the minimally required number for three smallest acceptable reliabilities $\alpha_{ok}$, four levels of statistical significance $p$, and several probabilities $P_{k=min}$, also expressed in terms of the numbers $v$ of equally likely values. Researchers may be able to extrapolate sample sizes whose individual requirements are not listed.

| Smallest acceptable $\alpha_{ok}$: | .667 | | | | .800 | | | | .900 | | | |
|---|---|---|---|---|---|---|---|---|---|---|---|---|
| Significance level $p$ (in $z_p$): | .100 | .050 | .010 | .005 | .100 | .050 | .010 | .005 | .100 | .050 | .010 | .005 |
| $P_k$ = .500; $v$ = 2 values | 36 | 60 | 119 | 146 | 62 | 103 | 206 | 252 | 128 | 211 | 422 | 518 |
| = .333; = 3 | 41 | 67 | 135 | 165 | 71 | 116 | 233 | 285 | 144 | 238 | 477 | 584 |
| = .250; = 4 | 49 | 81 | 161 | 198 | 84 | 139 | 277 | 340 | 172 | 283 | 566 | 694 |
| = .200; = 5 | 58 | 95 | 190 | 233 | 99 | 163 | 326 | 400 | 202 | 332 | 667 | 815 |
| = .167; = 6 | 67 | 110 | 220 | 270 | 114 | 189 | 377 | 462 | 233 | 384 | 768 | 941 |
| = .143; = 7 | 76 | 125 | 251 | 307 | 130 | 214 | 429 | 526 | 265 | 436 | 872 | 1069 |
| = .125; = 8 | 85 | 141 | 281 | 345 | 146 | 241 | 481 | 590 | 297 | 489 | 980 | 1198 |
| = .111; = 9 | 95 | 156 | 312 | 383 | 162 | 267 | 534 | 654 | 329 | 542 | 1083 | 1328 |
| = .100; = 10 | 104 | 172 | 344 | 421 | 178 | 293 | 587 | 719 | 361 | 595 | 1190 | 1459 |
| = .050; = 20 | 200 | 329 | 657 | 806 | 340 | 560 | 1119 | 1372 | 657 | 1131 | 2263 | 2775 |
| = .020; = 50 | 487 | 802 | 1604 | 1966 | 825 | 1360 | 2719 | 3334 | 1675 | 2759 | 5534 | 6765 |
| = .010; =100 | 966 | 1591 | 3182 | 3901 | 1640 | 2701 | 5403 | 6624 | 3307 | 5447 | 10896 | 13359 |

**TABLE 14.1**
The Number $T(P_{min}, \alpha_{ok}, p)$ of Pairable Values Minimally Required for $\alpha$ to be Credibly Above $\alpha_{ok}$ Regarding the Reliability of Distinctions between Any One $k$ and All Not-$k$s

Regarding the probability $P_{min}$, it is important to distinguish three separate aims of reliability tests which entail different conceptions of $P_{min}$:

- The *reliability of the data* to be analyzed. According to the above discussion, the value $k$ with the smallest frequency or proportion $P_{min}$ in (14.3) is the bottleneck for the size of adequate reliability sample sizes.
- The *reliability of a coding instrument*. When a variable is defined by $v$ categories with unknown frequencies, reliability data should include enough cases for each available category. In this case, the adequate size of reliable data cannot be linked to the probability of values on the population of data, rather to the number of values available for coding. Ideally, each of the $v$ categories or values of a variable should occur in sufficient frequency, in (14.3) with $P_{min} = 1/v$ and cover all $v$ values.
- When co mputing the agreement coefficient ${}_D\alpha$, $P_{min}$ should be ½.

The implications of the first two aims will be made clearer in Section 14.2.2.

To generate reliability data that enable statistically significant inferences about the reliability of the data to be analyzed, it is the rare phenomena that drive the required sample sizes up. However, when the categories or values are of unequal relevance to a research question, the category with the smallest probability may not matter. Instead, one may want to take the smallest probability among relevant categories as the basis for estimating adequate reliability sample sizes. Consider the probabilities of a population of five kinds of voters.

| | | | |
|---|---|---|---|
| Registered Independent | $P_i = 0.10$ | 0.23 | 0.33 |
| Registered Republican | $P_r = 0.15$ | 0.34 | 0.33 |
| Registered Democrat | $P_d = 0.19$ | 0.43 | 0.33 |
| Not voting | $P_n = 0.52$ | | |
| Unknown | $P_u = 0.04$ | | |

This example tabulates the voting records of $v = 5$ kinds of citizens. In the first column, the smallest proportion $P_{min} = P_u = 0.04$ is that of the default category "Unknown." As default categories attest to the lack of researchers' knowledge or interest in all categories of a variable, which provides no information about party affiliations, the proportion $P_u$ would most likely not be a good candidate for determining the required reliability sample size.

The next smallest proportion is $P_{min} = P_i = 0.10$, the proportion of independent voters. If this category of voters were of analytical interest, assuming the political scientist hopes to achieve $\alpha \geq \alpha_{ok} = 0.800$ at a $p = 0.05$ level of significance, (14.3) as tabulated in Table 14.1 would suggest the need for at least 293 pairable values. Coding 147 units by two observers or 98 by three, does not guarantee, of course, any agreement, only that a measured $\alpha \geq 0.800$ can serve as a statistically significant indicator of the reliability of the data.

Suppose the analyst is interested only in registered voters and excludes from the analysis those not voting and unknown. This would yield very different proportions, listed in the second column in the above example. The smallest proportion is 0.23. Trying to achieve the same data quality, according to (14.3) and Table 14.1, the researcher would need to generate 147 pairable values, half of the number required in the previous estimate.

If the value with the smallest frequency cannot be estimated in advance of generating the data whose reliability is to be evaluated, a researcher may want to start with a sample of the size needed to test the reliability of the coding instructions and adjust it as observed frequencies emerge.

For a variable with $v$ alternative codes, aiming at equal frequencies for each, one could consult Table 14.1 with $P_{min} = {}^1/_v$. For our example of coding three kinds of registered voters $P_{min} = {}^1/_3$. With the same statistical parameters, $\alpha_{ok} = 0.800$ at $p = 0.05$, according to Table 14.1: $T({}^1/_v, \alpha_{ok}, p) = T({}^1/_3, 0.800, 0.05) = 116$. This would suggest that the researcher asks three coders to identify the party affiliation of 39 registered voters. As it is unlikely that they turn up in a uniform distribution, it would therefore be advisable to continue sampling until the value with the smallest frequency satisfies the chosen statistical parameters. In our hypothetical example, finding 23% independent registered voters would allow the researcher to stop at 147 pairable values or 49 voters identified by three coders.

It is important to recognize that the definition of a statistically adequate sample size in (14.3), conveniently tabulated in Table 14.1, concerns $\alpha$ in its role as a reliability measure. As mentioned above, the required number of pairs of categories or values for alpha to be taken seriously can be achieved variously. It leaves researchers choices that need to satisfy additional criteria, not included in the $\alpha$-agreement measure. The following sections will elaborate on the criteria for the number of observers, coders, or judges, i.e., replications, on the volume of the data whose reliability is in question, and on the information about relevant phenomena data are meant to provide about the data and about the coding instructions respectively.

### 14.2.1 Regarding the Number of Replications

In many content analyses, reliability data are obtained by asking two individuals to independently replicate the coding of a set of units of analysis or segments of textual continua. Using only two replications may well suffice when the process of generating data is mechanically realized. By design, mechanical instruments respond causally to the phenomena to be recoded. As such, they respond to the physical qualities of the objects measured, ignore their contexts, and measurement errors tend to be systematic; for example, by always measuring a couple of degrees from the true temperature or failing to recognize special characters in texts. But in social research, in content analysis in particular, the phenomena of interest tend to be context dependent, symbolic, culturally significant, and afford multiple readings that only competent human beings can sort out. The mindless preference for generating data for content analyses by employing only two observers, coders, judges, or annotators is encouraged by at least three factors, none methodologically sound:

- *The false analogy between interpreting textual phenomena and applying mechanical measuring devices.* Reading texts is far more complex than identifying character strings in texts, which is what word counts indicate and search engines identify. Disagreements among human observers are rarely due to omitting objectively measurable features of phenomena. Rather, they tend to be due to lacking the literary competencies, intellectual backgrounds, open-mindedness and membership in communities with vested interest in particular research results. In addition, and unlike mechanical devices, humans can hardly avoid learning during repetitions, creating efficient shortcuts which introduce unnoticed drifts away from given coding instructions.
- *Available software.* Most software to calculate percent agreements, consistencies, associations, co-variations, and correlations tends to be limited to two observers or variables. Calculating agreement between 2 judgments or codes is conceptually obvious and softwares that provide numerical summaries are easy to understand. Their popular use gives the impression that there is no need for more than two replications.
- *Misguided attributions of reliabilities.* Calling agreement coefficients "inter-observer," "inter-coder," or "inter-rater" reliability measures naturally associates reliability with the relationships between human observers, coder, or raters. It is a truism that humans do not work like mechanical measuring instruments do. Zhao et al. (2012) and Feng (2013) go so far as to psychologize their conception of reliability by equating the lack of it with the difficulties that coders experience in achieving consensus and avoiding errors. Their conception diverts attention from the reliability of data to human competencies, just as Cohen's $\kappa$ is measured relative to the independence of two coders with their own predilections. If agreement measures are to be indicative of the extent to which data can be trusted as analyzable proxies for phenomena of analytical interest, the psychology of human coders has no place in that conception. I suggest the notion of inter-coder reliability is mistaken. Reliability is not located between coders. Agreement observed among independent replication is a means to infer the reliability of data.

To determine the number of replications, the purpose of reliability assessments needs to be kept in mind. Reliability tests are conducted to reassure the analysts that their data represent the phenomena they are concerned with, and are understandable by all who handle them including the conclusions of interest to the stakeholders. All of these assurances are embraced by the requirement that the process of generating data be replicable elsewhere, by equally instructed but different observers, chosen with the same criteria from a common pool of potential observers. The adequate number of replications depends on how certain one wants to be that they could be reproduced elsewhere. A researcher's assistants, good friends, or unique specialists are unlikely to be found in different social environments, and hence are unacceptable candidates for generating reliable data. The set of observers hired to generate the data for an analysis ought to be minimally representative of all the stakeholders in the research undertaken. Two replications rarely are.

The radical opposite of hiring and training two sympathetic coders is crowd coding, using software such as Amazon's Mechanical Turk to recruit many coders on the internet. The reliability problem faced by Julien Cornebise, discussed in Chapter 6, was due to an extraordinary number of up to 56 coders recording their assessment of 2,653,322 units of analysis, which resulted in 11,289,371 binary judgments by 3,712 coders. The great number of replications served Cornebise's project well. By selecting among the most trustworthy coders, he could report decent agreement measures.

Crowd coding outsources the process of coding to generally anonymous volunteers with internet connections. Its returns rely almost exclusively on written coding instructions, and do not involve training sessions which set scientific observers, coders and judges apart from untrained ordinary folks. This is not to say that crowd coding is without methodological problems. Except for being competent internet users and having time to volunteer their participation in a researcher's project, the coders' qualifications remain essentially unknown. The complexity of the coding tasks cannot exceed the capabilities of volunteers. Researchers who see benefits in crowd-coded data have difficulties controlling the knowledge, motivations, and social backgrounds of their coders. To prevent unreliabilities from polluting the crowd-sourced data due to incompetencies, disinterests, and personal biases, Section 12.7 already discussed the practice of adding test questions to their coding instructions for which the researchers know the answers. Such tests may well have methodological benefits as they standardize participation and ensure the ability of understanding what is to be done. However, crowd coding is known for other biases. Coders self-select their participation and may do so only if they are in line with questions posed by researchers. Nevertheless, crowd coding is a promising way of getting to many replications speedily and with low costs.

To repeat, while researchers do have options for how to satisfy the requirement of statistically significant reliability tests according to (14.3) and Table 14.1, to ensure the replicability of generating data calls for employing observers, coders, or competent judges whose qualifications are available elsewhere, and can represent the stakeholders in a research who need to be able to make sense of its results. Two coders rarely are enough.

The required number of pairable values also needs to capture the diversity of phenomena of concern for a research project. This will be the topic of the following section.

### 14.2.2 Regarding the Volumes of Recording Units and Texts

To answer the question of how many phenomena to include in a reliability test – observations, events, segments of texts or quotes, Regarding Information About Relevant Phenomena
or measurements – one has to distinguish four nested sets of phenomena.

(a) The universe of phenomena of ultimate analytical interest

(b) Samples of recorded phenomena in (a) to be analyzed in lieu of that universe

(c) Reliability data, that are part of (b), replicated to test (b)'s reliability

(d) Reliability data that include (c) but also other recordable phenomena needed to test the reliability of the instrument intended to generate (b) and (c)

The universe of (a) is conceivable but not knowable without data in (b). The data of (b), (c), and (d) respond to phenomena selected from (a). Inasmuch as reliability data (c) are concerned with the reliability of (b), (c) should be a subset of or equal to (b), and the reliability data (d) must intersect with (c) but may also include samples from the conceived universe (a).

Sampling (b) from the universe of phenomena (a) is the subject of well-established statistical sampling theories. It is not of concern here.

Sampling (c) from (b) is not too different from sampling (b) from (a). However, because (b) and (c) are data, whether the reliability data (c) are large enough to ensure that measures of agreement among its replications can indicate whether the data (b) can be taken as reliable surrogates of phenomena of interest is more easily established.

Selecting observers or instruments for replicating the process of generating analyzable data was the topic of the previous section. This section addresses mainly two issues.

- The agreement among replications in (c) to infer the reliability of data in (b) about (a).
- The agreement among replications in (d) to infer the reliability of the coding instructions, applied to generate data in (b) and (c) plus categories or values of yet unobserved phenomena in (a).

Regarding the first issue, randomly sampling units from a larger population of units is widely considered the gold standard. It ensures that all units of a population of units have an equal chance of being included in the sample. True random sampling requires that the units of a population be enumerable, can be assigned a random number, which in turn determines their inclusion in the sample. Other than the boundary of the population of interest, random sampling assumes no knowledge of the composition of that population. Analyzing a randomly drawn sample is to discover differences within that population – sometimes considered a fishing expedition. In much of social research, the boundary of the universe of analytical interest tends to be easily articulated but remains numerically fuzzy. Usually researchers do have some inclinations of the diversity within that universe that they can take advantage of.

Unlike the universe of phenomena of ultimate analytical interest (a), the analyzable data (b), sampled from and analyzed in place of (a), are finite and enumerable. To select (c) from (b) random sampling is recommended but may not be feasible when the universe (a) is non-enumerable and (b) is not yet established. For these reasons, I already recommended sampling data for (c) according to (14.3) by assuming uniform probabilities for each defined value of a variable and continuing sampling until the required sample size for unequal frequencies of values is reached.

Regarding the second issue, testing the reliability of a coding instrument, the anticipated distribution of phenomena in (a) and the actual distribution of data in (b) is less important than that all categories in a coding instrument are covered. The test of the reliability of (d) calls for a stratified sample that includes all conceived categories of phenomena more than once, whether the universe of phenomena includes all category of a coding instrument or not. Rare or absent categories need to be sampled more selectively or have to be artificially constructed to make sure that observers recognize the qualities that the coding instructions define. Making up examples of rare phenomena or phenomena that a variable includes without evidence that they occur in (a) sounds unscientific, but it is not. In the social sciences, most researchers start out with theoretical conceptions of the universe of phenomena of their interest not knowing which turn out to be absent. If the variables of a recording instrument are published and made available to other studies, their reliabilities must hold for all possible data in (b) to which that instrument is applied.

For example, before designing a questionnaire, public opinion researchers tend to decide the populations of interest in terms of easily identifiable variables. People are of a certain age, have incomes, achieved distinct levels of education, and live in particular locales.

But they also have their clients' questions and predefined answers in mind. Typically, pollsters interview not all but well-delineated populations of people of interest to their clients who are expected to be able to answer their questions. Sampling for select groups of people, stratified sampling, amounts to random sampling within particular categories or subgroups of people (within which variance is assumed to be relatively low), making sure that each category or subgroup of units (between which variance is assumed to be relatively high) is represented equally or at least fairly in the sample. Stratified sampling does not aim at representing a whole population proportionally, only within predefined strata.

In assessing the reliability of a coding or measuring instrument, the categories or values it consists of define the potential strata of a population of phenomena to be coded. It follows that the reliability data (d) need not refer to the actual distribution of units in (a) or in (b) but include adequate numbers of phenomena of each category or value that an instrument distinguishes. By comparison of how (c) is sampled, (d) may well oversample rare phenomena, undersample common ones, and make up examples that are expected but have not been found. Ideally, reliability data (d) are uniformly distributed over all available categories and numerically satisfy (14.3).

The reliability of a coding instrument is needed when its variables govern numerous research projects. If a coding instrument is established, it should apply to all situations regardless of their empirical distributions.

By contrast, tests of the reliability of data that lead a researcher to published conclusions do not need to care for categories conceived but never observed. However, the measured agreement can be considered a valid indicator of the reliability of data in (b) only when the reliability data (c) is a representative sample of data in (b). The replicability of data always refers to the ability to reproduce the data from the very same set of phenomena elsewhere and by different observers.

In social scientific practices replicating the data is rare. What is not so rare is that researchers utilize the coding categories of one study and continue exploring how they relate to other phenomena. To be able to claim that the categories or values used in one study would be reliable in other studies, requires that all categories that the coding instructions spell out are actually observed with sufficient frequencies.

These two reliabilities should not be confused and they may require different sample sizes.

### 14.2.3 Regarding Information About Relevant Phenomena

Section 14.1.2 showed how the $\alpha$-coefficient distinguished two kinds of variance within the variance of the replicated data: variance that conveys information about phenomena of interest to an analyst and variance of irrelevant origins or noise which is extraneous to how phenomena are defined. Figure 14.2 moreover exemplified that reliability data with identical $\alpha$-coefficients could vary greatly regarding their variance. For example, while D and E in Figure 14.2 measure $\alpha = 1.000$, D exhibits the smallest and E the largest possible variance within the constraints of these examples. It is apparent that D is most and E least vulnerable to small disagreements. Section 14.2 suggested that reliability data like D require sample sizes larger than E in order to ensure that $\alpha$ can be taken as a measure of the reliability of data. This was already anticipated in expression (14.3) which suggested that the ability to infer the reliability of data is loosely related to the product of the agreement and variance measures.

I once devised two simple variance indices (Krippendorff, 2011) which relate observed distributions of values in any one variable to the potential information about the phenomena attended to. Both make use of (14.3) and are of the general form:

$$0 \le i = min\left(1, \frac{\text{Number of pairable values observed}}{\text{Number of pairable values required}}\right) \le 1 \tag{14.4}$$

The proportion $i$ is capped not to exceed 1, at which point the reliability data exceed being informationally adequate and the α-coefficient can be said to indicate the reliability of the data without doubts. When the variance is zero, all values are identical, $P_{min} = 1$, $T(P_{min}, \alpha_{ok}, p) = \infty$ and $i = 0$.

The following section presents the information index $i$ for analyzable data and Section 14.2.3.2 for coding instruments – in both cases for any one variable.

#### 14.2.3.1 In Analyzable Data

Following the estimation of adequate sample sizes in (14.3), the proportion of adequate numbers of observed values available in the data is:

$$i_{data} = min\left(1, \frac{n_{..}}{T(P_{min}, \alpha_{ok}, p)}\right) \tag{14.5}$$

Where:

$T(P_{min}, \alpha_{ok}, p)$ = the required number of pairable values, sets, or distinctions from (14.3)

$n_{..}$ = the actual number of pairable values whether reliability data consist of single- or multi-valued predefined units or identified segments in partitions of continua

= the actual number of pairable sets of values when reliability data are analyzed with standardized multi-valued difference functions or consist of contiguous segments in continua

= the actual number of distinctions in reliability data when ${}_{D}\alpha$ is computed

For example, suppose the value used least often occurs 12 times out of a total of 446, then $P_{min} = {}^{12}/_{446} = 0.027$ or, to be conservative, about 0.025, which is found in Table 14.1. Suppose one aims for $\alpha_{ok} = 0.800$ at $p = 0.050$, the table shows $T(P_{min}, \alpha_{ok}, p) = 1095$ as the required number of values. Therefore, with a total of 446 available values, $i_{data} = {}^{446}/_{1095} = 0.407 < 1.000$ for this variable.

The quantities that define $i_{data}$ may also be used to determine how many values would have to be added to the data for the α-coefficient to indicate their reliability with the chosen level $p$ of statistical significance:

$$n_{\text{missing values}} = \left(1 - i_{data}\right) T(P_{\text{min}}, \alpha_{ok}, p) \tag{14.6}$$

(14.6) assumes that sampling would continue with the same population characteristics. In the above example, there would be a need to add a total of $n_{\text{missing values}} = (1-0.407)1095 = 638$ values for reliability data to be adequate indicators of the reliability of data as depicted in Figure 14.4.

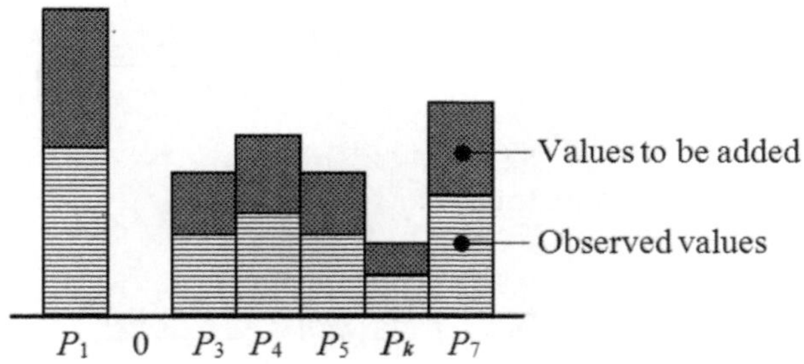


**FIGURE 14.4**
Observed Values in Data and Values to Be Added When Sampling Continues

Graphically:

When adding pairable values to a sample of reliability data, even when drawing them from the same source, it is quite common that the population characteristics shift. Even new categories of phenomena show up, perhaps in smaller numbers than the previous minimum. If this happens one may have to sample repeatedly until the desired numbers have been attended to, proceeding by analogy to the iterative process described at the beginning of this Section 14.2.

#### *14.2.3.2 In Coding Instruments*

Sampling so that the reliability of analyzable data can be inferred from $\alpha$ refers to the sampled proportions of values, sets of values or intervals between distinctions made. When categories or values of a coding instrument do not occur in the reality of data to be analyzed, they do not matter. There would be no need to sample what is not observable and codable.

However, as already mentioned, a coding instrument needs to be demonstratively reliable in the face of all possible phenomena to which it is designed to apply. It follows that in the reliability data that are adequate to confer reliability to the coding instrument, distinctions among all values, sets of values, that an instrument provides need to be demonstrably reliable, and hence present for a reliability test. There is no reason to consider the distribution of phenomena. The frequency of any one category or value $k$ of a variable needs to have or exceed the same minimum number of values whose sum equals $N_{k|not\text{-}k} = T(P_k, \alpha_{ok}, p)$ where $P_k$ is the proportion of any one of $v$ categories $k$ in the reliability data.

$$i_{\text{instrument}} = \sum_{k=1}^{v} min\left(\frac{1}{v}, \frac{n_k}{T(^1/_v, \alpha_{ok}, p)}\right) \tag{14.7}$$

Where: $v$ = The number of categories or values $k$ that the instrument makes available for coding given phenomena

$n_k$ = The number of pairable values $k$ in the reliability data, including $n_k = 0$

$T(^1/_v, \alpha_{ok}, p)$ = The minimally number of pairable values required for a uniform distribution for $v$ values as defined by (14.3)

To achieve or exceed that ideally uniform distribution of pairable values, one would have to add:

$$n_{\text{missing values k}} = \left(\frac{1}{v} - min\left(\frac{1}{v}, \frac{n_k}{T(^1/_v, \alpha_{min}, p)}\right)\right) T(^1/_v, \alpha_{min}, p) \tag{14.8}$$

to the reliability data and a total of:

$$n_{\text{missing values}} = \sum_{k=1}^{v}\left(\frac{1}{v} - min\left(\frac{1}{v}, \frac{n_k}{T(^1/_v, \alpha_{min}, p)}\right)\right) T(^1/_v, \alpha_{min}, p) \tag{14.9}$$

Graphically, Figure 14.5 depicts how many pairable values would have to be added to the example of reliability data depicted in Figure 14.4:

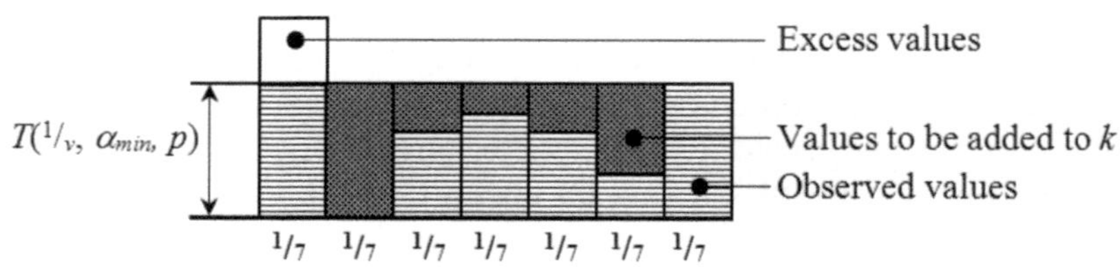


**FIGURE 14.5**
Observed Values in Instrument and Values to be Added to Each $k$ Specifically

Generally, uniform distributions call for fewer numbers of pairable values than distributions that require a minimum number for their least likely value. This can be seen in comparing Figures 14.4 and 14.5. Testing the reliability of an instrument gives the impression of being easier, even if a value was not observed as depicted in the second value from the left of the two figures. However, its requirement that all values are observed with a minimum frequency poses an empirical problem, already noted in Section 14.2.2 regarding the minimum sample sizes of units. To selectively sample units to satisfy the quota for individual categories may be even more difficult when phenomena of the kind needed are rare or not yet observed. In case of the latter, it is perfectly legitimate to purposively select missing phenomena from diverse sources, even construct them for this purpose as long as they are realistic to test the abilities of observers to distinguish among them.

The above did not mention the reliability of instruments for generating multi-valued data. In such instruments, the definitions of categories or values is independent of how coders group them into sets. Therefore, this should not present problems for finding the adequate sample sizes for categories or values according to Table 14.1.

Note also the omission of references to the reliabilities of drawing distinctions in continua of finite lengths, mentioned as a third alternative in Section 14.2. Because distinctions do not address what is distinguished and why, the sample sizes of distinguishable phenomena play no role here.

## 14.3 Alpha's Distribution

Section 14.2 discussed adequate sample sizes for αs obtained from replications of coding subsamples of the larger data whose reliability is in question. It quantified the theoretically adequate number of pairable values (14.3) and tabulated them in Figure 14.1 but added three dimensions that need to be satisfied to consider data to be considered reliable: adequate numbers of replications, samples of phenomena that are representative of the data whose reliability is in question, and the diversity of the phenomena to be turned into informative data. Section 14.2 was intended to facilitate the design of reliability tests, adjust to situations when reliability data are too small to serve their intended use, and evaluate whether a reliability test performed on a sample of phenomena could be generalized to a larger population of data.

There is another, perhaps more informative statistical property of α, tangentially mentioned in comparing the examples of reliability data depicted in Figure 14.2; its robustness or vulnerability to small errors and losses of information about the phenomena in question. This quality is manifest when seemingly minor disagreements in the reliability

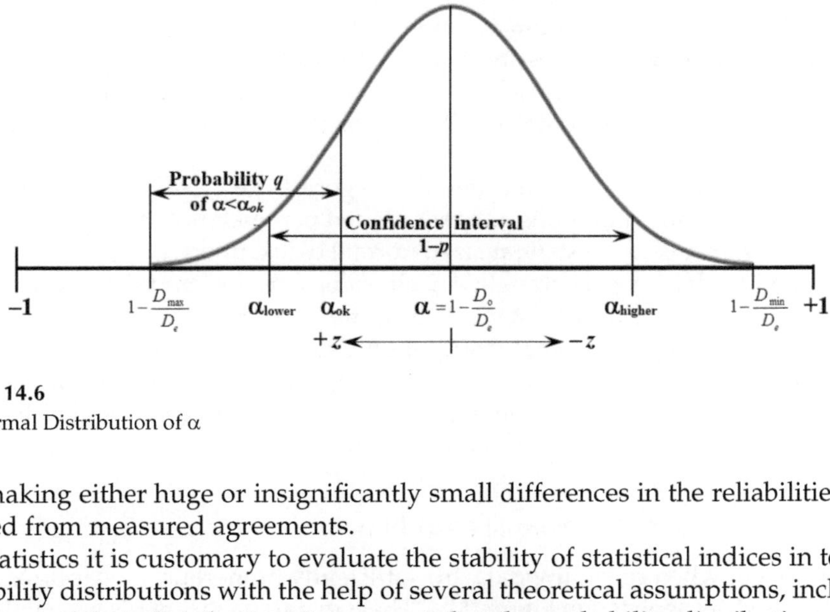


**FIGURE 14.6**
The Normal Distribution of α

data making either huge or insignificantly small differences in the reliabilities that can be inferred from measured agreements.

In statistics it is customary to evaluate the stability of statistical indices in terms of their probability distributions with the help of several theoretical assumptions, including about the nature of their distribution. It turns out that the probability distribution of α over the number of possible αs tends to be normally distributed, especially when the number of pairable values is large.

(14.1) considered the range of α's valid reliability interpretations but made clear that an observed α can range between −1 to +1. In the idealized distribution of α in Figure 14.6, the observed α occupies the most probable α among all possible αs and serves as the mean of its hypothetical normal distribution.

The width of the distribution depends on two factors: The size of the sample from which α is computed and how far the observed α deviates from measuring +1.

For inferring the reliability of data from the agreements measured on a sample, two parameters of α's distribution are important; both are identified in Figure 14.6.

- The *confidence interval* brackets the area under the normal curve which has the observed or true $\alpha$ at its center and estimates within which limits $\alpha$ is likely to vary if obtained from samples of replications similar to the reliability data evaluated. The width of this interval is a function of the chosen level of statistical significance $p$. For example, the choice of $p = 0.05$ (0.025 on either side of the distribution) would give a researcher the assurance that in 95% of all alternative samples from the same set of phenomena it would yield $\alpha$-agreements within these limits.
- The *probability* $q$ of accepting the data as reliable when there is a chance that $\alpha$ could be below a chosen minimum $\alpha_{ok}$. This error, sometimes called the Type I Error of accepting a hypothesis as true when there is a hopefully only small probability of being false.

The probability $q$ deviates from the Type I Error of testing the Null Hypothesis of something having no effect on a chosen variable. There have been content analysts who considered it sufficient to demonstrate that their coders do not work randomly. This is the lowest possible criterion for data to possess a hint of reliability. Researchers who want to know the reliability of their data are concerned with the other extreme, that their data deviate only minimally from being in perfect agreement and they need to spell out how much error variance they are willing to tolerate. The probability $q$ of accepting data as reliable when they could be below a chosen minimum $\alpha_{ok}$ should always be small.

$\alpha$'s probability distribution provides another path toward deciding whether the size and diversity of the reliability data from which $\alpha$ is computed would be adequate to infer the reliability of a population of data. We have considered two ways to obtain such a distribution for $\alpha$.

When reliability data consist of unordered categories, i.e., exhibit nominal metric relations and are sufficiently numerous, the two needed parameters of $\alpha$'s distribution can be approximated by means of the standardized normal distribution.

All other reliability data, especially when their values are ordered, i.e., have ordinal, interval, ratio, polar, circular or any other kind of metric properties and/or are not too numerous, the assumption of a normal distribution becomes questionable and computational estimates of $\alpha$'s probability distribution are complicated. In such cases we abandon assumptions of $\alpha$'s distribution and bootstrap a distribution by repeated resampling from the very data that gave rise to the computed $\alpha$-agreement.

### 14.3.1 By Approximation to the Normal Distribution

As above mentioned, when reliability data are sufficiently numerous, $\alpha$ is approximately normally distributed, as depicted in Figure 14.6. The two parameters of $\alpha$'s probability distribution that are most important for judging whether a measured $\alpha$ can indicate data to be sufficiently reliable – $\alpha$'s confidence limits at a chosen level $p$ of statistical significance and the probability $q$ of $\alpha$ not to exceed a chosen minimum $\alpha_{ok}$ – can be estimated from such a distribution. Section 14.3.2 outlines a path to obtain such a distribution by bootstrapping from the existing coincidence or contingency matrices. This approach has no limitation regarding the metric qualities of the data but consumes a lot of computer time when data are large. This section estimates the two parameters with the help of the known properties of the normal distribution. Generally, this approach is limited to data that do not have complicated metrics. This section concerns nominal data.

To estimate these two qualities, we consider the marginal sums of Table 6.8's coincidence or Table 6.9's contingency matrices as the best estimates of the distribution of values in the population of data whose reliability is in question. The expected disagreements $D_e, D_e^s, D_e^{sm}$, or $D_e^m$ make the same assumption. And for the sake of consistency this approach has considerable virtues.

While all estimates are coefficient-specific, most of them have the same form. Where the same expressions pertain to several coefficients, the superscripts that distinguish among them, $\alpha$, $\alpha^s$, $\alpha^{sm}$, $\alpha^m$ and their corresponding disagreement measure, are replaced by an asterisk when the expressions refer to more than one. Similarly, because most of the distributions are based on the same form but applied to $\alpha^*$-specific coincidence or contingency matrices, their cell entries $n_{ck}, x_{ck}, x_{\bar{c}k}$, and $x_{c\bar{c}}$ will be referred to by the generic $n_{ck}$, and so are their marginal sums $n_{c\cdot}$, $n_{\cdot k}$ and their totals $n_{\cdot\cdot}$.

With reference to Figure 14.6 and taking the marginal sums of the coincidence and contingency matrices as fixed, we have to be concerned with variations between extremes:

$$D^*_{min} \leq \begin{Bmatrix} D^*_o \\ D^*_e \end{Bmatrix} \leq D^*_{max} \tag{14.10}$$

Within the constraints of the marginal distribution, of the four disagreements shown in (14.10) $D^*_e$ remains fixed. The measured $D^*_o$ is in the center of a probability distribution between the two extremes of $D^*_{min}$ and of $D^*_{max}$. We define these disagreements as follows:

The *observed* and *expected disagreements* $D^*_o$ and $D^*_e$ are found in:

$$_c\alpha^*_{\text{nominal}} = 1 - \frac{_cD^*_o}{_cD^*_e}$$

The *smallest disagreements* $_cD^*_{min}$ within the constraints of their marginal sums of the observed coincidences or contingencies are:

$$_cD^*_{min} = n_{\cdot\cdot} - \sum_{c=1}^{v} min(n_{c\cdot}, n_{\cdot c}) \tag{14.11}$$

The *largest disagreements* depend on the nature of the coefficient.

- For decisiveness:

$$_cD^m_{max} = {}_cD^m_e \tag{14.12}$$

- For replicability, accuracy, and surrogacy $_cD^*_{max}$ depends on the number of values in the data. For $v = 2$ or binary data, for example valued 0 or 1:

$$_cD^*_{max} = min(n_{0\cdot}, n_{\cdot 1}) + min(n_{\cdot 0}, n_{1\cdot}) \tag{14.13}$$

- For nominal data with $v > 2$ categories, the largest disagreement within the constraints of the marginal sums of $v$-by-$v$ matrices and agreement measures other than decisiveness, $_cD^*_{max}$ is obtainable by iteratively manipulating a replica of the applicable $v$-by-$v$ matrices, initially containing the frequencies $n_{ck}$. It calls for:

two cells $a \neq b \begin{cases} n_{cc} > 0 \\ n_{a \neq c \neq b} > 0 \end{cases}$ and adding / subtracting $\begin{cases} n'_{cc} = n_{cc} - min(n_{cc}, n_{ab}) \\ n'_{ac} = n_{ac} + min(n_{cc}, n_{ab}) \\ n'_{cb} = n_{cb} + min(n_{cc}, n_{ab}) \\ n'_{ab} = n_{ab} - min(n_{cc}, n_{ab}) \end{cases}$

until all frequencies $n_{cc}$ in the resulting matrix are either 0 or can no longer be reduced.

Then: $$_{c}D^{*}_{max} = n.. - \sum_{c=1}^{v} n_{cc} \quad (14.14)$$

The working of this algorithm can be illustrated by means of the numerical example from Section 3.3, with categories $a$ and $b$ lumped into one category. Starting from its $D_{min} = 0$, it takes four steps to come to its $D_{max}$. The pair of cells selected before each iterative step are encircled, and the consequential modification is show in the following matrix:

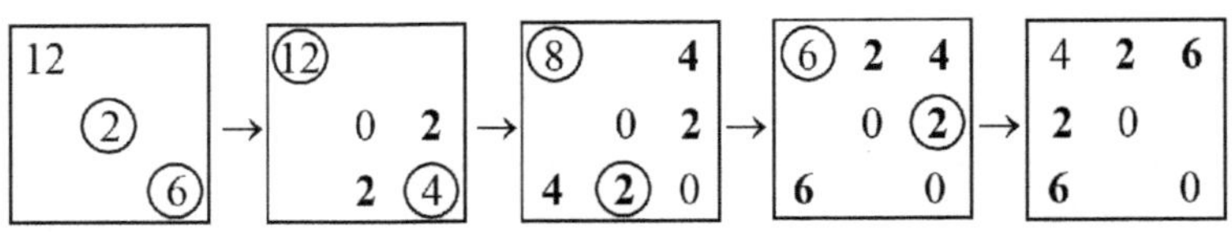


- ***The Confidence Limits of α***

The confidence limits $\alpha^{*}_{lower}$ and $\alpha^{*}_{higher}$ of $_{c}\alpha$ are located within the range of all possible $_{c}\alpha^{*}$s:

$$1 - \frac{_{c}D^{*}_{max}}{_{c}D^{*}_{e}} \leq \alpha^{*}_{lower} \leq {}_{c}\alpha^{*}_{\text{nominal}} \leq \alpha^{*}_{higher} \leq 1 - \frac{_{c}D^{*}_{min}}{_{c}D^{*}_{e}}$$

The degrees of $_{c}\alpha^{*}$'s freedom between which disagreements may vary are:

$$N^{*} = 1 + {}_{c}D^{*}_{max} - {}_{c}D^{*}_{min} \quad (14.15)$$

I am aware of the fact that for binary data, the number of possible matrices with different $_{c}\alpha^{*}$s is $N^{*}_{\text{binary}} = 1 + ({}_{c}D^{*}_{max} - {}_{c}D^{*}_{min})/2$, see (3.1). However, the following estimates turn out to be better when their range is taken as in (14.15). When $N^{*} < 6$, the following estimates are potentially uninterpretable and should not be attempted.

The range $N^{*}$ of possible $_{c}\alpha$s can be divided into one part $\geq D^{*}_{o}$ and another $\leq D^{*}_{o}$. They can be stated as proportions of $N^{*}$:

$$P^{*}_{\geq D^{*}_{o}} = \frac{0.5 + {}_{c}D^{*}_{max} - {}_{c}D^{*}_{o}}{N^{*}}$$

$$(14.16)$$

$$Q^{*}_{\leq D^{*}_{o}} = \frac{0.5 + D^{*}_{o} - D^{*}_{min}}{N^{*}} = 1 - P^{*}_{\geq D^{*}_{o}}$$

$N^{*}$, $P^{*}_{\geq D^{*}_{o}}$, and $Q^{*}_{\leq D^{*}_{o}}$ are needed to obtain the standard deviation of α's normal distribution, algebraically simplified:

$$\sigma^* = \sqrt{N^* P^*_{\geq {}_cD^*_o} Q^*_{\leq {}_cD^*_o}} = \sqrt{\frac{(0.5 + {}_cD^*_{max} - {}_cD^*_o)(0.5 + {}_cD^*_o - {}_cD^*_{min})}{1 + {}_cD^*_{max} - {}_cD^*_{min}}} \quad (14.17)$$

The $z$-values for a chosen level $p$ of statistical significance (two-tailed test), are as in the following familiar table:

| Level of significance $p$ | 0.000 | 0.001 | 0.002 | 0.005 | 0.01 | 0.02 | 0.05 | 0.1 | 0.2 | 0.5 |
|---|---|---|---|---|---|---|---|---|---|---|
| $z$-values (two-tailed) | 4.00 | 3.28 | 3.08 | 2.81 | 2.58 | 2.33 | 1.96 | 1.645 | 1.28 | 0.675 |

The estimated confidence limits of the computed ${}_c\alpha^*$ are obtained and capped by its extremes:

$$max\begin{Bmatrix} 1 - \dfrac{{}_cD^*_o + z \cdot \sigma^*}{{}_cD^*_e} \\ 1 - \dfrac{{}_cD^*_{max}}{{}_cD^*_e} \end{Bmatrix} = \alpha^*_{\text{lower}} \leq \alpha^*_{\text{nominal}} \leq \alpha^*_{\text{higher}} = min\begin{Bmatrix} 1 - \dfrac{{}_cD^*_o - z \cdot \sigma^*}{{}_cD^*_e} \\ 1 - \dfrac{{}_cD^*_{min}}{{}_cD^*_e} \end{Bmatrix} \quad (14.18)$$

- ***The probability q of data with the measured*** $\alpha$ ***failing to exceed a chosen minimum*** $\alpha_{ok}$
  It is obtained from the $z$-values:

$$\text{From } \alpha_{ok} = 1 - \frac{{}_cD^*_{ok}}{{}_cD^*_e}$$

$$\text{follows } {}_cD^*_{ok} = (1 - \alpha_{ok})\,{}_cD^*_e \text{ and } z^*_{ok} = \left| \frac{{}_cD^*_o - {}_cD^*_{ok}}{\sigma^*} \right| \quad (14.19)$$

By reference to a table of $z_{ok} \rightarrow p(z_{ok})$-correspondences[1] for one-tailed tests, with the addition of the rule governing extremes for $z^*_{\text{ok}} \geq 4.1 \rightarrow p(z^*_{ok}) = 0.5$, the probability $q^*$ of ${}_c\alpha^*$ failing to exceed the chosen minimum $\alpha_{ok}$ is:

$$\begin{aligned} {}_c\alpha^* \geq \alpha_{ok} \text{ then } q^* = 0.5 + p(z^*_{ok}) \\ {}_c\alpha^* \leq \alpha_{ok} \text{ then } q^* = 0.5 + p(z^*_{ok}) \end{aligned} \quad (14.20)$$

### 14.3.2 By Bootstrapping

In most practical endeavors, reliability data are small, have diverse metrics, include missing values and are irregularly distributed. These conditions do not lend themselves to easy computational approximations to known probability distributions. The few efforts to derive probability distributions for reliability coefficients I know of divert widely. This led me to the conclusion that probability distributions of $\alpha$ be best approximated by bootstrapping.

1 E.g.: http://stats.seandolinar.com/calculating-z-scores-with-r/

Bootstrapping is a statistical method for estimating the distribution of a statistic by frequent resampling the data from which it is obtained. In the case of α, one can bootstrap its distribution by repeatedly drawing samples from the original reliability data, allowing duplications and omissions to occur randomly, calculating the hypothetical αs for each resample, and thereby create a distribution of likely α-values. The bootstrapped distribution of the reliability data A in Figure 14.2.

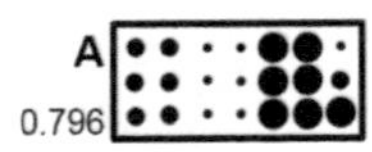


yields the following distribution involving 13 out of 22 possible probabilities $P_\alpha$:

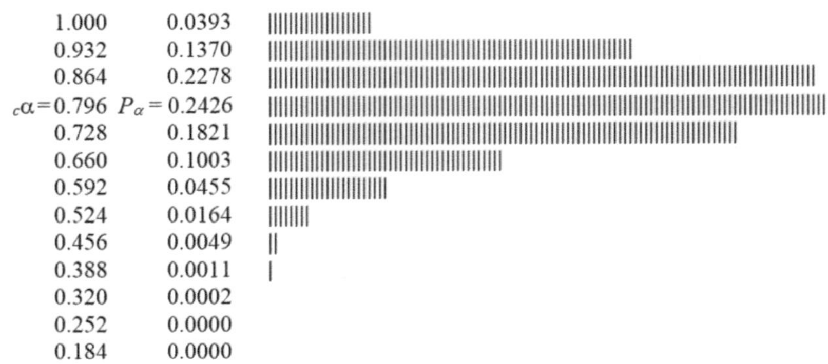


**FIGURE 14.7**
The Accurate Distribution of α for Reliability Data A of Figure 14.2

Note that all examples of reliability data in Figure 14.2 are numerically small. Data in A consist of three uniformly distributed categories. Within its marginal constraints there are only 12 combinatorically possible $_c\alpha$s. Without marginal constraints, which our bootstrapping algorithm, for simplicity's sake, does not honor, there are 22 $_c\alpha$s, of which Figure 14.7 reports the first 13 with their frequencies. In either case, here, $_c\alpha$-agreements presents itself as an equal interval scale while the distribution is skewed.

Examples D and E in Figure 14.2 are both perfect. But the lower variance of D makes these data more vulnerable to disagreements. Within its marginal constraints, there are only four combinatorically possible $_c\alpha$s, measuring between $_c\alpha$ = 1.000 and $_c\alpha$ = –0.111. By comparison the higher variance in example E makes these data more resistant to disagreements. The reliability data E yield 11 $_c\alpha$s, ranging between $_c\alpha$ = 1.000 and $_c\alpha$ = –0.058. As argued in Section 14.1.2: variance matters.

The following outlines an efficient algorithm to bootstrap distributions of α from its reliability data. Inasmuch as the previous chapters defined numerous versions of α, each of which incorporates features that allow α to cope with diverse reliability data, I will limit the description of the bootstrapping algorithm to its most common form: bootstrapping the agreement of replications of coding predefined units by various single-valued disagreement functions as implemented in the widely used KALPHA software (Hayes & Krippendorff, 2007).

Bootstrapping is not applicable when α = 1.000 in which case a probability distribution has no spread, and when reliability data exhibit no variance, i.e., all values are identical, and no spread can be obtained either.

I shall spell out the motivations and assumptions for each step in the hope that they may serve as a guide to readers who wish to bootstrap the distributions of agreement coefficients for other than predefined units of phenomena.

All bootstrapping has to start with the reliability data in their canonical form of an $m$-by-$N_u$ matrix containing values $c_{iu}$ and $k_{ju}$, as in Table 5.1 which lead to matrices of observed coincidences $n_{ck}$ as in Table 5.2 in whose terms the observed disagreement ${}_cD_o$, the expected disagreement ${}_cD_e$, and the agreement ${}_c\alpha_{\text{metric}}$ for a chosen metric difference function ${}_{\text{metric}}\delta^2_{ck}$ are defined.

The previous Section 14.3.1 and Figure 14.7 considered all possible agreement coefficients within the marginal constraints of the original data, according to (14.10) between $D_{max}$ and $D_{min}$ to which $n..$ pairable values contributed, albeit in fractions when paired with a number $m_u - 1$ of other values in units $u$. Within given marginal constraints, the expected disagreement ${}_cD_e$ is a constant.

In contrast, resampling from coincidence or contingency matrices may well generate marginal sums and, hence, expected disagreements other than the original. However, in practice, the calculated expected disagreements vary little relative to the resampled observed disagreements, which are the main target of bootstrapping. For efficiency, the following algorithm assumes ${}_cD_e$ to be given. The consequence of this assumption is that the bootstrapped distributions are a little wider, hence more conservative, than when estimated by means of the normal curve and within the observed marginal sums as in Section 14.3.1.

As the proportion ${}_cD_o/{}_cD_e$ in the definition of ${}_c\alpha$ subtracts from ${}_c\alpha = 1$, with the expected disagreement ${}_cD_e$ here assumed identical for all resamples, resampling has to focus on the observed disagreement ${}_cD_o$. They are the coincidences defined in terms of the original reliability data in Tables 2.4 and 5.1 according to (2.10) and (5.1):

$$n_{ck} = \sum_{u=1}^{N_u} \frac{1}{m_u - 1} \sum_{i=1}^{m_u} \sum_{j\neq 1}^{m_u} 1 \text{ iff } c_{iu} \text{ and } k_{ju}$$

They total $n.. = \sum_c^v \sum_k^v n_{ck}$ and stem from:

$$N_o = \sum_{u=1}^{N_u} \frac{m_u(m_u - 1)}{2} \text{ pair comparisons}$$

To resample possible alternatives of observed disagreement within the probabilities of $n_{ck}/n..$, we need to express $\alpha$ in terms of its differences, the $N_o$ pairs of values added to ${}_cD_o$ and subtracted from $\alpha$. This amounts to decomposing ${}_cD_o$ of (5.6) in terms of the sum of its differences between all pairs of values:

$${}_cD_o = \sum_{c=1}^{v} \sum_{k=1}^{v} n_{ck} \,{}_{\text{metric}}\delta^2_{ck} = \sum_{u=1}^{N_u} \frac{2}{m_u - 1} \sum_{i=1}^{m_u} \sum_{j>i}^{m_u} {}_{\text{metric}}\delta^2_{c_{iu}k_{ju}} \tag{14.21}$$

Entering the decomposed observed disagreement ${}_cD_o$ of (14.21) and the original ${}_cD_e$ of (5.7) into the definition of ${}_c\alpha_{\text{metric}}$ yields an expression in terms of all distractions $E_r$ from the perfect ${}_c\alpha_{\text{metric}}$ among $m_u(m_u - 1)/2$ observed pairs of values associated with each unit $u$:

$$
\begin{aligned}
{}_c\alpha_{metric} &= 1-\frac{{}_cD_o}{{}_cD_e} = 1-\frac{1}{{}_cD_e}\sum_{u=1}^{N_u}\frac{1}{m_u-1}\sum_{i=1}^{m_u}\sum_{j>i}^{m_u} 2\,{}_{\text{metric}}\delta^2_{c_{iu}k_{ju}} \\
&= 1-\sum_{u=1}^{N_u}\frac{1}{m_u-1}\sum_{x=1}^{(m_u-1)m_u} E_{ux}
\end{aligned}
\qquad (14.22)
$$

Define an array of $N_o$ values $1, \ldots, r, \ldots, N_o$ associated with

- All contributions $E_{ux} = 2\frac{{}_{\text{metric}}\delta^2_{c_{iu}k_{ju}}}{{}_cD_e}$ in consecutive order, relabeled: $E_r$
- And the pairs of values of ${}_{\text{metric}}\delta^2_{c_{iu}k_{ju}}$, relabeled: $c_r = c_{iu}$ and $k_r = k_{ju}$ (14.23)

With $E_r$'s and the two values $c_r$, $k_r$ to be randomly chosen from $1, \ldots, r, \ldots, N_o$; and the list of $N_u$ pairable values $m_u$ in $u$ from the reliability data, the bootstrapping algorithm can proceed in these steps:

Set the integer array $n_\alpha = 0$. $-1 \le \alpha \le +1$
Do $X$ times $X = 10{,}000$ recommended
  $\alpha = 1$
  Do $u = 1, N_u$
    Do $\frac{(m_u-1)m_u}{2}$ times (14.24)
      Pick a random number $1 \le r \le N_o$ (uniform distribution)
      If the 2nd pick shares $c_r$ or $k_r$ with any of the 1st pick
      $\alpha \Leftarrow \alpha - \frac{E(r)}{m_u - 1}$
  If $\alpha < -1$: $n_{-1} \Leftarrow n_{-1} + 1$
  If $\alpha \ge -1$: $n_\alpha \Leftarrow n_\alpha + 1$

The injunction against picking the same two values in the second random picks is to prevent resampling indeterminate αs.

For reliability considerations, the resulting distribution of α's probabilities $p_\alpha = \frac{n_\alpha}{X}$ is most appropriately characterized in terms of the same two parameters discussed in the previous Section 14.3.1:

- The *confidence limits* of ${}_c\alpha$ at a chosen level $p$ of statistical significance (two-tailed):

$$-1 \le \alpha_{\text{lower}} \le {}_c\alpha_{\text{metric}} \le \alpha_{\text{higher}} \le 1$$

Where: $\alpha_{\text{lower}}$ = the smallest α for which $max\left(\sum_{g=-1}^{g<\alpha_{\text{lower}}}\frac{n_g}{X}\right) \le \frac{p}{2}$

(14.25)

$\alpha_{\text{higher}}$ = the largest α for which $max\left(\sum_{g>\alpha_{\text{higher}}}^{g=1}\frac{n_g}{X}\right) \le \frac{p}{2}$

- The *probability q* of ${}_c\alpha_{\text{metric}}$ failing to exceed the required minimum $\alpha_{\text{ok}}$:

$$q = \sum_{g=-1}^{g<\alpha_{ok}} \frac{n_g}{X} \tag{14.26}$$

(14.24) is stated for bootstrapping the replicability of single-valued coding of predefined units using any of the difference functions defined in Chapter 5. For developing variations of this bootstrapping algorithm for other α-coefficients, I am suggesting to follow the logic preparatory to the use of the above algorithm, starting with creating a list of all observed differences among all pairs of values that enter the observed disagreement ${}_cD_o$, including any unequal weights – in the above case weighted by $(m_u-1)$.
Algorithms that do not consider the expected disagreements as fixed and those that resample within the confines of observed marginal distributions (as in the case of the approximation of the normal curve) are attractive possibilities but computationally far more inefficient than the above. However, the logic of decomposing the proportion ${}_cD_o/{}_cD_e$ into resampleable pair comparisons applies to all algorithms for bootstrapping αs.

The virtue of obtaining α's confidence limits and the probability of not exceeding a chosen minimum will be part of the following Chapter 15.

# 15

## *Reliability Standards*

Standards for deciding on whether data deserve subsequent analytical attention can be discussed only in conjunction with properly generated reliability data, appropriate agreement measures, and in view of the analytical consequences of using imperfect data.

For the use of data with uncertain reliabilities, this chapter discusses standards for two stages in a research project at which reliability assurances are needed: At the point where data are generated and after an analysis of data with somewhat compromised data is completed. Previous chapters developed variable-specific α-agreement measures, all of which pertain to the data-generating stage, leaving uncertain how any of their imperfections affect the results. The last section of this chapter concerns assessments of the reliability of research results, which is ultimately what researchers need to justify their conclusions. However, the latter is far more difficult to assess. This puts the burden squarely on the assurances researchers can derive from testing the reliability of their data. There is hardly any literature on how unreliabilities in the data affect the results of analyzing them largely because of the freedom scholars have in devising their own methods. This makes stating reliability standards for individual variables difficult to justify and hence often arbitrary – except for the desirable extreme of perfectly reliable data.

After a research project is conceptualized, has reached the stage at which the generation of suitable data begins and well before they are analyzed, assessing their reliability is the only safeguard against irregularities, misjudgments, and uncertainties entering the data, often unnoticed, and hence a primary concern. Data are the ground on which empirical research grows and inferring their reliability from agreements among replications has been the focus of most previous chapters. Obviously, if the data represent the phenomena of interest unambiguously and completely, their analysis has the best chance of yielding valid results – notwithstanding the possibility that the subsequent choices of analysts may not lead to where they want to be. Testing the reliability of data preceding their analysis is justified by the experience that even minimally unreliable data may introduce uncertainty in the analytical process and may lead researchers to wrong conclusions.

Measures of agreement are intended to enable researchers to decide whether to proceed with processing their data whose relationship with the phenomena of analytical interest may well be somewhat compromised. However, agreement measures cannot predict their effects. This is because analytical processes vary widely regarding how they transform data into conclusions. The effects of imperfect data on the conclusions they support tend to be specific to the analysis performed. Section 14.1.4 exemplified how data, even of a generally accepted level of measured α-agreement, introduce variability in the simple frequencies that a researcher may want to report or consider in deciding subsequent analytical steps. Correlations found between unreliable variables tend to be weaker than between reliable ones.

The following section considers conditions for generating reliability data whose agreement measures are unbiased, afford reliability interpretations, and enable the application of standards. Section 15.2 considers criteria for choosing standards for one variable. Section 15.3 extends standards for single variables to standards for multi-variable data structures. Section 15.4 will address what finally matters: the consequences of unreliabilities in whichever variables they occur on the analytical results, computing distributions of results heretofore not conceivable.

## 15.1 Conditions for Agreement Measures and Standards to Apply

Talking of standards based on variable-specific measures of agreement presupposes agreement measures that have appropriate mathematical structures and are applied to reliability data generated under conditions that do not bias the measured agreements.

Many textbooks discuss alternative agreement statistics with words that have conventional meanings which may have little to do with what they actually measure. Chapter 4 already examined the mathematical structure of several largely nominal agreement measures and their consequences. Table 4.1 compared the most common agreement measures regarding what they respond to and the ranges of their applicability. Calling an agreement measure a reliability coefficient does not make it so and suggesting options among various coefficients can invite uninformed readers to shop for a statistic that shows their data in the most favorable light without considering what they actually respond to and what they omit.

For example, Lombard et al. (2002) distinguishes between liberal and conservative coefficients, implicitly suggesting researchers' choices depend on their academic philosophy. The distinction is based on the numerical values different coefficients yield when applied to the same reliability data. Liberal coefficients produce larger measures than conservative ones. Undoubtedly, the most liberal measure is the measure of percent agreement. While noting that percent agreement has been criticized without explaining why, keeping it as one of the liberal options is seriously misleading (Section 4.1).

Also, calling agreement measures reliability coefficients ignores the crucial difference between measuring agreement and inferring the reliability of data from the lack of disagreement. Engineering, insurance industries, and pharmaceutical companies pursue vastly different conceptions of reliability. Wordings matter.

Section 4.2 points out that the indiscriminate use of the word "chance," in so-called "chance-corrected" agreement measures, does not have a single meaning. $\alpha$ corrects agreement by dividing its observed disagreement by the hypothetical disagreements observable in the absence of any statistical relationship between the variance in data and the diversity of phenomena that observers have been exposed to and coded. Cohen's (1960) $\kappa$, in contrast, takes chance to mean the statistical independence of two observers' categorizations, a conception at home in measuring correlations and associations. This conception is also at home in Cronbach's (1951) perfectly justifiable measure of the consistency of psychological tests. His consistency does not suggest data are reliable representations of what they are expected to be about. Failing to recognize the difference in conceptions of chance and calling all of them reliability coefficients hides why they differ. In effect, Cohen's $\kappa$ mistakes observers' unequal proclivities for categories as agreement. Craig

(1981) relates his definition of chance to majority judgments. Aikin (1990) acknowledges that raters face easy and hard decisions. Easy ones are those on which raters agree, and hard ones on which they don't. His percentage-like coefficient applies chance considerations only to difficult decisions. For Zhao, Liu, and Deng (2012) and Feng (2013), chance should reflect the difficulties of coding. While it is an empirical fact that the experiences of coding difficulties tend to cause data to be unreliable, avoiding such difficulties is a challenge to the designers of coding instructions, not to be given credit for in assessing the reliability of data.

The fact that humans bring diverse backgrounds, preferences, and literary abilities to the task of generating data, which are very different from how mechanical measuring devices vary or fail, has little to do with whether or the extent to which researchers can trust data they generate. What matters is the uncertainties they introduce in the data.

Labeling agreement measures "inter-observer," "inter-coder," or "inter-rater reliability coefficients" invites another misconception, of believing that only the agreement between coders matter. This has given some researchers the idea of allowing coders to talk among each other and reach agreements on their coding task. However, what matters is that the relationship between the phenomena of interest and the data meant to represent them is to be reproducible elsewhere. It follows that written coding instructions need to be applied by independently working coders, so that other researchers who may want to critically examine whether the published coding instruction works come to the same results as for replicability measures or conform to accepted conceptions as measured by accuracy measures. Reliability does not reside between observers; it is an inferred attribute of the ability of data to represent the target of research.

Examinations of how agreement coefficients handle observed disagreements among coded phenomena of interest ought to precede decisions of their usability and the adoption of numerical standards. I have suggested that agreement coefficients capable of supporting inferences about the reliability of data should exhibit two clearly defined reference points on a scale: They should measure 1 when independent replications agree without exceptions, and 0 when these replications are demonstrably unrelated to the differences in the phenomena of analytical interest at which point data cannot possibly be explained in terms of what observers have been exposed to. Chapter 3 explored the qualities of $\alpha$ in these terms and Chapter 4 evaluated the extent to which alternative coefficients claim to be reliability coefficients but are influenced by a variety of factors that are unrelated to the data whose qualities are in question. How close an acceptable agreement coefficient is to being perfect should not be confused with what they measure, what they respond to and what they inflate or omit.

However, while the mathematical properties of the scale by which agreement is measured are essential for inferring the reliability of variable-specific data from measured agreement, this condition can easily be undermined by inappropriate ways reliability data are generated and is not sufficient to interpret measured agreements as indicators of the ability to rely on the generated data in subsequent analyses. Section 1.7 already reviewed the following six preconditions that must be satisfied for agreement measures to be interpretable as indicators of the reliability of the resulting data. They are:

- Reliability data need to be statistically representative of the data whose reliability is in question. Section 14.2 addressed these conditions. Pretesting a coding instrument before generating the needed data for a research project is common. However, representativeness must be guaranteed for all subsequently coded data and may have to be checked in intervals throughout an extended period generating the data for a research project.

- To ensure replicability as well as understanding the generated data, coding instructions, aptitude tests for coders' qualifications, training manuals, the specifications of measuring devices, and everything that accounts for what coders do has to be unambiguously communicable (published, understandable, and hence independently executable). Despite its difficulties, crowd coding works entirely from written instructions to which competence tests may be added.
- Observers, coders, and judges, have to generate independent replications. Besides shared linguistic qualifications, the coding instructions and instruments they are asked to apply to the same phenomena should be the only explanation of agreements observed among resulting replications. When observers discuss their understandings of the given instructions and help each other to resolve difficulties of describing phenomena to be coded, agreement measures become polluted, generally inflated by noncommunicable consensus and the data generated are no longer unambiguously related to what the data are claimed to be about.
- Observers, coders, and judges, including the instruments they are asked to use, have to be widely available for other researchers to draw from. This speaks against researchers doing their own coding, hiring research assistants familiar with the research, close acquaintances or using secret tests as other researchers are likely to have their own. The qualifications of observers need to be explicit and sufficiently common.
- Agreement measures need to be sensitive to all relevant features of the phenomena to be analyzed. For example, if units of analysis are not predefined and coders are asked to identify the location and length of relevant segments in textual continua, agreement measures for predefined units of analysis would omit all disagreements on unreliable intersections among identified segments. Omissions of relevant features also include using agreement measures that do not respond to the ordering of data when that ordering is crucial for their analysis.
- All data need to provide to an analysis sufficient amounts of information about the phenomena of analytical interest in order to answer the guiding research questions about these phenomena. Data are proxies for analytical concerns, not facts. This requirement calls for matching the diversity of chosen phenomena by adequate numbers of coding options. Section 14.1 elaborated on the role of variance and interpreted the α-coefficients as the proportion of information in the data about the phenomena of interest. One cannot separate information from noise in the data without the replications that reliability data provide.

These six conditions specify the context in which reliability data need to be generated for agreements to be interpretable in terms of the reliability of data. The degree to which they are not satisfied is the degree to which even perfect agreements cannot be trusted. They are also a precondition to talk meaningfully about benchmarks for acceptable level of agreement. For example, if observers talk of or exchange experiences while doing their work, agreement may be high but reflects the social accommodations that observers make to each other more than being about the data they are meant to generate independently of each other. Or, when observers know that they are rewarded by achieving a high percentage of agreement, this amounts to an incentive to agree to code all units alike. This would deprive the researcher of the needed information to answer the controlling research objectives.

The forgoing, drawing on earlier work (Krippendorff, 2011) and visualized in Figure 1.4, is meant to be a reminder that reliability is not measured but inferred from observed agreement among replications, provided generating these replications satisfies the six preconditions. The epistemological problem of these requirements is that their violations are not measurable as such. They must be ensured separately from measuring agreements within reliability data. Because ignoring them influences measured agreements surreptitiously, they are often violated – unknowingly or deliberately. Violations of these preconditions tends to inflate agreement measures and put their interpretation as indicators of the reliability of data in question. This makes efforts to comply with these preconditions and assuring that they are satisfied fundamental to all reliability considerations.

## 15.2 Reliability Standards for Single Variables: Alpha$_{ok}$

Facing difficulties of achieving near-perfect agreement naturally raises the question of what an acceptable level of agreement might be to proceed with analyzing imperfect data. Should a chance-corrected agreement coefficient measure at least 0.900 or 0.800? Is 0.700 still acceptable? How low can the inferred reliability of data be for research results to be publishable in a respectable academic journal? We have a numerical scale whose values between 0 and 1 afford reliability interpretations. We talk about trusting data in natural language. However, translating this scale into everyday language can easily be misleading without understanding the consequences of the mathematical structure of the coefficients used. For example, Landis and Koch (1977) propose the following adjectives:

| κ (kappa) | Strength of agreement |
|---:|---|
| < 0 | Poor |
| 0.00 to 0.20 | Slight |
| 0.21 to 0.20 | Fair |
| 0.41 to o.60 | Moderate |
| 0.61 to 0.80 | Substantial |
| 0.81 to 1.00 | Almost perfect |

Attributing "Poor" to negative κ-values certainly misrepresents what κ indicates. With zero indicating the statistical independence of two coders, $\kappa < 0$ is not poor, it is evidence of systematic disagreements. The adjectives of the strength of agreement represent an ordinal scale that certainly does not correlate what in this case κ provides.

Fleiss (1981) proposed a slightly more realistic benchmark for his $K$ and by implication for Scott's π as well:

| $K$ (and π) | Strength of agreement |
|:---:|---|
| < 0.40 | Poor |
| 0.40 to 0.75 | Intermediate to Good |
| > 0.75 | Excellent |

However, even Fleiss' substitutions of quantitative agreement measures by easily understandable verbal attributes give the impression that using ordinary language to describe the quality of data is sufficient. It too fails to refer to the risks of drawing wrong conclusions from imperfect data. Once the label of "intermediate to good" has entered the conversation or a published report, the more important questions of its implications seem to vanish. I am suggesting that translating quantitative agreement measures into simple and fuzzy attributes is grossly inadequate.

One way to find suitable standards is to inquire about the extent to which analysts – and I would like to include all stakeholders in the results of research – are correct in interpreting what the analyzed data are about.

To explore how different levels of reliability can be interpreted, Marten Brouwer, a colleague of mine from the Netherlands in the 1960s, designed an experiment. He gave English-only speaking coders a set of complicated Dutch words and asked them to describe U.S. television characters in their terms. The Dutch categories did not resemble English ones, and their pronunciation was difficult for English speakers. However, these terms must have invoked some associations with the personalities of television characters because the measured agreement came to ${}_c\alpha = 0.44$. Knowing the coders' unfamiliarity with these words, even Dutch-speaking researchers would not be able to infer what coders might have seen on U.S. television. And processing these data would result in something totally uninterpretable. Indeed, nobody could possibly have relied on such data. However, by Fleiss' verbalization of the values of $\kappa$ – assuming $\alpha$ to be similar – this would call ${}_c\alpha = 0.44$ "intermediate to good" which would signal to researchers that the data are not ideal but ok. Knowing how these reliability data were obtained, there could be no doubt that their relations to the qualities of TV-characters are simply vacuous. In order to justify drawing meaningful conclusions from data, the agreement measures for replicability must be significantly higher.

Experiments like Brouwer's can provide a sense of the relationship between measured agreements and the ability of their analysts to interpret the data in terms of the phenomena they intend to analyze. Further experiments led us to adopt the following policy:

- Aim at reliabilities that deviate only minimally from $\alpha = 1.000$
- Rely only on variables with reliabilities above $\alpha_{ok} \geq 0.800$ (15.1)
- Consider variables with reliabilities betwee $0.667 \leq \alpha_{ok} < 0.800$ only for drawing tentative conclusions

Perfect reliability would mean that analysts are able to decode their data by applying the coding instructions that generated them in reverse and infer from them the phenomena of interest that observers had read, rated, judged, interpreted, and coded. Imperfect data would make decoding the data difficult. If the analysts of such data have difficulties in relating them to the kind of phenomena they are concerned with, I suggest they have even more difficulties interpreting their results.

I once observed such a test of the reliability of data in court. Unfortunately, the court proceeding is not public. The principal investigator of a huge content analysis, commissioned by the plaintiff in a legal proceeding, had devised instructions to identify qualities of advertisements under contention. A service was hired to train quite a number of coders and generate the needed data. Already the way the reliability of these data was estimated was questionable. Coders coded batches of 100 advertisements. Then they were informed about their disagreements and invited to discuss why they deviated from each other. After each batch of coding 100 advertisements, the measured agreement improved, in the end satisfying the principal investigator. Obviously, the agreement they achieved over time was influenced by their increasingly consensual interpretations of the coding categories. As time progressed the independence of coders diminished and the increased agreement the plaintiff presented to the court was polluted by the consensus coders surreptitiously developed in the course of their deliberations. This was one problem the proceedings revealed.

Given the contentious features that coders were asked to identify in advertisements, during preliminary proceedings, the principal investigator was asked to apply his own coding categories to a few advertisements. At first he refused, saying, correctly as it were, that he was not present during coder training, unintentionally admitting that coding had become no longer explainable by the coding instructions he had devised. However, he could not deny being the analyst who presented his conclusions about these advertisements as evidence in court. Therefore, he was shown several advertisements to identify the presence or absence of relevant qualities, which had been part of his coding instructions. He was pleased to match some of the coders' identifications, but not to a degree better than chance. If the author of the research could not reliably identify what his report was alleging, how could anyone trust his findings? This was the end of trusting his conclusions.

Standards like (15.1) have been cited and adopted in numerous content analyses, in the social sciences, and other disciplines. They do not deviate too much from other proposals for academic research (Fleiss, 1981; Popping, 1988; Riffe, Lacy & Fico, 1998; Neuendorf, 2002), and might continue to serve as guidelines in scholarly pursuits. However, I am hesitant to support the use of single standards for individual variables for three reasons; the last one will be the topic of Section 15.4.

The first source of hesitation is grounded in the statistical nature of agreement coefficients. Section 14.3 made clear that an $\alpha$-agreement is more justifiable when replications are many as opposed to the minimum of two, when sample sizes of phenomena to be coded are large and representative of all data to be analyzed as opposed to few, and when the diversity of phenomena is matched by a sufficient number of values in the coding instructions. Figure 14.6 depicted measured $\alpha$-agreement as the means of a distribution of possible $\alpha$-agreements associated with various probabilities. When reliability data are large, this distribution tends to be narrow and ${}_c\alpha$'s confidence interval tends to be small. The converse holds when reliability data are small. Comparing a single observed $\alpha$-agreement with a single numerical standard ignores whether this difference is significant or not. Any standard adopted should always be outside the *confidence intervals of an observed agreement* at the chosen level of significance (see Section 14.3).

Confidence intervals are well-established properties of any statistic. For considerations of whether a measured ${}_c\alpha$ is acceptable, the more important measure is the one-tail testing of the *probability q of accepting data as reliable when there is a chance that they measure below an acceptable level* $\alpha_{ok}$. Sections 14.3.1 and 14.3.2 are devoted to its estimation. This probability $q$ expresses the risk of failure – not in terms of verbal qualifiers but in terms of the chances of the data leading to mistaken conclusions. $1-\alpha_{ok}$ defines the range of acceptable agreement measures. The probability $q$ of $\alpha$ to be outside this range depends not only on the observed $\alpha$, but also on the sample size of the reliability data. In the academic literature, acceptable probabilities range between 0.001 and 0.100. Merely reporting that $\alpha \geq \alpha_{ok}$ would amount to being satisfied with $q \leq 0.5$, which provides hardly any assurances of the reliability of the data.

The second ground of my hesitation to recommend a single standard $\alpha_{ok}$ for inferring data to be reliable stems from asking what is at stake when unreliable data introduce uncertainties into research results. My suggestion has been to take a larger perspective and derive standards from the costs of drawing invalid conclusions from data, even if their unreliability is only small. Mistaken research results that lead to the launching of commercial products, signing contracts, adopting policies, or investing in businesses incur costs for someone. The magnitude of possible losses should determine the reliability standard $\alpha_{\text{ok}}$ employed for the data on which such decisions rest. When human lives are at stake, whether due to medical diagnoses, legal decisions in a court of law, or intelligence reports leading to war, standards for the reliability of data should be very high. For academic researchers, the costs of publishing wrong conclusions from unreliable data may well involve the loss of future funding, scholarly reputation, accreditation and employment. The $\alpha_{\text{ok}}$-agreements suggested in (15.1) are customary in the academic world.

In reality, much academic research is exploratory, and scholars should not be afraid to venture into uncertain territory. However, when scientific findings feed prejudices, incur individual hardship, or have policy implications affecting the lives of many, reliability standards of the data on which these consequences are based need to be considered carefully. Admittedly, the actual costs of potential failures may not be easily quantifiable, nor are the probabilities of mistaken conclusions. However, it would be a mistake to settle on a single number for all reliability tests without considering the impact of unreliable data on the validity of the published results and the costs of being mistaken.

## 15.3 Reliability Standards for Complex Multi-Variable Data

$\alpha$-coefficients are defined for single variables – except for Sections 12.5, 12.10, and 12.11, which concerned sets of values whether from one or more variables assigned to unitized phenomena not treated multi-dimensionally. However, almost all research projects generate data in many dimensions, relying on systems of variables whose interrelations support particular research goals. Two values are needed when the aim is to make simple comparisons. Two variables are needed to calculate correlations and associations. Clustering, factor analyses, multi-variate regressions, log-linear models, and network analyses require very many variables, posing the question of standards for larger data systems.

To report the reliability of larger data structures, with numerous variables and associated agreement coefficients, researchers have published average reliabilities of all the variables a research project depends on. However, sums do not preserve variances between the values summed. Reporting averages, by their very nature, hide small values behind big ones. In the case of agreement measures, averages hide unreliable variables behind reliable ones. In the reality of research projects, most larger date structures include unproblematic clerical variables which are likely to be perfect such as time stamps, sources of publications, socio-economic data culled from quantitative records, or measures provided by mechanical devices. Data in variables that pertain to observations, categorizations, interpretations, judgments, or annotations made by humans tend to be less reliable but typically make the more important contributions to settling social research questions rather than purely clerical ones. Whenever averages of agreement measures seem to be numerically acceptable, they may give the false impression of all variables being acceptable as well, which may not be the case. The more variables are averaged the more likely it is that they include variables far below the minimum acceptable level $\alpha_{ok}$. In other words, reliability measures should not be averaged or summed. But what would serve as the reliability of a system of variables?

When unreliable data are encountered early enough to still improve the process of making data methods to locate problems was discussed in the first part of Chapter 12. However, when improvements of their definitions are no longer practical, it is widely accepted to drop variables below acceptable levels of reliability from subsequent considerations as they could derail the whole process and lead to mistaken conclusions. Therefore, facing complex data in the form of many variables, two rules apply. The first is:

*Drop from further considerations any variable that cannot be improved upon above a chosen minimum reliability.* (15.2)

The best assurance against losing information that is essential to the outcome of a study is to introduce redundancy into the coding instructions. It means conceiving of multiple ways to code phenomena in dimensions deemed important, not only to cross-validate how they refer to the phenomena of interest, but also not to lose needed information when forced to abandon variables for failing to exceed the minimum standard. This is what Shannon's information theory has taught us. Instead of defining a data language in which every variable counts and having to abandon the research project if any one fails, it may save a project to build into that language alternative paths should some fail. It simply means thinking of several ways motivations could be characterized, not one; measuring socio-economic status by multiple indicators, or using several symptoms of a disease in the hope at least one succeeds.

Rigorously applied, (15.2) ensures that all variables on which a research project relies are above the accepted minimum. Some may be better if not perfectly reliable. Reporting averages being misleading, short of listing the agreement coefficients of all variables involved, the smallest α-agreement among all variables would count for the reliability of a large system of variables. Therefore, I have been arguing for a second rule:

*The reliability of a whole data system is the smallest reliability among its variables.* (15.3)

I have heard researcher objecting to this suggestion, arguing that it does not give credit to variables with high or perfect reliability among them. True. However, any one variable may be the flipping point, leading researchers to wrong conclusions and they may not know which variable caused it and how they were misled. Rules (15.2) and (15.3) ensure that the reliability of a larger system of variables will always be equal to or larger than the adopted minimum.

Researchers can always report the range of agreement coefficients found as long as the lowest one is arguably acceptable.

## 15.4 The Reliability of the Results of Analyzing Imperfect Multi-Variable Data

When all individual variables of a multi-variable set of data are perfectly reliable, short of serious conceptual missteps by their analyst, there is no doubt that all analytical transformations applied to such data remain related to the phenomena of interest.

Practicing rule (15.2) – deleting variables whose agreement measures below a chosen minimum from subsequent analyses – ensures that the reliability of multi-variable data is at least not below the chosen standard but gives analysts no assurances how, to what extent, and which variables will determine the outcomes of an intended analysis, and worse, how much an intended analysis will be directed by the noise in the data or remain representative of the information data contain about the phenomena of analytical concerns.

At the beginning of any research project, applying the same reliability standard to each variable of a larger system of variables is recommended for it is rarely possible to know which variables determine the direction an analysis of all variables will be taking and whether and how this direction may be compromised. This is so because analyses of complex data rarely are "democratic" in the sense that each variable impacts the conclusion based on them alike. It is also futile to make general claims about how the disagreements in the data affect analytical efforts – except for fairly simple analyses, exemplified regarding frequency accounts of data in Section 14.1.4 – calculating correlations, networks, or clusters with known software. Analysts design their own methods from readily available components which can differ quite drastically in how unreliabilities affect their intermediate and final results. For example, when a number of variables are bundled into one index, a common analytical practice to gain a level of abstraction, the effects of disagreements in the reliability data among them is likely reduced in favor of their common property which that index ends up representing. The regression lines of correlations tend to be shaped by the information in the data whereas their noise renders them weaker or less significant statistically – whereby it is far from clear whether the uncorrelated variance is due to coding errors or due to the indeterminacy of phenomena observed and recorded. By analyzing one set of data, analysts are unable to know whether the disagreements in their data lead them astray or not. Short of having validating evidence at their disposal, the answer lies in finding comparative results.

Note that fundamental to all reliability concerns is the agreement among replications that should ideally be identical. This is so whether the focus is on replicability, accuracy, surrogacy, or decisiveness. However, once the reliability of each variable of multi-variable data is considered sufficiently reliable to proceed, the common practice is to select any one replication for analyzing all variables, either as the only data or a statistically significant portion of them, and to discard all other replications, having made their contributions and being now redundant and dispensable. However, even if the most accurate, the best surrogate or the most decisive replication was chosen for analysis among available replication, analyzing any one replication hides its disagreements with all other disregarded replications.

Note also that all computational methods are determinist by design, i.e., their repeated applications to the same data should yield identical results. Although data analysts tend to exercise their choices among the method at their disposal and how they network them into an analysis, the standards of scientific publishing requires reporting how reported findings came about, in replicable step-by-step details so that colleagues, critics, users, even the analysts themselves can reproduce them, leaving nothing open to chance or to unspecifiable individual preferences. After trying out different methods and formalizing their decision criteria for networking the chosen analytical methods, they should be replicable as often as needed by all stakeholders in a research and yield the same results when applied to the same data.

Considering the common practice of analyzing any one replication of the reliability data or its extension beyond them once the agreement measures are above the chosen minimum for each variable, the choice of that replication is more or less arbitrary and so are the results of their analysis. This fact is rarely acknowledged as such, except by assurances that the chosen replication was part of others which jointly were considered sufficiently reliable. In the absence of validating evidence with which analytical results could be compared, there is no way to evaluate these results other than by going back to the replications that provided the comparisons for reliability assessments and recover the disagreements that any one set of data can no longer represent.

If the generation of data is perfectly reliable, given the deterministic nature of their analysis, there can be only one analytical result. Selecting any one replication of multi-variable data, as implied in Figure 1.1, also leads to only one analytical result. The reliability of these singular results is actually not the same. The assumption of the latter practice is that the data from which the results were obtained either are perfectly reliable or their unreliabilities are insignificant. This assumption may not be justified, even if the analyzed multi-variable data contain relatively small imperfections.

To test this assumption requires the construction of a set of possible data which include the arbitrarily chosen one from the combinations of mismatching values in the reliability data. Figure 15.1 extends Figure 1.1 (shaded) by adding this possibility.

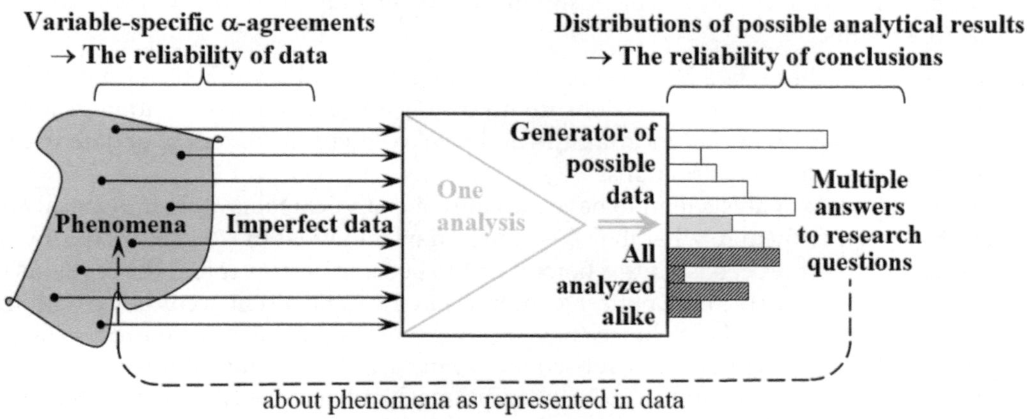


**FIGURE 15.1**
The Distribution of Possible Results as a Result of Unreliable Data

It charts the possibility of generating possible data that lead to a distribution of possible results, which is far more conclusive than measuring the variable-specific agreements at the point at which data enter a research project. It focuses on where reliability actually matters, namely where research results are published and have theoretical or practical implications. Even from only slightly compromised data the distributions may be narrow or wide. Wide distributions indicate uncertainties that no analysis of one set of data can reveal. This calls for benchmarks that spell out how stable findings have to be depending on the implications of acting on published findings; see Section 15.2, for example, in terms of the probabilities of alternative answers to research questions instead of one, or the uncertainty of distributions of possible results instead of a single score.

The shift from measuring the agreements of individual variables to revealing the reliability of the results of analyzing multi-variable data is more than a simple extension of the idea of the reliability of data to the conclusions of their analysis. For one, we obtain distributions of possible findings, not agreements on replications. For another, the mismatches of replications are not summed to disagreements associated with individual variables; they are given the very weight their analytical procedure grants them. And finally, these mismatching values remain no longer associated with individual variables. Their possible impact on the interrelations among variables is accounted for if – which is common to most empirical research – variables are considered constitutive of a system of interconnections from which research conclusions are derived.

Let me outline what is needed to establish the reliability of analytical results from imperfect data:

1. Reliability data as in Table 6.7, generated under conditions outlined in Section 15.1, have to include the data on ideally all ultimately analyzed units or at least a large number, minimally a statistically representative sample of them.

Instead of dismissing all replications used to establishing the reliability of data but one as redundant and dispensable, when it comes to the analysis of data, all mismatching values in these replications are indications of alternative interpretations and give way to the possible data. The more replications are available the more can be said about the reliability of the analytical consequences of compromised data.

2. The reliability data must include all variables that contribute to a particular result. Whether individual variables are perfectly reliable or exhibit minor disagreements, all variables that lead to a reportable result need to be included in efforts of generating all possible results.
3. An analysis of these data has to consist of an algorithmically worked out network of analytical methods and the decisions involved in directing the process to its result. It must be replicable without uncertainties and able to accept the diversity of data within the governing coding instructions.

   Certainly, data analysts tend to have a wealth of analytical tools at their disposal and selectively apply them to what they see in their data and what they wish to explore. As a research project proceeds, these choices tend to become narrower and the methods of analysis are fine tuned to the patterns observed in the data and in accordance with the aims of a study. After these choices have been exhausted, the analytical process ought to be fixed and be replicable in subsequent applications. The algorithmization of the analytical process might appear to be an additional burden to the analyst; however, this should not be entirely strange as all authors of scientific research results are required to describe their methods of analysis in such detail that it can be replicated by colleagues, critics, and stakeholders.
4. Generating combinatorically possible data from mismatching values in the reliability data.

   When reliability data exhibit no disagreements in any of the variables of a study, data imply no alternative possibilities and analyzing perfectly reliable data yields a single perfect result.

   By contrast, observed disagreements always signal alternative interpretations of the phenomena under scrutiny. According to Figure 15.1 the one arbitrarily selected replication used in customary analyses of multi-variable data has to be included as merely one of several possible variations of data for separate analysis. These variations have to be constructed by combining ideally all mismatching values in the given reliability data – regardless of the variable-specific agreement measures. Let me describe three methods for constructing possible data from the mismatches in replications.

   First, *generating all combinatorically possible data.* Such data consist of the values of all units of analysis on which observers agree plus all combinations of values on which they do not agree. This method acknowledges that all mismatching values belong to alternative conceptions of the phenomena coded, which needs to be recognized as such. In variable-specific measures of agreements, mismatches are summed into disagreements and related to what is expected if coded data were unrelated to the phenomena as coded. Here we bypass such summations by constructing all possible combinations of mismatching values, acknowledging that each represents a different conception of the phenomena under consideration.

   For example, the following reliability data consist of three replications generated by three observers, A, B, and C, each coding six units in terms of three variables. Regarding unit 3 of the 1st variable, all three observers disagree. Regarding unit 1 of the 2nd variable, observers B and C agree on valuing that unit 2 from which A deviates, valuing it 3. They agree on all other units, most obviously on coding the 3rd variable. With agreement on all other units being unanimous there are 3 × 2 = 6 possible data implied by these mismatches:

**Reliability data for three variables**

| Variables: | | 1st $_c\alpha$=0.734 | | | | | | 2nd $_c\alpha$=0.709 | | | | | | 3rd $_c\alpha$=1.000 | | | | | |
|---|---|---|---|---|---|---|---|---|---|---|---|---|---|---|---|---|---|---|---|
| Units: | | 1 | 2 | 3 | 4 | 5 | 6 | 1 | 2 | 3 | 4 | 5 | 6 | 1 | 2 | 3 | 4 | 5 | 6 |
| Observers: | A | a | a | **b** | b | c | a | **3** | 3 | 1 | 1 | 2 | 4 | 0 | 0 | 1 | 0 | 1 | 1 |
| | B | a | a | **a** | b | c | a | **2** | 3 | 2 | 1 | 2 | 4 | 0 | 0 | 1 | 0 | 1 | 1 |
| | C | a | a | **c** | b | c | a | **2** | 3 | 2 | 1 | 2 | 4 | 0 | 0 | 1 | 0 | 1 | 1 |

**6 possible data** = **3** mismatches × **2** mismatches

| Possible data | Their frequencies |
|---|---|
| <a a **b** b c a><**3** 3 1 1 2 4><0 0 1 0 1 1> | **1** |
| <a a **b** b c a><**2** 3 1 1 2 4><0 0 1 0 1 1> | **2** |
| <a a **a** b c a><**3** 3 1 1 2 4><0 0 1 0 1 1> | **1** |
| <a a **a** b c a><**2** 3 1 1 2 4><0 0 1 0 1 1> | **2** |
| <a a **c** b c a><**3** 3 1 1 2 4><0 0 1 0 1 1> | **1** |
| <a a **c** b c a><**2** 3 1 1 2 4><0 0 1 0 1 1> | **2** |
| | **9 Results** |

Analyzing these six data may lead to between one to six different conclusions.

Real data rarely are that simple and enumeration of all possible data implied by large numbers of replications and substantial disagreements can easily exceed computational limits. Therefore the two other approaches may offer different approximations to the combinatorial construction of all possibilities that unreliable replications imply.

The second alternative for generating possible data involves *bootstrapping possible data from all mismatching values*. It leaves the coded units on which all observers agree as is and inserts randomly resampled values from observed mismatches. Therefore, each resampled bootstrap constructs one possible data set from all those implied by observed mismatches. The larger the number of bootstraps the more do the resampled data sets approximate the combinatorically generated possibilities described in the first approach. The advantages of bootstrapping possible data over combinatorically generating them are two. First, as the number of bootstraps is arbitrary and can be decided by the analyst, it can be kept within computational limits and selected according how important the findings of a research project are. Second, randomly selecting among mismatching values will select among mismatching values regardless of whether the agreement is small or large. Values in the majority will be selected with the same probability in which they are observed. It follows that bootstrapping does not need to keep track of the number of identical data as shown in the complete combinatorial approach above.

A third alternative, not recommended, is to *analyze each replication separately* and gain one result for each. In the above example it would mean computing the results from the data generated by each of the three observers, A, B, and C. This alternative is the easiest one to compute as replications are already available when measuring the agreement of individual variables. It may serve as a preliminary check. However, it is not recommended because it ignores all intra-replication combinations of mismatches. Note in the above example, while the coding by A, B. and C is part of the six combinatorial possibilities, they do not capture all six of them. Not only does computing possible data from the contributions by individual observers exclude all intra-observer combinations, it requires far more replications for distributions of possible results to become meaningful.

I discussed this third possibility merely to discourage researchers from relying on it exclusively. Calculating the results of different observers' data may be interesting but should not be taken as evidence of the reliability of the research results of a multi-variabled study.

One limitation of all three approaches is that one can compute possible data only for the portion of data for which replications are available. If assurances of the reliability of published results are important, there is considerable virtue in replicating all data ultimately analyzed. A statistically unrepresentative sample of replications can yield only statistically unrepresentative distributions of results.

5. The reliability of analytical results is a function of the widths of the distribution of possible results obtained from the analyses of the possible data extracted from imperfect reliability data:

   The reliability of analytical results from imperfect data can be considered:

   - **Perfect** when all possible data yield the same result.
   - **Tolerable** when all results remain within what is claimed to be the result.
   - **Uncertain** when the distribution of possible results crosses the boundaries of one answer to research questions to its opposite with probabilities associated with each alternative answer.
   - **Absent** when the distribution of possible results prevents decisions on the answer to any of the research questions.

Note that the reliability of individual variables does not enter the reliability of their analyses. In fact they are surrendered to their joint analytical consequences. This might well be construed as an argument against pursuing uniform agreement standards for data in all variables – advocated in the previous sections. However, it is not unusual that some unreliable variables have little influence on the results while other almost perfectly reliable variables include statistically minor mismatches that constitute the tipping point for results to go in another direction. Therefore, calculating the distribution of possible results from the possible data is the reliability analysis method preferred to measuring the reliabilities of individual variables.

However, as already mentioned, there are several reasons not to abandon the practice of assuring that all variables are well above the minimum $\alpha_{ok}$. The first and most obvious one is that it is difficult to know the impact of any one variable on the research results before settling on an appropriate method of analysis. Absent this knowledge, the safest strategy is to ensure that all variables are nearly perfectly reliable. The second reason is one of costs involved. To establish the reliability of research results calls for more data than for individual variables, ideally replicating all data twice, and also ideally by multiple observers, interpreters, or measurements. These coding efforts are not normally undertaken but may have to be if the results really matter. A third reason is the unavailability of sufficiently general software to generate possible data from mismatches in the reliability data. Softwares for computing all combinatorically possible data or bootstrapping representative samples are not complicated but so far not too common. A fourth reason is that it may not always be within the skillset of researchers to translate their analytical method into a replicable algorithm. The use of software for standard analytical methods, like for sentiment analyses, tracing citations or infections, or performing variance, factor, or network analyses already is replicable. However, when unusual analyses are developed, turning them into replicable analytical processes is not that easy but it may have to be done if the possible results become public, legal, or economic issues.

Coming from a different research tradition, Gustafson (2004) is the only researcher I know who explored how measurement errors – the disagreements in α's definition – affect the outcomes of diverse analytical techniques used in epidemiology. He did not mention the word reliability as he was not concerned with what data represented. Rather, he started, for example, from imperfect correlations in cross-tabulations and traced these imperfections through specific analytical processes to diverse conclusions that could emerge from them. His generalizations stayed within the analytical techniques he investigated. However, he also showed, just as I argued above, that measurement errors affect the results of an analysis rather unequally. Some made little difference; others derailed the process and yielded invalid conclusions. Some results are close to a tipping point into their opposites. If research is limited to well-defined and widely available analytical methods, as Gustafson explored, they are always replicable, produce the same results from the same data, and yield the needed distributions when applied to all or a subsample of possible data.

I wish to make clear that the reliability of analytical results obtained by combinatorial or bootstrapping methods should not to be confused with an indication of the validity of research results – unless the reliability data involve a standard of factual evidence. While the dispersion of possible results can be traced to the unreliabilities or noise in multi-variable data, their majorities point back to the phenomena seen, interpreted, and recorded by observers. This is so whether the mismatching values concern the replicability of the process of generating data, the accuracy of the generated data, the surrogacy of a substitute for human coding, or for the decisiveness of human judgments. However, inasmuch as analytical results also depend on the analysts' choices of analytical methods and analytical preferences, the resulting distributions may no longer be exclusively explainable by the nature of the phenomena that gave rise to the research.

While knowing the reliability of research results is methodologically feasible and of practical value, it is for the four reasons mentioned above that I consider the variable-specific α-agreement measures indispensable for assessing the reliability of data. Except for very simple analyses and obvious steps, exemplified by frequency counts in Section 14.1.4, there can be no generalizable formulae by which one could estimate the impact of variable-specific disagreements on the analytical results of multi-variable data before their analysis is worked out and sufficiently formalized. In the absence knowing how coded data will affect the research results, assuring high reliabilities with α-agreement measures developed in previous chapters provides the arguably best assurances that data are worth taking seriously.

# 16

## *Toward a General Calculus of Differences and Agreements*

As I detailed in Chapter 0, I started my journey into issues of data reliability with my dissertation on content analysis (Krippendorff, 1967), knowledge of Claude Shannon's (Shannon & Weaver, 1949) information theory and Ross Ashby's (1969) multi-variable extension, paralleling variance analysis (Fisher, 1925; Garner & McGill, 1956; Krippendorff, 1986). Unexpectedly, it turned out that the information-theoretical $\log_2$ function affected the granularity of its entropy measures as a function of the size of data to which it was applied, making comparisons of different partitions of a variable – needed to test the reliability of data – difficult. I then focused on what variance analysis had to offer (Rajaratnam, 1960; Krippendorff, 1970a) but was stymied by its limitations, being applicable only to interval data whose values could be summed, subtracted, and transformed. These metric qualities are absent when texts are categorized, and hence unordered.

### 16.1 Its Difference Functions (Reiterated)

Then I discovered that the standard deviation *SS* of the analysis of variances could be equated to the sum of differences between all pairs of values (0.3).

$$SS = \sum_{i=1}^{n}(x_i - \bar{x})^2 = \frac{1}{n}\sum_{i=1}^{n}\sum_{j>i}^{n}(x_i - x_j)^2$$

In variance analysis, this difference, $(x_i - x_j)^2$ ties variances to interval data. I saw the possibility of replacing it by one that accommodated nominal data, which arise whenever coding involves categorizations of which two are either the same or different. Defined by (0.4) and again in Chapters 3 and 4:

$$_{\text{nominal}}\delta_{ck}^2 = \begin{cases} 1 \text{ iff } c \neq k \\ 0 \text{ iff } c = k \end{cases} \tag{16.1}$$

With (16.1) in place of the interval differences this opened the path to a first calculus of disagreements among unordered categories (Krippendorff, 1971b).

To standardize the range of all difference functions to be between 0 and 1, the interval difference had to be stated relative to the largest possible difference, yielding the interval difference function of (0.5):

$$0 \leq {}_{\text{interval}}\delta^2_{c_i c_j} = \left(\frac{c_i - c_j}{c_{max} - c_{min}}\right)^2 \leq 1 \tag{16.2}$$

Inasmuch as a expresses all observed disagreements relative to the expected disagreements, the standardizing range $c_{max} - c_{min}$ is a constant in both and these cancel each other, which renders their proportion equivalent to the proportion of two variances.

The extension of (16.2) to the difference function for ordinal data, to rankings, was introduced as (5.12a) for symmetrical coincidences occurring with frequencies $n_{ck}$:

$$_{\text{ordinal}}\delta^2_{ck} = \left(\frac{\sum_{b=c}^{b=k} n_{b\cdot} - \frac{n_{c\cdot} + n_{k\cdot}}{2}}{n_{\cdot\cdot} - \frac{n_{\text{first}\cdot} + n_{\text{last}\cdot}}{2}}\right)^2 \tag{16.3a}$$

and as (5.12b) for asymmetrical contingencies for accuracy, surrogacy, and decisiveness – for generality referred to here by $n$:

$$_{\text{ordinal}}\delta^2_{ck} = \left(\frac{\sum_{b=\text{first}}^{b=c} n_{b\cdot} - \frac{n_{c\cdot}}{2} - \sum_{b=\text{first}}^{b=k} n_{\cdot b} + \frac{n_{\cdot k}}{2}}{n_{\cdot\cdot} - \frac{min(n_{\text{first}\cdot}, n_{\cdot\text{first}}) + min(n_{\text{last}\cdot}, n_{\cdot\text{last}})}{2}}\right)^2 \tag{16.3b}$$

It expanded the scope of this calculus further. To these four well-known metrics of data comes the difference function for ratio data, e.g., proportions, which have absolute zero-points and no negative values:

$$_{\text{ratio}}\delta^2_{ck} = \left(\frac{c-k}{c+k}\right)^2 \tag{16.4}$$

The difference function for polar scales, whose values are limited by defined maximum and minimum values, the polar opposites:

$$_{\text{polar}}\delta^2_{ck} = \frac{(c-k)^2}{(c+k-2c_{min})(2c_{max}-c-k)} \tag{16.5}$$

And the difference function for circular data; complying with the modular arithmetic, circular metric differences can be defined in two mathematically equivalent ways:

$$_{\text{circular}}\delta^2_{ck} = \left(\sin\left[180\frac{c-k}{U}\right]\right)^2 \quad \text{in degrees}$$

$$_{\text{circular}}\delta^2_{ck} = \left(\sin\left[\pi\frac{c-k}{U}\right]\right)^2 \quad \text{in radians} \tag{16.6}$$

According to (5.3) to (5.5) the above difference functions are:

symmetrical: $$_{\text{metric}}\delta^2_{ck} = {}_{\text{metric}}\delta^2_{kc} \tag{16.7}$$

measure: $$_{\text{metric}}\delta^2_{c=k} = 0 \tag{16.8}$$

and are standardized: $$0 \le {}_{\text{metric}}\delta^2_{ck} \le 1 \text{ (comparatively convenient)} \tag{16.9}$$

All other difference functions that anyone may want to introduce to cope with different situations should conform to these three properties. Standardization in (16.9) is not required but offers convenient comparisons as demonstrated in Chapter 5.

## 16.2 Multi-Variable Analysis of Differences

For assessing the replicability of data generated from predefined units, Table 5.1 depicts their canonical forms in terms of an $m$-by-$N_u$ matrix of $m$ independent replications by $N_u$ discrete units of analysis, containing the values or multi-valued sets of values assigned by one of $m$ replications to one of $N_u$ units. In effect, replicability is obtained from three-dimensional data in the product set of: $R \times U \times C$. They serve different purposes. The set $R$ contains the $m$ replications and the set $U$ the $N_u$ units of analysis. Both are explanatory variables that account for the differences between pairable values in two partitions of the set $C$ of values. The disagreements $D_o$ among the values $c \in_u C$ assigned to units $u \in U$ and the disagreements $D_e$ among all values $c \in C$.

This chapter seeks to overcome the limitation to the canonical form of reliability data, defined in Table 5.1 essential for assessing replicability and decisiveness, part of Table 6.7 of the quadrilogy, making the above difference functions available to other forms of analysis. Previous chapters already deviated from traditional reliability analyses by allowing some cells $\langle r, u, c\rangle \in R \times U \times C$ to be absent. The frequency $n_{ruc}$ associated with each cell $\langle r, u, c\rangle$ could be 0 or 1. This chapter relaxes this requirement in two respects. First, any number of explanatory variables can be considered, not just two. Second, it allows each cell in a multi-dimensional space to have frequencies of any positive number, including 0.

### 16.2.1 Frequencies for Multi-Variable Differences in *C*

For convenience we will continue to describe the data with the three familiar variables with the understanding that these could stand for more. This expands my earlier effort (Krippendorff, 1971b) but incorporates the insights since acquired.

The needed frequencies within two variables, here $R \times U$, explanatory of one, here $C$:

$$n_{ruc} = \text{the frequency of the triple } \langle r, u, c\rangle \in R \times U \times C$$

$$n_{ruC} = \sum_{c \in C} n_{ruc} = \text{the number of values } c \in_{ru} C \text{ associated with the pair } \langle r, u\rangle \in R \times U$$

$$n_{uc} = \sum_{r \in R} n_{ruc} = \text{the frequency of the pair } \langle u, c\rangle \in U \times C$$

$$n_{uC} = \sum_{c \in C} n_{uc} = \text{the number of values } c \in_u C \text{ associated with the value } u \in U \tag{16.10}$$

$$n_c = \sum_{r\in R}\sum_{u\in U} n_{ruc} = \text{the frequency of the value } c\in C$$

$$n_C = \sum_{c\in C} n_c = \text{the total number of values } c\in C$$

Frequencies for other foci and more variables can be obtained by analogy.

### 16.2.2 Primary Definitions of Differences in C

In terms of the above sums, (16.10), the following defines the several primary differences in C:

$$\text{Unconstrained}: D(C) = \frac{\sum_{c\in C} n_c \sum_{k\in C}^{n} n_k \,{}_{\text{metric}}\delta_{ck}^2}{n_C - 1} \tag{16.11}$$

$$\text{Conditional on } u\in U:\ D_U(C) = \sum_{u\in U} \frac{\sum_{c\in C} n_{uc} \sum_{k\in C}^{n} n_{uk} \,{}_{\text{metric}}\delta_{ck}^2}{n_{uC} - 1} \tag{16.12}$$

$$\text{Conditional on } r\in R: D_R(C) = \sum_{r\in R} \frac{\sum_{c\in C} n_{rc} \sum_{k\in C}^{n} n_{rk} \,{}_{\text{metric}}\delta_{ck}^2}{n_{rC} - 1} \tag{16.13}$$

$$\text{Conditional on } \langle r,\ u\rangle \in R\times U: D_{RU}(C) = \sum_{r\in R}\sum_{u\in U} \frac{\sum_{c\in C} n_{ruc} \sum_{k\in C}^{n} n_{ruk} \,{}_{\text{metric}}\delta_{ck}^2}{n_{ruC} - 1} \tag{16.14}$$

### 16.2.3 Secondary Definitions of Differences in C: Agreements and Accounting Equations for Differences

The agreement among values $c\in C$ partitioned by $u\in U$:

$$D(U : C) = D(C) - D_U(C) \tag{16.15}$$

The agreement among values $c\in C$ partitioned by $u\in U$ and conditional on $r\in R$

$$D_R(U : C) = D_R(C) - D_{RU}(C) \tag{16.16}$$

The agreement among values $c\in C$ partitioned by $r\in R$ and conditional on $u\in U$

$$D_U(R : C) = D_U(C) - D_{RU}(C) \tag{16.17}$$

The agreement among values $c\in C$ partitioned by $\langle r, u\rangle \in R\times U$

$$D(R,U : C) = D(C) - D_{RU}(C) \tag{16.18}$$

The interaction of agreements among values $c \in C$ due to the unique combinations $<r,u> \in R \times U$

$$D(\overline{ru} : C) = D_R(U : C) - D(U : C) \tag{16.19}$$

### 16.2.4 Accounting Equations for Partitions of *C*

The first explains the agreements among values *c* of $_{ru}C$ in the cells $<r, u> \in R \times U$:

$$D(R,U : C) = D(U : C) + D(R : C) + D(\overline{ru} : C) \tag{16.20}$$

where $D(U{:}C)$ is the agreement among values *c* of $_uC$ by values $u \in U$, $D(R : C)$ is the agreement among values *c* of $_rC$ by values $r \in_u R$, and $D(\overline{ru} : C)$ is the interaction between *r* and *u* regarding *C*.

The second explains unconstrained differences *D*(*C*)among all *C*:

$$D(C) = D(U : C) + D(R : C) + D(\overline{ru} : C) + D_{RU}(C) \tag{16.21}$$

The third expresses (16.21) in terms of the proportions of unconstrained differences *D*(*C*):

$$1 = \frac{D(U : C)}{D(C)} + \frac{D(R : C)}{D(C)} + \frac{D(\overline{ru} : C)}{D(C)} + \frac{D_{RU}(C)}{D(C)} \tag{16.22}$$

When the conditions of reliability data are met, and only then, α is the proportion of agreement in *C* partitioned by $u \in U$ (the units of analysis) – measured by the deviating from 1:

$$\alpha = 1 - \frac{D_o}{D_e} = 1 - \frac{D_U(C)}{D(C)} = \frac{D(U{:}C)}{D(C)} \tag{16.23}$$

Under these conditions, one could define the proportion β of agreement within replications – ideally measuring 0 – by:

$$\beta = 1 - \frac{D_R(C)}{D(C)} = \frac{D(R : C)}{D(C)} \tag{16.24}$$

and the proportion of agreement in interactions by:

$$\iota = \frac{D(\overline{ru} : C)}{D(C)} \tag{16.25}$$

Because reliability data contain at most one value in the $<r, u>$-cells of their canonical form, $D_{RU}(C) = 0$ in (16.21) enables expressing (16.23) through (16.25) as the sum of the three coefficients:

$$1 = \alpha + \beta + \iota \tag{16.26}$$

In (16.26), $\alpha$ is the extent to which data are replicable; $\beta$ is the extent to which they are due to coder predilections, and $\iota$ is the extent to which both interact.

### 16.2.5 A Numerical Example

Note that the following analysis deviates slightly from a previous publication (Krippendorff, 1971b:198) at which time I defined disagreements as proportions of $n_C$ and made a distinction between measured and estimated disagreements. The current quantities differ from the earlier definitions mainly by the factor of $n_C$. Their logic seamlessly maps into the current and more advanced accounting equations.

The numerical example consists of five unordered values $c \in C$, $n_c = 112$ values in total which are partitioned into four rows $R$, three columns $U$, and twelve cells of its $R \times U$ matrix. These values are distributed over the cells of this 4-by-3 matrix. The distribution of its values in rows, columns, and totals is expressed in terms of their frequencies.

As the values in this example consist of alphabetical characters, only the nominal difference function ${}_{\text{nominal}}\delta^2_{ck}$ is applicable.

Finally, although the terms used to designate the three dimensions of this example are taken from typical reliability considerations of the previous chapters, by not restricting the frequencies of cell contents to one or less, this example would not lend itself to the computation of valid reliability coefficients. The purpose of this example is to demonstrate that the calculus is far more general than depicted in Table 16.1.

| *R* \ *U* | 1 | 2 | 3 | | | | | | |
|---|---|---|---|---|---|---|---|---|---|
| *v* | dd d b | bb d e | cccc dd dd | $n_{va}=0$ | $n_{vb}=3$ | $n_{vc}=4$ | $n_{vd}=8$ | $n_{ve}=1$ | $n_v=16$ |
| *w* | dd b d<br>dd b d | bb b b<br>bb b b | cccc bb ae<br>cccc bb cc | $n_{wa}=1$ | $n_{wb}=14$ | $n_{wc}=10$ | $n_{wd}=6$ | $n_{we}=1$ | $n_w=32$ |
| *x* | dd c e<br>dd c d<br>dd c b | bb c b<br>bb c b<br>bb c a | cccc cc ca<br>cccc cc ca<br>cccc cc bd | $n_{xa}=2$ | $n_{xb}=9$ | $n_{xc}=26$ | $n_{xd}=9$ | $n_{xe}=2$ | $n_x=48$ |
| *y* | dd a c | bb a a | cccc aa ea | $n_{ya}=6$ | $n_{yb}=2$ | $n_{yc}=5$ | $n_{yd}=2$ | $n_{ye}=1$ | $n_y=16$ |
| | $n_{1a}=1$ | $n_{2a}=3$ | $n_{3a}=5$ | $n_a=9$ | | | | | |
| | $n_{1b}=4$ | $n_{2b}=19$ | $n_{3b}=5$ | | $n_b=28$ | | | | |
| | $n_{1c}=4$ | $n_{2c}=3$ | $n_{3c}=38$ | | | $n_c=45$ | | | |
| | $n_{1d}=18$ | $n_{2d}=2$ | $n_{3d}=5$ | | | | $n_d=25$ | | |
| | $n_{1e}=1$ | $n_{2e}=1$ | $n_{3e}=3$ | | | | | $n_e=5$ | |
| | $n_1=28$ | $n_2=28$ | $n_3=56$ | | | | | | $n_C=112$ |

Table 16.1
Literal and Numerical Exemplification of Three-Dimensional Nominal Data

The confinement to the nominal difference function $_{\text{nominal}}\delta^2_{ck}$ allows for computational simplifications of the primary differences, here taken advantage of.

In terms of Table 16.1, The unconstrained differences within $C$ are obtained by (16.11):

$$D(C) = \frac{\sum_{c\in C} n_c \sum_{k\in C}^{n} n_k \ {}_{\text{nominal}}\delta^2_{ck}}{n_C - 1} = n_C - \frac{\sum_{c\in C} n_c(n_c - 1)}{n_C - 1}$$
$$= 112 - \frac{9(9-1)+28(28-1)+45(45-1)+25(25-1)+5(5-1)}{112-1} = 81.117$$

According to (16.12), the differences in $C$ conditional on $u\in U$, i.e., in three columns of Table 16.1 are:

$$D_U(C) = \sum_{u\in U} \frac{\sum_{c\in C} n_{uc} \sum_{k\in C}^{n} n_{uk} \ {}_{\text{nominal}}\delta^2_{ck}}{n_{uC} - 1} = n_C - \sum_{u\in U} \frac{\sum_{c\in C} n_{uc}(n_{uc} - 1)}{n_{uC} - 1}$$
$$= 112 - \frac{0+4(3)+4(3)+18(17)+0}{28-1} - \frac{3(2)+19(18)+3(2)+2(1)+0}{28-1}$$
$$- \frac{5(4)+5(4)+38(37)+5(4)+3(2)}{56-1} = 59.829$$

According to (16.13) the differences in $C$ conditional on $r \in R$, i.e., in the four rows of Table 16.1 are:

$$D_R(C) = \sum_{u\in U} \frac{\sum_{c\in C} n_{rc} \sum_{k\in C}^{n} n_{rk} \ {}_{\text{nominal}}\delta^2_{ck}}{n_{rC} - 1} = n_C - \frac{\sum_{c\in C} n_{rc}(n_{rc} - 1)}{n_{rC} - 1}$$
$$= 112 - \frac{0+3(2)+4(3)+8(7)+0}{16-1} - \frac{0+14(13)+10(9)+6(5)+0}{32-1}$$
$$- \frac{2(1)+9(8)+26(25)+9(8)+2(1)}{48-1} - \frac{6(5)+2(1)+5(4)+2(1)+0}{16-1} = 76.746$$

According to (16.14) the difference in $C$ conditional on $\langle r, u\rangle \in R\times U$, i.e., in the 12 cells of Table 16.1 are:

$$D_{RU}(C) = \sum_{r\in R}\sum_{u\in U} \frac{\sum_{c\in C} n_{ruc} \sum_{k\in C}^{n} n_{ruk} \ {}_{\text{metric}}\delta^2_{ck}}{n_{ruC} - 1} = n_C - \sum_{r\in R}\sum_{u\in U} \frac{\sum_{c\in C} n_{ruc}(n_{ruc} - 1)}{n_{ruC} - 1}$$
$$= 112 - \frac{3(2)+0}{4-1} - \frac{2(1)+0+0}{4-1} - \frac{4(3)+4(3)}{8-1} - \frac{2(1)+6(5)}{8-1} - \frac{8(7)}{8-1}$$
$$- \frac{0+4(3)+10(9)+0}{16-1} - \frac{0+3(2)+7(6)+0}{12-1} - \frac{0+8(7)+3(2)}{12-1}$$
$$- \frac{2(1)+0+20(19)+0}{24-1} - \frac{0+0+2(1)}{4-1} - \frac{2(1)+2(1)}{4-1} - \frac{3(2)+4(3)+0}{8-1} = 55.353$$

The secondary differences, which are agreements among values $c \in C$ across diverse partitions are as follows, starting with (16.15), in the partition by $u \in U$:

$$D(U:C) = D(C) - D_U(C) = 81.117 - 59.829 = 21.288$$

The agreement (16.16) on values $c \in_r C$ partitioned by $r \in R$:

$$D(R:C) = D(C) - D_R(C) = 81.117 - 76.746 = 3.371$$

The agreement (16.17) among values $c \in_u C$ partitioned by $u \in U$ and conditional on $r \in R$:

$$D_R(U:C) = D_R(C) - D_{RU}(C) = 76.746 - 55.353 = 21.393$$

The agreement (16.18) among values $c \in_{ru} C$ partitioned by $<r, u> \in R \times U$, in the cells of the table:

$$D(R,U:C) = D(C) - D_{RU}(C) = 81.117 - 55.353 = 25.764$$

The interaction of agreement (16.19) among values $c \in C$ due to the combinations $<r, u> \in R \times U$:

$$D(\overline{ru}:C) = D_R(U:C) - D(U:C) = 21.415 - 21.288 = 0.105$$

Equation (16.20) which accounts for all agreements among values $c \in C$ would yield:

$$D(R,U:C) = D(U:C) + D(R:C) + D(\overline{ru}:C)$$

$$25.764 = 21.288 + 3.371 + 0.105$$

Equation (16.21) which accounts for the unconstrained differences $D(C)$ reveals:

$$D(C) = D(U:C) + D(R:C) + D(\overline{ru}:C) + D_{RU}(C)$$

$$81.117 = 21.288 + 3.371 + 0.105 + 55.353$$

And equation (16.22), expressing (16.21) as proportions of unconstrained differences $D(C)$, becomes:

$$\frac{D(C)}{D(C)} = \frac{D(U:C)}{D(C)} + \frac{D(R:C)}{D(C)} + \frac{D(\overline{ru}:C)}{D(C)} + \frac{D_{RU}(C)}{D(C)}$$

$$1 \quad = 0.262 + 0.054 + 0.001 + 0.683$$

Visual inspection of Table 16.1 should confirm the proportional accounts of (16.21). Obviously. the proportional agreement in these three columns is larger than in the four rows. In fact, the proportional agreement in $\frac{D(U:C)}{D(C)} = 0.262$ is nearly five times larger than

the proportional agreement $\frac{D(R:C)}{D(C)} = 0.054$ which measures near zero. The proportion of
the remaining differences in C, which none of the three other agreements can accounts for, the proportion of the differences in the twelve cells of Table 16.1, i.e., $\frac{D_{RU}(C)}{D(C)} = 0.683$ accounts for most of the differences $\frac{D(C)}{D(C)} = 1.000$ in these data.

## 16.3 Multi-Variable Analyses of Decisiveness as a Form of Agreement

Although this chapter goes beyond the reliability of data, the concept of decisiveness was introduced in Chapter 6 as one kind of agreement measure. There, replicability is defined as a measure of agreement among all pairs of replications or observers; accuracy as a measure of agreement between each observer and a trusted standard; surrogacy as a measure of the agreement between a potential substitute for what trusted observers have in common; and decisiveness as a measure of agreement between each observer and what all of them have in common. This chapter speaks for the latter form of agreement. For example, if the members of a group are unanimous regarding an issue they constitute the most decisive majority possible. If some members dissent, decisiveness suffers. If all were concerned with an issue not shared with anyone else, decisiveness would be absent or zero. Decisiveness is one form of agreement.

Chapter 0 noted the equivalence of the sum of all pairwise differences and the variance for interval scales. In (0.3):

$$SS = \sum\nolimits_{i=1}^{n}(x_i - \bar{x})^2 = \frac{1}{n}\sum\nolimits_{i=1}^{n}\sum\nolimits_{j>i}^{n}(x_i - x_j)^2$$

Clearly, variance is related to decisiveness by being affected by the deviation of a set of values from their mean. To express variance in terms of the interval difference function ${}_{\text{interval}}\delta^2_{ck}$ defined in Section 16.1 merely requires replacing the value $k$ with the mean $\bar{c}$ of all values $c \in C$ and acknowledging its standardization:

$$\left(\frac{c - \bar{c}}{c_{max} - c_{min}}\right)^2 = {}_{\text{interval}}\delta^2_{c\bar{c}}$$

Then:

$$SS = \sum\nolimits_{c \in C}(c - \bar{c})^2 = (c_{max} - c_{min})^2 \sum\nolimits_{c \in C} {}_{\text{interval}}\delta^2_{c\bar{c}}$$

And the relation between the two expressions is a simple constant, a function of all data being analyzed.

Generalizing ${}_{\text{interval}}\delta^2_{c\bar{c}}$ to other metrics ${}_{\text{metric}}\delta^2_{c\bar{c}}$ effectively generalized the conception of Fisher (1925) of variance to data other than interval scales of measurements. This requires close attention to the analogues of averages in different metric qualities of data, which is the topic of the following section.

### 16.3.1 Means, Medians, Modes or Majorities

The following states the conceptual analogues of averages or means, developed in Section 6.6.3, but in terms of this chapter. All commonalities refer to $n_{wC}$ values $c$ in a set ${}_wC$, defined conditional on the context of qualities $w$, which are logically independent explanatory variables of $c \in {}_wC$. With $n_{wc}$ as the multi-valued frequency of pairs $<w, c> \in W \times C - W$ standing in for the explanatory variables for values $c \in {}_wC$ – the number of values $c \in {}_wC$ is:

$$n_{wC} = \sum\nolimits_{c \in C} n_{wc} \qquad (16.27)$$

For interval, ratio and polar data, the mean $\bar{c}$ of a set of $n_{wC}$ of values $c \in {}_wC$ is their average (6.21):

$$\bar{c} = \frac{1}{n_{wC}} \sum\nolimits_{c=1}^{n_{wC}} c \in {}_wC \qquad (16.28)$$

For ordinal data or ranks, the analogue of a mean is the median rank, which is found in the center of the frequency distribution of $n_{wC}$ ranks between its $c_{lowest}$ and its $c_{highest}$ rank:

$$\bar{c} = \text{The} \left(\frac{m_u + 1}{2}\right)^{\text{th}} \text{rank between the } c_{\text{lowest}} \text{ and } c_{\text{highest}} \text{ rank} \qquad (16.29)$$

When the number $m_u$ of ranks is uneven, $\bar{c}$ is an observed rank. When $m_u$ is even, $\bar{c}$ may not be an observed rank. If it falls between two different ranks, $\bar{c}$ is the mean of the two neighboring ranks.

For nominal data, the analogue of a mean is the majority or mode of a distribution of $n_{wc}$ categories $c \in {}_wC$. In the absence of ties, a single mode $\bar{c}$ is:

$$\bar{c} \mid max(n_{wc \in {}_wC}) \qquad (16.30a)$$

However, it may happen that $n_{wC}$ values $c$ in ${}_wC$ exhibit two or more tied modes. Ties exist when $h > 1$ different categories $c$ share the highest frequencies ${}_w\{n_{c_1} = n_{c_2} = ... = n_{c_h}\}$ in the context of $w$. To embrace the possibility of $h$ modes characterizing the set of values ${}_wC$, (16.30a) is generalized to (16.30b):

$${}_w\left\{ \bar{c}_1, ...\bar{c}_g, ...\bar{c}_h \right\} \mid max(n_{wc_1} = \cdots = n_{wc_g} = \cdots = n_{wc_h}) \qquad (16.30b)$$

While ties become less likely as the number $n_{wC}$ of categories becomes large, their possibility cannot be ignored.

### 16.3.2 Frequencies for the Multi-Variable Decisiveness in *C*

This section follows Section 16.2.1 with frequencies in three dimensions to which the reference to their appropriate $\bar{c}$ is added as follows:

$n_{ruc}$ = the frequency of the triple $<r, u, c> \in R \times U \times C$

$n_{ruC} = \sum_{c \in C} n_{ruc}$ = the number of values $c \in_{ru} C$ in the above triple

In the absence of ties, the reference to $\bar{c}$ is added to each $n_{ruc \in_{ru} C}$ to yield $n_{ruc\bar{c}_{g=1=h}}$

In case of $h$ ties $_{ru}\{\bar{c}_1, \ldots \bar{c}_g, \ldots \bar{c}_h\}$ in $_{ru}C$, each $n_{ruc \in_{ru} C}$ is replaced by $h$ frequencies $n_{ruc\bar{c}_g}$

$n_{uc} = \sum_{r \in R} n_{ruc}$ $\quad n_{uC} = \sum_{c \in C} n_{uc}$ $\quad$ each $n_{uc \in_u C}$ is replaced by $h$ frequencies $n_{uc\bar{c}_g}$ (16.31)

$n_{rc} = \sum_{u \in U} n_{ruc}$ $\quad n_{rC} = \sum_{c \in C} n_{rc}$ $\quad$ each $n_{rc \in_u C}$ is replaced by $h$ frequencies $n_{rc\bar{c}_g}$

$n_{c} = \sum_{r \in R} \sum_{u \in U} n_{ruc}$ $\quad n_{C} = \sum_{c \in C} n_{c}$ $\quad$ each $n_{c \in_u C}$ is replaced by $h$ frequencies $n_{c\bar{c}_g}$

### 16.3.3 Primary Definitions of Decisiveness in *C*

$$\text{Unconstrained: } D(C)^m = \sum_{c \in C} \frac{1}{h} \sum_{g=1}^{h} n_{c\bar{c}_g} \text{ metric } \delta^2_{c\bar{c}_g} \tag{16.32}$$

$$\text{Conditional on } U\text{: } D_U(C)^m = \sum_{u \in U} \sum_{c \in C} \frac{1}{h} \sum_{g=1}^{h} n_{uc\bar{c}_g} \text{ metric } \delta^2_{c\bar{c}_g} \tag{16.33}$$

$$\text{Conditional on } R\text{: } D_R(C)^m = \sum_{r \in R} \sum_{c \in C} \frac{1}{h} \sum_{g=1}^{h} n_{rc\bar{c}_g} \text{ metric } \delta^2_{c\bar{c}_g} \tag{16.34}$$

$$\text{Conditional on } R \times U\text{: } D_{RU}(C)^m = \sum_{r \in R} \sum_{u \in C} \sum_{c \in C} \frac{1}{h} \sum_{g=1}^{h} n_{ruc\bar{c}_g} \text{ metric } \delta^2_{c\bar{c}_g} \tag{163.5}$$

### 16.3.4 Secondary Definition of Decisiveness in *C*: Accounting Equations for Agreements and Decisiveness

The decisiveness of values $c \in C$ in relation to their respective modes, medians, or means $\bar{c}$ within the partitions of $C$ of analytical interest and in the context of other variables follows Section 16.2.2 as well as the accounting equations in Section 16.2.3. The numerical example in Section 16.2.5 illustrated its computations, but only for nominal differences.

## 16.4 Concluding Remarks

As mentioned at the beginning of this chapter, defining this calculus of differences and agreements for three variables, any two to explain the third, was merely convenient and it evolved from the analysis of reliability data in the previous chapters.

There are no restrictions on turning one's analytical lenses to the two explanatory variables and analyzing $D(R)$ or $D(U)$ in view of $R \times C$ or $U \times C$ instead of $D(C)$ in terms of $R \times U$, using difference functions suitable to the metric of their values $r \in R$ and $u \in U$. It means merely rotating the names of the variables in the definitions of differences and agreements and proceeding accordingly.

There also are no restrictions on using fewer or larger numbers of variables. The frequencies and primary and secondary definitions presented here should be easily extendable to numbers of explanatory variables larger than the three elaborated above.

Although the values of the explanatory variables used in the example of Table 16.1 are categorial, if the explanatory variables have other metrics, this should not prevent them from distinguishing the various versions of differences, decisiveness, and agreements, and there is no reason to limit the explanatory variables to nominal data.

I have toyed with developing this calculus for considerations of accuracy and surrogacy. When applied to reliability data which cannot have more than one value (or sets of values together with the multi-valued difference function as defined in Chapter 7) in their canonical form, evaluating the accuracy of coding, and evaluating the surrogacy of potential substitutes for coding are extremely useful in reliability contexts of large data but probably less so in other situations.

I suggest that decisiveness as developed in Chapter 6 and the calculus for decisiveness and agreements developed here, have significant futures, not least because they most closely resemble measures of variance in terms of which many analytical techniques have been developed.

# Appendix

## Terms for Alpha

**In general:**

$\alpha = 1 - \frac{D_o}{D_e}$ => the variable-specific coefficient Alpha (0.2) ff., wherein:

$D_o$ => the observed disagreement in replicated the records of phenomena.

$D_e$ => the expected disagreement, absent of any relationship between records and the phenomena of interest.

**The quadrilogy of reliabilities** (distinguished by superscripts), defined in Chapter 6:

$\alpha$ => measures the replicability of the process of generating data. Started as (0.2), defined as (2.5) in Chapter 2, in Chapters 3, 4, and 5, and explicitly distinguished by (6.2) in Section 6.2.1 and (6.11) in Section 6.3 and understood thereafter.

$\alpha^s$ => measures the accuracy of data. (6.4) in Section 6.2.2 and (6.14) in Section 6.3. Repeatedly used in other chapters.

$\alpha^{sm}$ => measures the surrogacy of an alternative (algorithm, coder, or theory), the degree to which it can serve as a substitute for (human) coders. (6.6) in Section 6.2.3 and (6.20) in Section 6.3.

$\alpha^m$ => measures the decisiveness of majorities, modes, medians, and averages. (6.8) in Section 6.2.4 and (6.23) in Section 6.3. Generally the degree to which a set of values conform to what they have in common.

**$\alpha$ accommodates different orderings in the phenomena represented in data**. Chapter 5:

$\alpha_{\text{metric}}$ => "metric" in subscript refers to any of the following metrics matching the ordering of the phenomena to be coded. (5.8) and (5.11) in Chapter 5.

$\alpha_{\text{binary}}$ => for dichotomous distinctions, a special case of all $\alpha_{\text{metric}}$. (2.9) in Sections 2.2 and extensively explored in Chapter 3.

$\alpha_{\text{nominal}}$ => for unordered data, containing mutually exclusive categories, implied in Chapters 2 to 4 and explicitly defined by (5.8) and (5.10) in Section 5.2.

$\alpha_{\text{ordinal}}$ => for phenomena that possess an ordering describable as ranked, defined by (5.8) and (5.12) in Section 5.3.2.

$\alpha_{\text{interval}}$ => for phenomena whose linear ordering is quantifiable, scales of measurement. (5.8) and (5.13) Section 5.3.3.

$\alpha_{\text{ratio}}$ => for phenomena that can be expressed in terms of proportions, e.g., observed frequencies, which have absolute zero points. (5.8) and (5.14) in Section 5.3.4.

$\alpha_{\text{polar}}$ => for phenomena best described in polar opposite terms like degrees to which something is good or bad, slow or fast. (5.8) and (5.15) in Section 5.3.5.

$\alpha_{\text{circular}}$ => for phenomena in recursive loops. (5.8) and (5.16) in Section 5.3.6.

**The $\alpha$s for the single-valued coding of predefined units of phenomena:**

$_c\alpha$ => This $\alpha$ is assumed in Chapters 0 through 4. Explicated in Chapters 5 and 6.

$_{(k)c}\alpha$ => The $\alpha$ for a designated value $k$ among all single values used. (12.8) in Section 12.5.

$_{c\text{-}k}\alpha$ => The $\alpha$ for a $c$–$k$ pair of values among all single-valued pairs of values. (12.11) in Section 12.5.

$_{(u)c}\alpha$ => The $\alpha$ for a designated unit $u$ among $N_u$ single-valued units. (12.2) in Section 12.2.

$_{(u\text{-specific})c}\alpha$ => The $\alpha$ for the reliability of selecting unit-specific single categories. (13.6) and (13.7) in Section 13.2.

**The $\alpha$s for the multi-valued coding of predefined units of phenomena:**

$_C\alpha$ => The $\alpha$ for the multi-valued set of values assigned to units. (7.8) in Section 7.3.

$_{c \in C}\alpha$ => The $\alpha$ for a single value $c$ in multi-valued sets of values. (7.10) in Section 7.4.

$_{(u)C}\alpha$ => The $\alpha$ for the multi-valued set of values assigned to a single unit $u$. (13.16) in Section 13.4.

$_{(K)C}\alpha^{sm}$ => The $\alpha$ for measuring the surrogacy of any one group's set $K$ of evaluations of given units relative to what all groups' multi-valued evaluations of these units have in common. (13.19) in Section 13.4.

$_{\bar{C}}\alpha$ => The $\alpha$ for standardized multi-valued set of values, ignoring the sizes of the set $C$ of values. (7.20) in Section 7.6.

**The $\alpha$s for freely unitized and selectively coded segments of continua:**

$_u\alpha$ => The family of $\alpha$s for identifying and coding segments in the partition of a continuum. (8.5) in Section 8.1.

$_{\phi|u}\alpha$ => The $\alpha$ for the binary distinction between relevant and identified and irrelevant and ignored segments in the partition of a continuum. (8.8) in Section 8.2.1.

$_{cu}\alpha$ => The $\alpha$ for the single-valued coding of identified segments in the partition of a continuum. (8.13) in Section 8.2.3.

$_{Cu}\alpha$ => The $\alpha$ for multi-valued coding of identified segments in the partition of a continuum. (8.18) in Section 8.4.

$_U\alpha$ => The $\alpha$ for the single-valued coding of identified segments, preserving their contiguities within a continuum. (9.5) in Chapter 9.

$_D\alpha$ => The $\alpha$ for distinctions drawn within a continuum. (10.5) in Chapter 10.

**The $\alpha$s for special applications**

$_c\alpha^{IR}$ => The $\alpha$ for information retrieved. (11.5) in Section 11.3.

$_c\alpha^R$ => The $\alpha$ for the reliability of search engines, part of $_c\alpha^{IR}$. (11.5) in Section 11.4.

$_c\alpha^I$ => The $\alpha$ for the reliability of analysts handling the searches and identifying what is relevant, part of $_c\alpha^{IR}$. (11.6) in Section 11.5.

$_P\alpha$ => The $\alpha$ for ranked units of analysis. (13.10) in Section 13.3.

$_{[\text{metric}]}\alpha$ => The $\alpha$ for circumscribing [××] and single-valued identifications of two-dimensional images. (13.23) in Section 13.5.

**Agreement measures not for assessing reliability:**

$_{Cu}\hat{\alpha}$ => An approximation of $_{cu}\alpha$ when single values are not mutually exclusive in a variable or semantic domain. (13.4) in Section 13.1.

$_{(\bar{k})c}\alpha^{=}$ => A variance-independent $\alpha$ obtained by averaging all $_{(k)c}\alpha$. (13.25) in Section 13.6.

$A_{\text{metric}}$ => The % agreement coefficient generalized to multiple contributors and any metric judgments. (13.28) in Section 13.7.

# References


Aickin, Mikel. 1990. Maximum likelihood estimation of agreement in the constant predictive probability model, and its relation to Cohen's Kappa. *Biometrics* 46: 293–302.

Alfonso, Herminia, C. M. 1999. *Socially Shared Inquiry*. Quezon City, Philippines: Great Books Trading.

Altheide, David L. 1987. Ethnographic content analysis. *Qualitative Sociology* 10: 65–77.

Armstrong, David, Ann Gosling, Josh Weinman, and Theresa Martaeu. 1997. The place of inter-rater reliability in qualitative research: an empirical study. *Sociology* 31: 597–598.

Artstein, Ron and Massimo Poesio. 2005. Kappa3=alpha (or beta). http://cswww.essex.ac.uk/technical-reports/2005/csm-437.pdf (accessed 2019.3.13).

Ashby, W. Ross. 1964. The set theory of mechanisms and homeostasis. *General Systems Yearbook* 9: 83–97.

Ashby, W. Ross. 1965. Measuring the internal information exchange in a system. *Cybernetics* 8: 5–22.

Ashby, W. Ross. 1969. Two tables of identities governing information flows within large systems. *Communications of the American Society for Cybernetics* 1: 3–8.

Banerjee, Mousumi, Michelle Capozzoli, Laura McSweeney, and Debajyoti Sinha. 1999. Beyond kappa: A review of interrater agreement measures. *The Canadian Journal of Statistics* 27: 3–23.

Benini, Rodolfo. 1901. *Principii di Demongraphia: Manuali Barbera Di Scienze Giuridiche Sociali e Politiche* (No. 29). Firenze, Italy: G. Barbera.

Bennett, Edward M., R. Alpert, and A. C. Goldstein. 1954. Communications through limited response questioning. *Public Opinion Quarterly* 18: 303–308.

Berelson, Bernard. 1952. *Content analysis in communications research*. New York: Free Press.

Berelson, Bernard and Paul F. Lazarsfeld. 1948. *The Analysis of Communication Content*. Chicago, IL: University of Chicago Press.

Berry, Charles C. 1992. The Kappa Statistic (letter to the editor). *Journal of American Medical Association* 268, 18: 2513.

Bloch, Daniel A. and Helena Chmura Kraemer. 1989. 2×2 kappa coefficients: Measures of agreement or association. *Biometrics* 45: 269–287.

Bollier, David. 2010. *The Premise and Perils of Big Data*. Washington, DC: Aspen Institute. https://dorutodpt4twd.cloudfront.net/content/uploads/files/content/docs/pubs/The_Promise_and_Peril_of_Big_Data.pdf (accessed 2020.5.4).

Boyd Danah and Kate Crawford. 2012. Critical questions for big data. *Information, Communication & Society* 15: 662–679. www.tandfonline.com/doi/pdf/10.1080/1369118X.2012.678878 (accessed 2020.5.4).

Brennan, Robert L. and Dale J. Prediger. 1981. Coefficient kappa: Some uses, misuses, and alternatives. *Educational and Psychological Measurement* 41: 687–699.

Broussard, Meredith. 2018. *Artificial unintelligence: How computers misunderstand the world*. Cambridge, MA: MIT Press.

Brouwer, Marten, Cedric C. Clark, Michael F. Eleey, George Gerbner, and Klaus Krippendorff. 1969. "The Television World of Violence." and "Content Analysis Procedures and Results," in *Mass*

*Media and Violence*, Vol. IX. A Report to the National Commission on the causes and prevention of violence, eds. Robert K. Baker and Sandra J. Ball, Washington, DC: U.S. Government Printing Office, November 1969, 311–339 and 519–591. http://repository.upenn.edu/asc_papers/214 (accessed 2020.5.4).
Byrt, Ted, Janet Bishop, and John B. Carlin. 1993. Bias, prevalence and kappa. *Journal of Clinical Epidemiology* 46: 423–429.
Carletta, Jean. 1996. Assessing agreement on classification tasks: The kappa statistic. *Computational Linguistics* 22: 249–254.
Carmines, Edward G. and Richard A. Zeller. 1979. *Reliability and validity assessment*. Beverly Hills, CA: Sage.
Carol, Sean. 2016. *The big picture: On the origins of life, meaning and the universe itself.* New York: Dutton.
Cicchetti, Domenic V. and Alvan R. Feinstein. 1990. High agreement but low kappa: II. Resolving the paradoxes. *Journal of Clinical Epidemiology* 43: 551–559.
Checkland, Peter and Sue Holwell. 1998. *Information, Systems, and Information Systems: Making Sense of the Field*. Chichester, West Sussex: John Wiley & Sons.
Cohen, Jacob 1960. A coefficient of agreement for nominal scales. *Educational and Psychological Measurement* 20: 37–46.
Cohen, Jacob. 1968. Weighed kappa: Nominal scale agreement with provision for scaled disagreement or partial credit. *Psychological Bulletin* 70: 213–220.
Coombs, Clyde H. 1967. *A Theory of Data*. New York: John Wiley & Sons.
Crawford, Kate, Kate Miltner, and Mary L. Gray. 2014. Critiquing Big Data: Politics, ethics, epistemology. *International Journal of Communication* 8: 1663–1672.
Craig, Robert T. 1981. Generalization of Scott's index of intercoder agreement. *Public Opinion Quarterly* 45: 260–264.
Cronbach, Lee J. 1951. Coefficient alpha and the internal structure of tests. *Psychometrika* 16: 297–334.
Cicchetti, Domenic V. and Alvan R. Feinstein. 1990. High agreement but low kappa: II. Resolving the paradoxes. *Journal of Clinical Epidemiology* 43: 551–559.
Dice, Lee R. 1945. Measures of the amount of ecologic association between species. *Ecology* 26: 297–302.
Ejima, Keisuke, Kazuyuki Aihara, and Hiroshi Nishiura. 2013. On the use of chance-adjusted agreement statistic to measure the assortative transmission of infectious diseases. *Computational & Applied Mathematics* 32: 303–313.
Eye, Alexander von. 2006. An alternative to Cohen's κ. *European Psychologist* 11: 12–24.
Eye, Alexander von and Eun Young Mun. 2005. *Analyzing Rater Agreement; Manifest Variable Methods*. Mahwah, NJ: Lawrence Erlbaum.
Feinstein, Alvan R. and Domenic V. Cicchetti. 1990. High agreement but low kappa: I. The problems of two paradoxes. *Journal of Clinical Epidemiology* 43: 543–549.
Feng, Guangchao Charles. 2013. Factors affecting intercoder reliability: A Monte Carlo experiment. *Quality and Quantity* 47: 2959–2982.
Feng, Guangchao Charles. 2013. Underlying determinants driving agreement among coders. *Quality & Quantity* 47: 2983–2997.
Feng, Guangchao Charles. 2014. Intercoder reliability indices: disuse, misuse, and abuse. *Quality & Quantity* 48: 1803–1815.
Feng, Guangchao Charles. 2015. Mistakes and how to avoid mistakes in using intercoder reliability indices. *Methodology: European Journal of Research Methods for the Behavioral and Social Sciences* 11: 13–22.
Fisher, Ronald A. 1925. *Statistical Methods for Research Workers*. Edinburgh: Oliver and Boyd.
Fleck, Ludwik. 1935/1979. *Genesis and Development of a Scientific Fact*. Chicago: University of Illinois Press. (Originally published 1935)
Fleiss, Joseph L. 1971. Measuring nominal scale agreement among many raters. *Psychological Bulletin* 76: 378–382.
Fleiss, Joseph L. 1981. *Statistical Methods for Rates and Proportions*. New York: John Wiley & Sons.
Frakes, William B. and Ricardo Baeza-Yates, eds. 1992. *Information Retrieval, Data Structure and Algorithms*. Englewood Cliffs, NJ: Prentice Hall.

Fretwurst, Benjamin. 2015. Lotus manual; reliability and accuracy with SPSS. http://www.iakom.ch/Lotus/LotusManualEng.pdf (accessed 2020.3.22).

Friese, Susanne. 2019. *Qualitative Data Analysis with* ATLAS.ti, *3rd Edition*. Thousand Oaks, CA: Sage.

Garfinkel, Harold. 1967. *Studies in Ethnomethodology*. Englewood Cliffs, NJ: Prentice Hall.

Garner, Wendel R. and William J. McGill. 1956. The relation between information and variance analyses. *Psychometrika* 21: 219–228.

Geertz, Clifford. 1973. *The Interpretation of Cultures: Selected Essays*. New York: Basic Books.

Goldman, Ronald L. 1992. In reply to Berry 1992. *Journal of American Medical Association* 268: 2513–4.

Goodman, Leo A. and William, H. Kruskal. 1954. Measures of association for cross classification. *Journal of the American Statistical Association* 49: 732–764.

Goodwin, Charles. 2018. *Co-operative Actions; Learning in Doing, Social, Cognitive & Computational Perspectives*. New York: Cambridge University Press.

Graham, Mark. 2012, March 9 Big data and the end of theory? The Guardian. Retrieved from www.theguardian.com/news/datablog/2012/mar/09/big-data-theory (accessed 2019.8.13).

Grove, William M., Nancy C. Andreasen, Patricia McDonald-Scott, Martin B. Keller, and Robert W. Shapiro. 1981. Reliability studies of psychiatric diagnosis. *Archives of General Psychiatry* 38: 408–413.

Guetzkow, Harold. 1950. Unitizing and categorizing problems in coding qualitative data. *Journal of Clinical Psychology* 6: 47–58.

Gustafson, Paul. 2004. *Measurement Error and Misclassification in Statistics and Epidemiology: Impacts and Bayesian Adjustments*. New York: Chapman & Hall/CRC press.

Guttman Louis. 1946. The test-retest reliability of qualitative data. *Psychometrika* 11: 81–95.

Gwet, Kilem Li. 2002. Kappa statistics is not satisfactory for assessing the extent of agreement between raters. *Statistical Methods for Inter-Rater Reliabiliy* 1: 1–5. http://agreestat.com/papers/papers/kappa_statistic_is_not_satisfactory.pdf (accessed 2020.5.4).

Gwet, Kilem Li. 2008. Computing inter-rater reliability and its variance in the presence of high agreement. *British Journal of Mathematical and Statistical Psychology* 61: 29–48.

Gwet, Kilem Li. 2010. *Handbook of Inter-Rater Reliability, 2nd ed.* Gaithersburg, MD: Advanced Analytics.

Hayes, Andrew F. and Klaus Krippendorff. 2007. Answering the call for a standard reliability measure for coding data. *Communication Methods and Measures* 1: 77–89. http://afhayes.com/public/kalpha.pdf (accessed 2020.5.4).

Holley, J. W. and J. P. Guilford. 1964. A note on the G-index of agreement. *Educational and Psychological Measurement* 24: 749–754.

Holsti, Ole K. 1969. *Content Analysis for the Social Sciences and Humanities*. Reading, MA: Addison-Wesley.

Hoyt, Cyril. 1941. Test reliability estimated by analysis of variance. *Psychometrika* 6: 153–160.

Hubbard, Ruth. 1989. Science, facts, and feminism, in *Feminism and Science*, ed. Nancy Tuana, 119–131. Bloomington and Indianapolis, IN: Indiana University Press.

Hubert, L. 1977. Kappa revisited. *Psychological Bulletin* 84: 289–297.

Janes, Cynthia L. 1979. An extension of the random error coefficient of agreement to N×N tables. *British Journal of Psychiatry* 134: 617–619.

Janson, Sacante and Jan Vegelius. 1979. On generalizations of the G index and the phi coefficient to nominal scales. *Multivariate Behavioral Research* 14: 255–269.

Kaplan, Abraham and Joseph M. Goldsen. 1965. The reliability of content analysis categories, in *Language of Politics: Studies in Quantitative Semantics*, ed. Harold D. Lasswell, Nathan Leites, and Associates, 83–112. Cambridge, MA: MIT Press.

Katz, Elihu, Christopher Ali, and Joohan Kim. 2014. *Echoes of Gabriel Tarde: What We Know Better or Different 100 Years Later*. Los Angeles, CA: USC Annenberg Press.

Kolbe, Richard H. and Melissa S. Burnett. 1991. Content-analysis research: An examination of applications with directives for improving research reliability and objectivity. *Journal of Consumer Research* 18: 243–250.

Kraemer, Helena Chmura. 1979. Ramifications of a population model for κ as a coefficient of reliability. *Psychometrika* 44: 461–472.

Krippendorff, Klaus. 1967. *An examination of content analysis: A proposal for a general framework and an information calculus for message analytic situations.* Ph.D. dissertation, University of Illinois, Urbana.
——— 1970a. Estimating the reliability, systematic error and random error of interval data. *Educational and Psychological Measurement* 30: 61–70.
——— 1970b. On generating data in communication research. *Journal of Communication* 20: 241–269, 1970. http://repository.upenn.edu/asc_papers/273 (accessed 2020.3.22).
——— 1971a. Reliability of recording instructions: Multivariate agreement for nominal data. *Behavioral Science* 16: 222–235.
——— 1971b. A calculus for disagreements: A categorical equivalence to variance analysis. *General Systems* 16: 187–203. http://repository.upenn.edu/asc_papers/219 (accessed 2020.3.22).
——— 1978. Reliability of binary attribute data. *Biometrics* 34: 142–144.
——— 1980. *Content Analysis: An Introduction to Its Methodology*. Beverly Hills, CA: Sage.
——— 1986. *Information Theory: Structural Models for Qualitative Data*. Beverly Hills, CA: Sage.
——— 1992, May, "Recent Developments in Reliability Analysis." Paper presented at the 42nd Annual Meeting of the International Communication Association, Miami, FL. http://repository.upenn.edu/asc_papers/44 On 2020.5.4.
——— 2004b. Reliability in content analysis: Some common misconceptions and recommendations. *Human Communication Research 30,* 3: 411–433. http://repository.upenn.edu/asc_papers/242/ (accessed 2020.5.4).
——— 2005. The social construction of public opinion, in *Kommunikation über Kommunikation. Theorie, Methoden und Praxis; Festschrift für Klaus Merten,* eds. Edith Wienand, Joachim Westerbarkey and Armin Scholl, 129–149. Wiesbaden, Germany: VS-Verlag. http://repository.upenn.edu/asc_papers/75/ (accessed 2020.5.4).
——— 2009. Pathology, power and emancipation, in his *On Communicating: Otherness, Meaning, and Information*, ed. Fernando Bermejo, 131–155. New York: Routledge.
——— 2011. Agreement and information in the reliability of coding. *Communication Measures and Methods* 5: 93–112. http://repository.upenn.edu/asc_papers/278 (accessed 2020.5.4).
——— 2012. Commentary: A dissenting view on so-called paradoxes of reliability coefficients, in *Communication Yearbook 36*, ed. Charles T. Salmon, 481–499. New York: Routledge.
——— 2013. *Content Analysis; An Introduction to its Methodology, 3rd Edition*. Thousand Oaks, CA: Sage. Replacement of Section 12.4 for its 2nd printing.
——— 2019. *Content Analysis; An Introduction to its Methodology, 4th Edition*. Thousand Oaks, CA: Sage.
——— and Mary Angela Bock. 2009. *The Content Analysis Reader*. Thousand Oaks, CA: Sage.
——— and Richard Craggs. 2016. The reliability of multi-valued coding of data. *Communication Methods and Measures* 10: 181–198. www.researchgate.net/publication/272581846_On_the_Reliability_of_Unitizing_Continuous_Data (accessed 2020.5.4).
———, Yann Mathet, Stéphane Bouvry, and Antoine Widlöcher. 2016. On the reliability of Unitizing Textual Continua: Further developments. *Quality & Quantity* 50: 2347–2364. Online since 2015.9.15 at https://link.springer.com/article/10.1007/s11135-015-0266-1 (accessed 2016.1.23).
Kuhn, Thomas S. 1962. *The Structure of Scientific Revolutions* (1st ed.). Chicago, IL: University of Chicago Press.
Kuder, G. Frederic, and M. W. Richardson. 1937. The theory of the estimation of test reliability.
Landis, J. Richaer and Gary G. Koch. 1977. The measurement of observer agreement for categorical data. *Biometrics* 33: 159–174.
Lasswell, Harold D. 1948. *The structure and function of communication in society*, in *The Communication of Ideas*, ed. Bryson, Lyman 117–129. New York: Institute for Religious and Social Studies.
Lawlis, G. Frank and Elba Lu. 1972. Judgment of counseling process: Reliability, agreement, and error. *Psychological Bulletin* 78: 17–20.
Lin, Lawrence, I. 1989. A concordance correlation coefficient to evaluate reproducibility. *Biometrics* 45: 255–268.
Lombard, Matthew, Jennifer Snyder-Duch, and Cheryl C. Bracken. 2002. Content analysis in mass communication research: An assessment and reporting of intercoder reliability.

*Human Communication Research* 28: 587–604. https://engagedscholarship.csuohio.edu/cgi/viewcontent.cgi?article=1049&context=clcom_facpub (accessed 2020.5.4).

Manning, Christopher D., Prabhakar Raghavan, and Hinrich Schütze. 2008. *Introduction to Information Retrieval*. Cambridge, MA: Cambridge University Press.

Maxwell, A. E. 1970. Comparing the classification of subjects by two independent judges. *British Journal of Psychiatry* 116: 651–655.

Maxwell, A. E. 1977. Coefficients of agreement between observers and their interpretation. *British Journal of Psychiatry* 130: 79–83.

Mayer-Schönberger, V. and Kenneth Cukier. 2013. *Big Data: A Revolution That Will Transform How We Live, Work, and Think*. Boston, MA: Houghton Mifflin.

McFarland, Daniel A. and H. Richard McFarland. 2015. Big data and the danger of being precisely inaccurate. *Big Data & Society* 1: 1–4. https://www.researchgate.net/publication/285544247_Big_Data_and_the_danger_of_being_precisely_inaccurate. (accessed 2020.5.4).

McGill, William J. 1954. Multivariate information transmission. *Psychometrika* 19: 97–116.

Miller, George A. 1956. The magical number seven, plus or minus two: Some limits on our capacity for processing information. *Psychological Review* 63: 81–97. https://psycnet.apa.org/fulltext/1957-02914-001.pdf (accessed 2020.5.11).

Neuendorf, Kimberly A. 2002. *The Content Analysis Guidebook*. Thousand Oaks, CA: Sage.

Neuendorf, Kimberly A. 2009. Reliability for content analysis, in *Media Messages and Public Health: A Decision Approach to Content Analysis*, eds. Amy B. Jordan, Dale Kunkel, Jennifer Manganello, and Martin Fishbein, 67–87. New York: Routledge.

NIST – National Institute of Standards and Technology 2012. *e-Handbook of Statistical Methods*, www.itl.nist.gov/div898/handbook/ (accessed 2020.5.4).

Noble, Safiya, Umoja. 2018. *Algorithms of Oppression; How Search Engines Reinforce Racism*. New York: NYU Press.

Nunnally, Jum C. and Ira H. Bernstein. 1994. *Psychometric Theory* (3rd ed.) New York: McGraw-Hill.

Ogilvie, Daniel M., Philip, J. Stone, and Erwin S. Shneidman. 2009. Some characteristics of genuine versus simulated suicide notes, in *The Content Analysis Reader*, eds. Klaus Krippendorff and Mary A. Bock, 404–409. Thousand Oaks, CA: Sage.

O'Neil, Cathy. 2017. *Weapons of Math Destruction: How Big Data Increase Inequality and Threaten Democracy*. New York: Broadway Books.

Osgood, Charles E. 1959. The representational model and relevant research methods, in *Trends in Content Analysis*, ed. Ithiel de Sola Pool, 33–88. Urbana: University of Illinois Press.

Pearson, Karl et al. 1901. Mathematical contributions to the theory of evolution: IX. On the principle of homotyposis and its relation to heredity, to variability of the individual, and to that of race. Part I: Homotyposis in the vegetable kingdom. *Philosophical Transactions of the Royal Society, 197* (Series A), 285–379.

Pearson, Karl. 1926. On the coefficient of racial likeness. *Biometrika 18*, 1–2, 105–117.

Perreault, William D. and Lawrence E. Leigh. 1989. Reliability of nominal data based on qualitative judgments. *Journal of Marketing Research* 26: 135–148.

Popping, Roel. 1988. On agreement indices for nominal data, in *Sociometric Research: Volume 1, Data Collection and Scaling*, eds. Willem E. Saris and Irmtraud N. Gallhofer, 90–105. New York: St. Martin's Press.

Potter, W. James and Deborah Levine-Donnerstein. 1999. Rethinking reliability and validity in content analysis. *Journal of Applied Communication Research* 27: 258–284.

Powers, David M. W. 2011. Evaluation: From precision, recall and F-measure to ROC, informedness, markedness & correlation. *Journal of Machine Learning Technologies* 2: 37–63.

Rajaratnam, Nageswari. 1960. Reliability formulas for independent decision data when reliability data are matched. *Psychometrika* 25: 261–271.

Riffe, Daniel, Stephen Lacy, and Frederick G. Fico. 1998. *Analyzing Media Messages: Using Quantitative Content Analysis in Research* (2nd ed.). Mahwah, NJ: Lawrence Erlbaum.

Riffe, Daniel and Alan Freitag. 1996. *Twenty-five years of content analyses*. Paper presented at the annual meeting of the Association for Education in Journalism and Mass Communication, Anaheim, CA. *Journalism & Mass Communication Quarterly* 74: 873–882, 1997.

Rogot, Eugene and Irving. D. Goldberg. 1966. A proposed index for measuring agreement in test-retest studies. *Journal of Chronic Diseases* 19: 991–1006.

Rorty, Richard. 1989. *Contingency, Irony, and Solidarity*. New York: Cambridge University Press.

Rosenberg, Daniel. 2013. Data before the fact, in *"Raw Data" is an Oxymoron*, ed. Lisa Gitelman, 15–40. Cambridge, MA: MIT Press.

Sargeant, Jan M. and S. Wayne Martin. 1998. The dependence of Kappa on attribute prevalence when assessing the repeatability of questionnaire data. *Preventive Veterinary Medicine, 34*, 2–3: 115–123.

Schuster, Christof. 2001. Kappa as a parameter of a symmetry model for rater agreement. *Journal of Educational and Behavioral Statistics* 26: 331–342.

Scott, William A. 1955. Reliability of content analysis: The case of nominal scale coding. *Public Opinion Quarterly* 19: 321–325.

Seung, Hui K. and Patric S. C. Lee. 1985. Effects of the use of percentage agreement on behavioral observation reliabilities: A reassessment. *Journal of Psychopathology and Behavioral Assessment* 7: 221–234.

Shneidman, Ervin S. 1966. *The Logics of Communication: A Manual for Analysis.* China Lake, CA: U.S. Naval Ordnance Test Station.

Siegel, Sidney and N. John Castellan. 1988. *Non-Parametric Statistics for the Behavioral Sciences*, (2nd ed.). New York: McGraw-Hill.

Shannon, Claude E. and Warren Weaver. 1949. *The Mathematical Theory of Communication.* Urbana: University of Illinois Press.

Shoukri, Mohamed M. 2011. *Measures of Interobserver Agreement and Reliability, 2nd ed.* Boca Raton, FL: Taylor&Francis/CRC Press.

Spearman, Charles. 1904. The proof and measurement of association between two things. *American Journal of Psychology* 15: 72–101.

Spiegelman, Marvin, Charles Terwilliger, and Franklin Fearing. 1953. The reliability of agreement in content analysis. *Journal of Social Psychology* 37: 175–187.

Stemler, Steven E. 2004. A comparison of consensus, consistency, and measurement approaches to estimating interrater reliability. *Practical Assessment, Research & Evaluation* 9:1–11. https://scholarworks.umass.edu/cgi/viewcontent.cgi?article=1137&context=pare (accessed 2020.6.20).

Stevens, Stanley S. 1946. On the theory of scales of measurement. *Science* 103, 2684: 677–680. https://science.sciencemag.org/content/103/2684/677 (accessed 2019.8.8).

Strauss, Anselm and Juliet Corbin. 1994. Grounded theory methodology, in *Handbook of Qualitative Research*, eds. Norman K. Denzin and Yvonna S. Lincoln, 217–285. Thousand Oaks, CA: Sage.

Surowiecki, James. 2005. *The Wisdom of Crowds; Why the Many Are Smarter than the Few and How Collective Wisdom Shapes Business, Economies, Societies, and Nations.* New York: Doubleday.

Tildesley, Michael L. 1921. A first study of the Burmese Skull. *Biometrika* 13: 2–3: 176–262, www.jstor.org/pss/2331753 (accessed 2019.8.11).

Tinsley, Howard E. A. and David J. Weiss. 2000. Interrater reliability and agreement, in *Handbook of Applied Multivariate Statistics and Mathematical Modeling*, eds. Howard E. A. Tinsley and Steven D. Brown, 95–124. New York: Academic Press.

Turow, Joseph, Joseph N. Cappella, and Kathleen Hall Jamieson. 1996. Call-in political talk radio: Background, content, audiences, portrayal in mainstream media. In Report Series No. 5, *The Annenberg Public Policy Center of the University of Pennsylvania:* https://repository.upenn.edu/cgi/viewcontent.cgi?referer=&httpsredir=1&article=1410&context=asc_papers (accessed 2020.5.4).

van Rijsbergen, Cornelis Joost. 1979. *Information Retrieval, 2nd ed.* London, UK; Boston, MA: Butterworth.

Voorhees, Ellen M. and Donna Harman. 2000. Overview of the Sixth Text REtrieval Conference (TREC-6). *Information Processing and Management* 36: 3–25.

Viera, Anthony J. and Joanne M. Garrett. 2005. Understanding interobserver agreement: The kappa statistic. *Family Medicine* 37: 360–363.

Wittgenstein, Ludwig. 1933. *Tractatus Logico-Philosophicus*. New York: Harcourt, Brace & Company.

Zhao, Xinshu, Jun S. Liu, and Ke Deng. 2012. Assumptions behind inter-coder reliability indices, in *Communication Yearbook 36*, ed. Charles T. Salmon, 419–480. New York: Routledge. https://repository.hkbu.edu.hk/cgi/viewcontent.cgi?article=7399&context=hkbu_staff_publication (accessed 2019.8.8 – with minor changes to the original).

Zwick, Rebecca. 1988. Another look at inter-rater agreement. *Psychological Bulletin, 103*: 374–378. https://files.eric.ed.gov/fulltext/ED279710.pdf (accessed 2020.5.4).

# *Index*

*Note*: Tables are shown in **bold type** and figures are in *italics*.

## N



## O



## P



## Q

## R

Taylor & Francis Group
an **informa** business

# Taylor & Francis eBooks

www.taylorfrancis.com

A single destination for eBooks from Taylor & Francis with increased functionality and an improved user experience to meet the needs of our customers.

90,000+ eBooks of award-winning academic content in Humanities, Social Science, Science, Technology, Engineering, and Medical written by a global network of editors and authors.

## TAYLOR & FRANCIS EBOOKS OFFERS:

- A streamlined experience for our library customers
- A single point of discovery for all of our eBook content
- Improved search and discovery of content at both book and chapter level

## REQUEST A FREE TRIAL

**support@taylorfrancis.com**

Routledge
Taylor & Francis Group

CRC Press
Taylor & Francis Group